ANNALS *of* THE NEW YORK ACADEMY OF SCIENCES

ISSUE
Reproductive Aging

ISSUE EDITORS
Maxine Weinstein and Kathleen O'Connor

This volume presents manuscripts stemming from "The Workshop on Reproductive Aging," held June 5–6, 2009 at Georgetown University. The workshop activities were supported by Grant K07-AG019165, "Infrastructure for Aging and Health," awarded to Georgetown University by the National Institute on Aging, and by additional funding from Georgetown University.

TABLE OF CONTENTS

1	Reproductive aging: theoretical perspectives, mechanisms, nonhuman models, and health correlates	
	John Haaga, Kathleen O'Connor, Maxine Weinstein, and Phyllis Wise	

Theoretical perspectives

11	Life historical perspectives on human reproductive aging	
	Peter T. Ellison	
21	The connections between general and reproductive senescence and the evolutionary basis of menopause	
	Thomas B. L. Kirkwood and Daryl P. Shanley	
30	Learning, menopause, and the human adaptive complex	
	Hillard Kaplan, Michael Gurven, Jeffrey Winking, Paul L. Hooper, and Jonathan Stieglitz	
43	Do women stop early? Similarities in fertility decline in humans and chimpanzees	
	Kristen Hawkes and Ken R. Smith	
54	An evolutionary and life history perspective on human male reproductive senescence	
	Richard G. Bribiescas	
65	Dynamic heterogeneity and life histories	
	Shripad Tuljapurkar and Ulrich K. Steiner	

Become a Member Today of the New York Academy of Sciences

The New York Academy of Sciences is dedicated to identifying the next frontiers in science and catalyzing key breakthroughs. As has been the case for 200 years, many of the leading scientific minds of our time rely on the Academy for key meetings and publications that serve as the crucial forum for a global community dedicated to scientific innovation.

Select one FREE *Annals* volume and up to five volumes for only $40 each.

Network and exchange ideas with the leaders of academia and industry.

Broaden your knowledge across many disciplines.

Gain access to exclusive online content.

Join Online at **www.nyas.org**

Or by phone at **800.344.6902** (516.576.2270 if outside the U.S.).

TABLE OF CONTENTS, CONTINUED

Mechanisms

73 Mechanisms of reproductive aging: conserved mechanisms and environmental factors
Mary Ann Ottinger

82 Ovarian aging in developmental and evolutionary contexts
Caleb E. Finch and Donna J. Holmes

95 Relating smoking, obesity, insulin resistance, and ovarian biomarker changes to the final menstrual period
MaryFran R. Sowers, Daniel McConnell, Matheos Yosef, Mary L. Jannausch, Sioban D. Harlow, and John F. Randolph, Jr.

104 Estrogen and the aging brain: an elixir for the weary cortical network
Dani Dumitriu, Peter R. Rapp, Bruce S. McEwen, and John H. Morrison

113 The hypothalamic median eminence and its role in reproductive aging
Weiling Yin and Andrea C. Gore

Nonhuman models

123 Animal models of reproductive aging: what can they tell us?
Steven N. Austad

127 Life history context of reproductive aging in a wild primate model
Jeanne Altmann, Laurence Gesquiere, Jordi Galbany, Patrick O. Onyango, and Susan C. Alberts

139 Reproductive aging in tephritid fruit flies
James R. Carey and Freerk Molleman

149 Reproductive aging in invertebrate genetic models
Marc Tatar

156 A pathway that links reproductive status to lifespan in *Caenorhabditis elegans*
Cynthia Kenyon

Health correlates

163 Health consequences of reproductive aging: a commentary
Siobán D. Harlow

TABLE OF CONTENTS, CONTINUED

169 Reproductive aging, menopause, and health outcomes
JoAnn V. Pinkerton and Dale W. Stovall

179 Reproductive aging and its consequences for general health
Michael L. Traub and Nanette Santoro

188 Longitudinal, epidemiologic studies of female reproductive aging
Rebecca J. Ferrell and MaryFran Sowers

198 Corrigendum for Ann. N. Y. Acad. Sci. 1173: 865–873

The New York Academy of Sciences believes it has a responsibility to provide an open forum for discussion of scientific questions. The positions taken by the authors and issue editors of the *Annals of the New York Academy of Sciences* are their own and not necessarily those of the Academy unless specifically stated. The Academy has no intent to influence legislation by providing such forums.

Reproductive aging: theoretical perspectives, mechanisms, nonhuman models, and health correlates

John Haaga,[1] Kathleen O'Connor,[2] Maxine Weinstein,[3] and Phyllis Wise[4]

[1]Division of Behavioral and Social Research, National Institute on Aging, Bethesda, Maryland. [2]Department of Anthropology and Center for Studies in Demography and Ecology, University of Washington, Seattle, Washington. [3]Graduate School of Arts and Sciences, Georgetown University, Washington, DC. [4]University of Washington, Seattle, Washington

Address for correspondence: Maxine Weinstein, Distinguished Professor of Population and Health, Graduate School of Arts and Sciences, Georgetown University, Washington, DC 20057. weinstma@georgetown.edu

Every 365.2421999 days—more or less—the earth rotates, revolves, wobbles, and precesses along an elliptical path around the sun. Those of us who survive the journey are a year older: we have aged one solar year. Some years we seem to age faster than other years; some people seem to age faster than other people; some systems seem to age faster than other systems. As we begin to mature, reach our middle years, and become elderly, reproductive changes are among the markers of aging that are most notable, particularly among women. What—if anything—can we learn about more general processes of aging from reproductive aging? Does our postreproductive survival contribute to our fitness, or is it just a chance event, a result of selection on other characteristics? Can our insights and research be translated into improved clinical practice? We explore reproductive aging with a wide-angle multidisciplinary lens that we use to focus on four articulating areas: theoretical perspectives, mechanisms, nonhuman models, and health correlates. We propose directions for future work.[a]

Keywords: reproductive; aging; fitness

The Workshop on Reproductive Aging, held at Georgetown University in June 2009, was organized with the idea of bringing together an intellectually diverse group of scholars and practitioners to take a look at wide-ranging questions related to somatic and reproductive aging. This volume, which grew out of the meeting, is divided into four sections, very much along the lines of the workshop: theoretical perspectives, mechanisms, nonhuman models, and health correlates. Each comprises a group of papers that collectively explores the theme of the section; a commentary at the beginning of each section pulls together the various approaches to the subject and brings the research of the individual commentator to bear.

Our introduction provides a pathway through each section of the volume; the commentaries draw together threads within each section. We also use this introduction to discuss a few themes that cut across the sections. We focus on three of those cross-cutting issues: study design; integration across systems, species, and sexes; and context: the joint ecology of reproduction and aging. The workshop discussions were lively, occasionally heated; our introduction draws on those conversations as well as the talks of our two valedictory speakers, John Haaga and Phyllis Wise. We close with a few thoughts about directions for future research.

Theoretical perspectives

Tom Kirkwood and Daryl Shanley[1] set the stage for our discussion of underlying theory. Their paper lays out a few fundamental questions that are addressed throughout the volume from various perspectives. First, what are the underlying factors that account for the general biology of aging, and more specifically, which of them has the greatest effect on reproductive aging? Second, what are the relationships among these factors? Third, how have evolutionary forces jointly shaped these factors?

---

[a]The views expressed here are those of the authors and are not attributable to the National Institute on Aging.

Humans have a notably long postreproductive lifespan. Among women, age at menopause occurs around age 50[2]; expectation of life at birth is currently about 69 years globally and about 77 years in more developed countries.[3] And, as noted by Kaplan and his colleagues[4] in this volume, modern populations of hunter gatherers and simple horticulturalists have an expectation of life at age 45 of about 20 years. Overall, then, the available evidence suggests that postreproductive life is a pervasive feature of humans.

Hawkes and Smith[5] examine postreproductive survival, focusing on females. From a simple accounting point of view, the extended postreproductive lifespan we currently see might be attributable to increases in expectation of life after menopause, to decreases in age at menopause, or to some combination of both. We know, of course, that expectation of life at birth has increased dramatically over the past century. Expectation of life at age 50 has also increased. But assuming age at menopause is around 50 years, even lower life expectancies—even substantially lower expectations of life than we observe today—would typically be associated with significant postreproductive years because age-specific mortality rates are bathtub shaped, with high mortality and generally rapidly declining mortality immediately after birth. For example, in a female population with expectation of life at birth of 45 years, with a West-model life table, a woman who survived to the age of childbearing (say 15) would expect to live another 45 years; a woman who survived to menopause (say 50), would have an expectation of life of another 20 years.[6] Hawkes and Smith[5] look at the complementary piece of the equation: do modern women reach the end of fertile life earlier than our ancestors? Using data from the Utah Population Database[7] and from published records of wild chimpanzees, they conclude that secondary sterility has not come earlier in modern women, a result consistent with a version of the grandmother hypothesis that proposes that age at onset of decline in fecundity has not markedly changed, whereas expectation of life has increased.

Bribiescas[8] brings us into the realm of the male contribution to fertility and how it changes with age; Kaplan and his colleagues[4] put the sexes together in constructing an explanation for the evolution of menopause. The processes of reproductive aging and decline in men differ from women in some fundamental ways: men do not run out of germ cells (sperm), do not experience an abrupt end to reproductive ability, and they can and do reproduce into elderly ages.[9] Nevertheless, spermatogenesis declines, spontaneous abortion and fetal abnormalities appear to rise, and coitus becomes compromised with increasing age.[10,11] As illustrated among the Tsimane in the chapter by Kaplan and colleagues,[4] despite the biological ability to reproduce at older ages, few men actually do. Kaplan *et al.*[4] argue that menopause evolved in the context of postreproductive investment by both females and males. It is maintained by selection because women who invested in existing offspring (and stopped reproducing) had greater net fertility, and it evolved in a setting in which cooperative division of labor between males and females was joined to similar (albeit not identical) age-specific declines in reproduction.

Variation in life histories among individuals and variation across a single individual's life are crucial factors in understanding reproductive aging. The importance of heterogeneity has long been recognized; the possibility of drawing incorrect conclusions by improperly accounting for it is well documented.[12,13] As noted by Trussell and Rodriguez,[14] the central point is to balance the tension between the search for commonalities and the need to account for differentials: "the goal in either case is to separate the population into the minimum number of groups sufficient to capture the important sources of variation" (p. 111).

The chapter by Tuljapurkar and Steiner[15] takes an important step forward in enriching our understanding of the potential effects of heterogeneity. Here, the notion of frailty— the idea that there are differentials in individual susceptibility to mortality, for example—is extended by relaxing the assumption that frailty is fixed over an individual's lifetime. Instead, their model allows individual frailty to vary across a range of values. Using longitudinal data on the acute swan, Tuljapurkar and Steiner[15] examine reproductive success and survival rates. The central idea is that it provides a platform for beginning to decompose intrinsic and exogenous factors affecting a given phenotype; thus, it has potential for identifying the factors that affect reproductive and somatic aging in concert or separately.

Mechanisms

In a simple sense, we know the underlying mechanism causing final reproductive cessation in human females: exhaustion of the pool of oocytes a woman is endowed with as a fetus. The mechanisms associated with the *process* of reproductive aging that begins up to a decade or more before menopause in the human female are less clear and involve multiple components of the reproductive hormone signaling axis: the hypothalamus, pituitary, and ovaries. Reproductive aging may occur along separate mechanistic paths in each of these organs, earlier in one component than others, or to receptors or hormones common to all three. Moreover, there are multiple times, some as early as *in utero*, when environmental and somatic aging mechanisms may intersect with the reproductive axis. The chapters and commentary in this section examine some of these relationships.

Finch and Holmes[16] focus on ovarian aging with the goals of assessing whether homologous patterns of ovarian decline are seen across taxa and whether these patterns are related to specific life histories. This information helps to identify the extent to which reproductive and somatic aging processes and mechanisms are conserved or derived in various species. They illustrate, for example, that postreproductive survival is not altogether unusual in the animal world, nor is it limited to long-lived species with kin networks. One implication of this observation is that mechanisms underlying both reproductive and somatic aging are not necessarily tightly linked; cross-taxa plasticity appears to be present in each.

Sowers and her colleagues[17] also focus on ovarian aging and investigate whether smoking, obesity, or insulin resistance—factors that advance somatic aging—mediate reproductive aging through their effects on ovarian reserve (as indicated by anti-Müllerian hormone (AMH), inhibin B, and follicle-stimulating hormone). They report that smoking is associated with accelerated reproductive aging partly through its effects on AMH and inhibin B, while insulin-resistance accelerates time to the final menstrual period through a pathway that does not involve indicators of ovarian reserve. These results suggest that interventions to reduce smoking and insulin resistance may help to reduce both early reproductive and somatic aging.

Dumitriu and colleagues[18] and Yin and Gore[19] examine structural changes within the hippocampus and hypothalamus, respectively. Reproductive aging in rodents and nonhuman primates is characterized by different endocrine milieus: aging rodents have elevated estrogen with declining gonadotropin-releasing hormone (GnRH), whereas aging primates have declining estrogen and elevated GnRH. Yin and Gore's[19] work shows that structural changes associated with GnRH neuroterminals within the hypothalamic median eminence degrade with age, physically impeding GnRH release in the rodent. Similar work has not been done in primates, but elevated gonadotropins in the perimenopause suggest upregulated GnRH. As postmenopause progresses, the primate endocrine milieu becomes more like that of the rodent. This work suggests that signaling axes downstream from GnRH would also be compromised, thus somatic—as well as reproductive—aging may proceed rapidly at late stages of reproductive aging in different species. Dumitriu and colleagues[18] examine the effects of estrogen on hippocampal spinogenesis and synapse formation in the context of examining the neuroprotective effects of estrogen on cognition. Tracking down underlying mechanisms is a challenge because of estrogen's multiple receptors and multiple nongenomic and genomic pathways. These authors elegantly show that reduced cognitive function in older nonhuman primates and rodents is a function of two distinct aging processes: loss of neuron spines and spine density results from both somatic aging and ovarian estrogen deficiency.

Nonhuman models

Our section on nonhuman models comprises four chapters: Altmann *et al.*,[20] Carey and Molleman,[21] Tatar,[22] and Kenyon.[23] It is rare to find data on reproductive senescence in nonhuman animals and most especially rare to find data on noncaptive populations. The contribution of Altmann and her colleagues,[20] using nearly 40 years of data from the Amboseli Baboon Research Project, provides a correspondingly rare window onto a close primate relative of humans. Her data show that among the Amboseli baboons, declines in reproduction occur only late in life; postreproductive survival among females is confined to only the very oldest baboons. Male reproductive decline in baboons occurs earlier than female. These trails appear to be sensitive to social context and to food availability.[24–26] The data support similar findings in chimpanzee female reproductive aging[27] and provide some additional evidence in support of the hypothesis that long human postreproductive survival may be attributable to improvements in mortality while the reproductive lifespan has remained stable.

What can worms and flies tell us about reproductive aging and somatic aging in humans? Three papers—those by Kenyon,[23] Tatar,[22] and Carey and Molleman[21]—suggest that, in fact, we can learn quite a bit from these nonmammalian animals. In contrast to mammals, in *Drosophila melanogaster* and *Caenorhabditis elegans*, oocytes are produced from stem cells in mature adults. Despite this difference, egg production declines with age, and females often survive after reproduction ceases. The papers in this section explore the signaling networks influencing reproductive aging and lifespan and identify some potentially conserved pathways influencing mammalian ovarian aging. As summarized by Tatar,[22] reproductive senescence encompasses two processes that are not unique to humans: first, a decline in egg production over time (in humans, we think that production ceases before birth); and second, reduction in egg quality over time (although in humans, it is not established whether this reduction is attributable to selective use of better eggs at earlier ages, to accumulation of damage to the reserved eggs, or both).

The common theme is that there is a reciprocal relationship between reproduction and lifespan, but the relationship does not appear to be a simple reduction in lifespan attributable to a reproductive burden: Tatar[22] found no tradeoff between reproduction and expectation of life in wild-type *C. elegans*; it was not failure to reproduce that affected aging, rather it was loss of the germline that added years of life. A complementary observation is made by Carey and Molleman[21] (with regard to tephritids), who note that the effects of reproduction on lifespan have been given substantial attention, but that the price of postponing reproduction in favor of lifespan is less examined.

The mechanisms that underlie the reciprocal relationship are not fully understood, but both Kenyon[23] and Tatar[22] suggest some pathways through which reproductive and nonreproductive tissues communicate. And how (indeed whether) these findings extend to other animals is also unclear: for example, loss of the germline in *Drosophila* is not associated with extended lifespan.[28] The work of Carey and his colleagues on mice (see, e.g., Cargill *et al.*[29]) is intriguing in this regard: it suggests that signals from reproductive tissue (specifically ovaries) can affect subsequent survival in mammals. Tatar[22] and Kenyon's[23] work suggests that in the fly and worm the germline and somatic gonad influence lifespan in counter-balancing ways, with the germline generally inhibiting, and the somatic gonad promoting lifespan extension. What are the consequences of reproductive aging for health and survival in humans? Intriguingly, in human females, hysterectomy with ovarian conservation is associated with decreased risk of all-cause mortality when compared with removal of both the uterus and ovaries.[30]

Health correlates

Clinically, menopause is customarily defined as occurring when a year elapses after the final menstrual period. Prospectively, of course, it is therefore well after an observable event that a woman would be able to say she has reached menopause. While the shift from menstrual bleeding to none is a highly noticeable marker, earlier changes are also notable and may serve as important sources of motivation for seeking medical care. The processes are complex, our understanding is at an early stage, and the appropriate intervention, if any, is often not clear.

The contributions from Traub and Santoro[31] and from Pinkerton and Stovall[32] are reminders of the importance of translational research in this area: cardiovascular disease, bone loss, cognition, and mood are among a few of the correlates of hormonal change. From a clinical perspective, however, the many gaps in our knowledge mean that the kind, timing, and duration of interventions are still not well established.

Better guidelines for appropriate interventions depend on the quality and breadth of the data that are available to us, as well, of course, as the quality of the analyses. The earliest path-breaking studies of menstrual cycles that were conducted by Treloar and his colleagues (e.g., Treloar *et al.*[33]) and Vollman (e.g., Vollman[34]) focused on documenting cycle characteristics across age, most notably central tendency and variability in cycle length.

Advances in our ability to collect and assay hormones from specimens collected from larger, more representative samples have allowed us to improve our picture of age-related changes in the hormonal milieu, mostly during the perimenopause. The chapter from Ferrell and Sowers[35] focuses on eight longitudinal studies that include data on how reproductive hormones change with age; they summarize some common findings, but their comparison also sheds some light on seemingly discrepant results that have emerged across the studies. It is a sobering reminder of the difficulties in amassing and analyzing the data. In particular, the authors sound an important caution regarding the kinds of health outcomes for which we might be able to find some links to reproductive aging: rare events are probably beyond our current reach. Advances in our ability to combine and analyze data across studies and harmonization of new efforts will be needed to explore any but the most common outcomes.

The recent reemergence of attention in the popular press (see, e.g., the article by Gorney in the *New York Times*, April 12, 2010[36]) to the effects of peri- and post-hormone replacement therapy is a timely example of how our research has real consequences for the health of women. And of course, the attention in the popular press reflects increased scholarly scrutiny, as well. Goldman's summary of the controversy surrounding the discrepancies between findings regarding hormone therapy from the Women's Health Initiative (a randomized controlled trial) and observational studies underscores its translational importance: starting in 2002 hormone therapy dropped sharply in the United States and elsewhere, and clinical guidelines were modified.[37] In short, our studies matter.

Study design

An unusual feature of the workshop was that it not only brought together investigators from diverse disciplinary backgrounds, but that both clinicians and nonmedical sorts attended. The exchange was particularly helpful in focusing attention on how study design constrains the applicability of research findings in clinical practice.

Many of our research protocols include only women who have intact reproductive organs (i.e., who have not had their uterus or ovaries removed) and who are not taking exogenous hormones. In the real world, of course, something like a third of all women are hysterectomized[38]—many before the menopausal transition—and at both ends of the reproductive age spectrum, women are using hormones either for contraception or for alleviating discomfort associated with the transition to menopause. And these factors (among others) are known to differ by ethnicity and by race.[39] The question, therefore, is the representativeness of our study populations: how far can we generalize any of our study findings?

Can we devise designs that will allow us to translate our findings and to generalize across subgroups? At the start of her talk, Siobán Harlow[40] amended it by adding "human." Peter Ellison[41] added "modern, western, industrial," and we have already—at least for studies in the United States—noted the selectivity with respect to hormone use and surgery. At the same time, the work of Bribiescas[8] and of Kaplan and his colleagues[4] implicitly makes a case for more attention to male reproductive aging. In short, our designs impose Thwackum-worthy restrictions.

These design considerations are important not only for clinical translation but for our fundamental understanding of the commonalities and differences between somatic and reproductive aging. The selective design of our studies may make it impossible to disentangle, for example, a link between early (or for that matter, late) onset of menopause and early risk of death from underlying health-related factors that may influence both age at menopause and age at death. A similar argument holds when we choose women on the basis of (some set of) cycle characteristics.

Good design needs also to account for the life course experience. As noted by Phyllis Wise, we often talk about "estrogen," "progesterone," and "testosterone," but the reality is complex: estrogen, for example, refers to a family of hormones. And we were reminded by Stovall and Ellison that the receptors may be up- or downregulated and that receptor set points may vary widely across populations. Sensitivity in these factors are dependent, perhaps, on chronic estrogen levels or on exposure at particular times in the life course. The response in Western women during the menopausal transition may be unusually amplified compared with populations with lower lifetime exposures.

A life course perspective is also important for examining the effects of early events (Pinkerton, for example, pointed to the effect of childhood trauma on vasomotor response during the menopausal transition) and for examining the effects of the duration of the transition on responses, such as cognitive function and bone loss. From a design standpoint, however, these are tall orders: we know that there can be significant measurement error associated with retrospective reporting and the costs of long-term longitudinal studies can be quite high. It may be that the perfect study design is a will-o'-the-wisp. Still, it is crucial that we have a heightened awareness of, and sensitivity to, the deficiencies in our current study designs, especially with an eye toward assessing how these vulnerabilities constrain our ability to generalize and interpret the results.

Integration across systems, species, and sexes

A recurring point in our discussions was how reproductive aging could be understood both as part of, and separate from, other aging processes. On the one hand, we know quite a bit about various changes that occur during aging, but not how they link up across systems. On the other hand, there are intrinsic difficulties in discriminating between the concurrent processes of somatic and reproductive aging.

As discussed by Harlow,[40] the interdependence of reproductive and somatic aging means that disentangling markers may be quite complex. As an example of what might be an approach, she cites the data cited by Ferrell and Sowers[35] that identify declines in estrogen as a late (relative to changes in other factors that occur as part of reproductive aging) occurrence, suggesting that laying out the sequence of changes more clearly may be of help in identifying the effects of the process.

The underlying questions are first, What, if any, underlying mechanisms do somatic and reproductive aging have in common? second, How have evolutionary pressures resulted in the combination we currently observe? and third, To what extent are these features preserved across species (or what do we learn from examining other species)?

There are many ways to think about the mechanisms that contribute to aging; Kirkwood and Shanley[1] suggest that we ask whether somatic aging is attributable to the accumulation of damage over time or whether it is caused by changes in (i.e.,

the deterioration of) our repair systems over time. For human females, reproductive life ends (completely) when insufficient eggs remain to maintain cycles, but, in fact, earlier cessation is almost universal. Menopause occurs at an average age of about 50 or 51 years, but fertility drops sharply—even in natural fertility populations—well before that time. And the results reported by Bribiescas,[8] by Kaplan and his colleagues,[4] and by Ellison[41] suggest a similar phenomenon for males as well. Few follicles (relative to the initial total endowment) are lost to ovulation. Unlike other cells, follicles are not replenished; are the causes of apoptotic follicular loss different from other apoptotic cell death?

Again, as Kirkwood and Shanley[1] point out, the germline ultimately can not survive if it reproduces accumulated damage. Thus, one would expect selective pressures on the system for the heightened levels of the maternity schedule to correspond to the use of the best—or least damaged—germ cells in combination with a low, nonreproductive force of mortality. Such a pattern is observed in humans, but what about other animals? And how unusual is it in the animal kingdom to live postreproductively, whether we imagine that question as an early end to reproduction or as an extended survival period postreproduction?

In humans, both parents make important contributions to the survival of their offspring to reproductive ages, and both maternal and paternal age matter for the likelihood of successful reproduction.[42] Tuljapurkar and colleagues in earlier work,[9] and Kaplan and colleagues[4] in this volume, break new ground by considering two sex models of the evolution of reproductive aging. What might we learn from considering these models for other monogamous species where paternal investment is important for survival of offspring? How are somatic or reproductive aging similar or different in timing and sequence in nonmonogamous social species? The longitudinal data on wild primates from Altmann and her colleagues[20] suggest that reproductive, behavioral, and somatic aging occur earlier in males than females—at least in a savannah setting. What can we learn regarding humans from the underlying mechanisms contributing to this pattern? But data from wild populations are rare, and Austad[43] rightly cautions us to be careful about making inferences about evolutionary processes based only on laboratory results.

Context: the joint ecology of reproduction and aging

The interaction between somatic aging and reproductive aging is central to research on reproductive aging. Several of the chapters emphasize that growth, development, reproduction, and survival are inextricably intertwined. It should not surprise us that hormones that are thought to be reproductive hormones also exert effects on somatic growth. Estrogens are not only reproductive hormones, but they are growth and trophic factors. Insulin growth factor is not only a growth and trophic factor, but it is also a reproductive hormone that acts upon GnRH and other aspects of the reproductive axis. But aging—reproductive or somatic—does not occur in a vacuum. Our environment has massive effects on both. This point came up repeatedly throughout our discussions—often in different guises—but over and over.

Environmental exposures and the timing of those exposures may have profound effects on both somatic and reproductive aging. The clinical context was one focus for the importance of environment. The workshop participants whose practice included a strong clinical component (JoAnn Pinkerton, Ricki Pollycove, James Simon, and Dale Stovall) raised concerns about the women they were seeing in their day-to-day practice, women who express distress about characteristics of the menopausal transition. The research on menopausal women in other settings may tell us that many of the "symptoms" experienced by Western women are strongly influenced by the environment, but it does not relieve those symptoms in a meaningful way for those Western women whose culture is the Western medical model and whose prior exposure is probably quite different from women in those other ecological settings. As stated by Hawkes during a discussion, "the important point about your patients is the take-home message we should hear: that we need to take seriously what women are experiencing in the socioecology."

The results of Sowers and her colleagues[17] show that smoking, obesity, and metabolic dysregulation (as indicated, for example, by insulin resistance) can affect the timing of reproductive aging as well as how it manifests. As discussed by Ottinger,[44] there is substantial individual variation in response to environmental factors. She points to increased recent attention to endocrine-disrupting chemicals, noting, by way of example, their different effects on the reproductive success of long-lived versus short-lived birds. Phyllis Wise reminded us that, "hormones are feathers—not hammers": it may not take a large stimulus for a noticeable change in outcome. Much remains unknown regarding the role of environment on hormonal set points, determinants of the sensitivity of response to environmental exposures, or the importance of the timing of exposure for subsequent response. Surprisingly—given the wide range of ecology, both physical and social—at least in the aggregate, cross-population variation in the timing of menopause is not great,[2,45] but variation among individuals is extensive. We have much to learn about the effects of environmental triggers.

Comparative studies across populations with longitudinal data at the individual level will be needed to understand interactions among individual endowment, environmental exposure, and culture. The data required for this kind of work is only one example of the importance of shared access to, and the development of, infrastructure for data resources. It would be hard to overstate this point. Two common threads were implicit throughout the workshop: the need for databases and the importance of sharing data. The complexity of data and how to manage data sets have become frequently voiced critical concerns of researchers around the world. We wonder how much funding is wasted—or if not wasted, at least underused—because we are unable (or sadly, sometimes unwilling) to share data. Clearly, collecting and managing data is only the first step to knowledge. We must be able to transform data into information, then transform that information into knowledge, and finally, transform that knowledge into understanding. Without the capacity to enter data into databases that can be easily managed and accessed by other investigators, information, knowledge, and understanding will never happen. We all appreciate the importance—as well as the difficulty—of managing the complex data that we are collecting.

Future directions

Future advances will depend on obtaining additional data, not simply—or, more correctly, not only—sharing the data we already have. A good example is the "grandmother hypothesis," which, in various forms, appeared numerous times throughout the workshop. The work and the progress being made on the grandmother hypotheses, and to what extent they explain the distinctive timing of the menopause, are fascinating. Additional work will be necessary to determine the contribution of the length and extent of parenting to the timing of the menopause. Is there a good correlation between the need for parenting and the importance of grandparenting to the timing of the menopause? What kinds of data are necessary to explore the hypothesis?

Some of the most unexpected findings in reproductive biology relate to GnRH. There are only 1,000 GnRH neurons in the hypothalamus of many mammalian species, there is very little modulation in the structure of this hormone, and GnRH plays a central role in the reproductive success of virtually all organisms. That this singularly simple decapeptide plays such a pivotal role in the survival of a species is counterintuitive. The redundancy and complexity of neurotransmitter and neuropeptide interactions that focus around this one neuronal phenotype are intriguing and continue to be the topic of numerous studies. Successful reproduction is so essential to the continuation of the species that one would think that there would be all sorts of hormones that could regulate gonadotropins and then the steroid hormones. Instead, there is ultimate simplicity: if this is an hourglass, GnRH stands at the constriction with all sorts of complexities in terms of regulation of GnRH.

The presentations at the workshop give us a good illustration of the potential of comparative research to address a fundamental question that has both clinical and behavioral applications. "Comparative" in this instance did not just have the usual biological meaning of cross-species but also across levels of organization (cellular to whole-organism and population health) and across human populations. A recurrent theme in aging research is the need for attention to variation among individuals rather than just average values. Examples discussed at the workshop and in the volume range from an especially frisky elderly nematode to the large differences in circulating testosterone levels for human males in different populations. The finding that testosterone levels are much lower in non-Western societies raises many questions about functional significance and evolutionary origins of variability. One came away from these presentations wanting more research on aging in the wild for other species and on the biodemography of aging in human populations across a wide range of social and physical environments. How much of what we consider normal for Americans in the 21st century is the result of long exposure to cheap calorie-dense foods, a lifetime of ease and cleanliness, and what may be an overload of social connections and demands? Looking across populations and species, looking back in evolutionary time, or looking from "bottom–up" and "top–down" in organization,[46] are all productive ways to approach what otherwise would be intractable.

But it is much easier to call for interdisciplinary and comparative research than to organize it and get it funded. There are very practical questions of career development. Young researchers who want to finish training in a finite period of time, work with a reasonable expectation of sustained funding and eventual professional autonomy, and maintain some balance between work and their own reproduction would have to think carefully before getting involved in projects involving regular travel to other countries and other subdisciplines. Peer review is also a thorny issue. One does not want to encourage second-rate science just because it is interdisciplinary, but there is a widespread feeling that it is especially hard to organize review of interdisciplinary and comparative projects. Review committees accustomed to the power of laboratory studies to isolate single sources of variation do not always welcome studies of field biology. Cohort studies conducted in non-Western societies will typically come up short on some aspect of measurement or analysis compared with a narrowly focused study of an easier-to-reach population; reviewers are usually quick to point out the shortcomings.

A workshop like the one resulting in this volume, taking on a central issue from several angles, can help overcome such obstacles. A creatively planned workshop can partially succeed (*ex post*) at what might have been an impossible task (*ex ante*), namely, putting together a team and a research plan to address reproduction and aging across such a wide frontier. Another benefit can be to sensitize young investigators, potential peer reviewers, and research funding agency staff to an area poised to produce useful results. A sequence of meetings, possibly with varied formats, including didactic sessions and some focus on data harmonization and sharing, or a research network with permeable membranes, could help sustain the momentum of discovery.

Conflicts of interest

The authors declare no conflicts of interest.

References

1. Kirkwood, T.B.L. & D.P. Shanley. 2010. The connections between general and reproductive senescence and the evolutionary basis of menopause. *Ann. N.Y. Acad. Sci.* **1204:** 21–29.
2. Thomas, F. *et al.* 2001. International variability of ages at menarche and menopause: patterns and main determinants. *Hum. Biol.* **73:** 271–290.
3. Population Reference Bureau. 2009. *2009 World Population Data Sheet*. Population Reference Bureau. Washington, DC.
4. Kaplan, H., M. Gurven, J. Winking, *et al.* 2010. Learning, menopause, and the human adaptive complex. *Ann. N.Y. Acad. Sci.* **1204:** 30–42.
5. Hawkes, K. & K.R. Smith. 2010. Do women stop early? Similarities in fertility decline in humans and chimpanzees. *Ann. N.Y. Acad. Sci.* **1204:** 43–53.
6. Coale, A.J., P. Demeny & B. Vaughan. 1983. *Regional Model Life Tables and Stable Populations*, 2nd Ed. Academic Press. New York.
7. Bean, L.L., G.P. Mineau & D.L. Anderton. 1990. *Fertility Change on the American Frontier: Adaptation and Innovation*. University of California Press. Berkeley.
8. Bribiescas, R.G. 2010. An evolutionary and life history perspective on human male reproductive senescence. *Ann. N.Y. Acad. Sci.* **1204:** 54–64.
9. Tuljapurkar, S.D., C.O. Puleston & M.D. Gurven. 2007. Why men matter: mating patterns drive evolution of human lifespan. *PLoS ONE* **2:** e785.
10. Sitzmann, B.D., H.F. Urbanski & M.A. Ottinger. 2008. Aging in male primates: reproductive decline, effects of calorie restriction and future research potential. *Age* (Dordr) **30:** 157–168.
11. ESHRE Capri Workshop Group. 2008. Fertility and ageing. *Hum. Reprod. Update* **11:** 261–276.
12. Vaupel, J.W., K.G. Manton & E. Stallard. 1979. The impact of heterogeneity in individual frailty on the dynamics of mortality. *Demography* **16:** 439–454.
13. Vaupel, J.W. & A.I. Yashin. 1985. Heterogeneity's ruses: some surprising effects of selection on population dynamics. *Am. Stat.* **39:** 176–185.
14. Trussell, J. & G. Rodriguez. 1990. Heterogeneity in demographic research. In *Convergent Issues in Genetics and Demography*. J. Adams, A. Hermalin, D. Lam & P. Smouse, Eds.: 111–132. Oxford University Press. New York.
15. Tuljapurkar, S. & U.K. Steiner. 2010. Dynamic heterogeneity and life histories. *Ann. N.Y. Acad. Sci.* **1204:** 65–72.
16. Finch, C.E. & D.J. Holmes. 2010. Ovarian aging in developmental and evolutionary contexts. *Ann. N.Y. Acad. Sci.* **1204:** 82–94.
17. Sowers, M.F.R., D. McConnell, M. Yosef, *et al.* 2010. Relating smoking, obesity, insulin resistance, and ovarian biomarker changes to the final menstrual period. *Ann. N.Y. Acad. Sci.* **1204:** 95–103.
18. Dumitriu, D., P.R. Rapp, B.S. McEwen & J.H. Morrison. 2010. Estrogen and the aging brain: an elixir for the weary cortical network? *Ann. N.Y. Acad. Sci.* **1204:** 104–112.
19. Yin, W. & A.C. Gore. 2010. The hypothalamic median eminence and its role in reproductive aging. *Ann. N.Y. Acad. Sci.* **1204:** 113–122.
20. Altmann, J., L. Gesquiere, J. Galbany, *et al.* 2010. Life history context of reproductive aging in a wild primate model. *Ann. N.Y. Acad. Sci.* **1204:** 127–138.
21. Carey, J.R. & F. Molleman. 2010. Reproductive aging in tephritid fruit flies. *Ann. N.Y. Acad. Sci.* **1204:** 139–148.
22. Tatar, M. 2010. Reproductive aging in invertebrate genetic models. *Ann. N.Y. Acad. Sci.* **1204:** 149–155.
23. Kenyon, C. 2010. A pathway that links reproductive status to lifespan in *Caenorhabditis elegans*. *Ann. N.Y. Acad. Sci.* **1204:** 156–162.
24. Hrdy, S.B. 1981. *The Woman That Never Evolved*. Harvard University Press. Cambridge, Massachusetts.
25. Beehner, J.C. *et al.* 2009. Testosterone related to age and life-history stages in male baboons and geladas. *Horm. Behav.* **56:** 472–480.
26. Wright, P., S. King, A. Baden & J. Jernvall. 2008. Aging in wild female lemurs: sustained fertility with increased infant mortality. In *Primate Reproductive Aging: Cross-Taxon Perspectives, Interdisciplinary Topics in Gerontology*, Vol. 36. S. Atsalis, S.W. Margulis & P.R. Hof, Eds.: 17–28. Karger. Basel, Switzerland.
27. Emery Thompson, M. *et al.* 2007. Aging and fertility in wild chimpanzees provide insights into the evolution of menopause. *Curr. Biol.* **17:** 2150–2156.
28. Barnes, A.I. *et al.* 2006. No extension of lifespan by ablation of germ line in *Drosophila*. *Proc. Biol. Sci.* **273:** 939–947.
29. Cargill, S.L., J.R. Carey, H.G. Müller & G. Anderson. 2003. Age of ovary determines remaining life expectancy in old ovariectomized mice. *Aging Cell* **2:** 185–190.
30. Parker, W.H., *et al.* 2009. Ovarian conservation at the time of hysterectomy and long-term health outcomes in the nurses' health study. *Obstet. Gynecol.* **113:** 1027–1037.
31. Traub, M.L. & N. Santoro. 2010. Reproductive aging and its consequences for general health. *Ann. N.Y. Acad. Sci.* **1204:** 179–187.
32. Pinkerton, J.V. & D.W. Stovall. 2010. Reproductive aging, menopause, and health outcomes. *Ann. N.Y. Acad. Sci.* **1204:** 169–178.
33. Treloar, A.E., R.E. Boynton, B.G. Behn & B.W. Brown. 1967. Variation of the human menstrual cycle through reproductive life. *Int. J. Fertil.* **12:** 77–126.
34. Vollman, R.F. 1977. The menstrual cycle. *Major Probl. Obstet. Gynecol.* **7:** 1–193.
35. Ferrell, R.J. & M.F. Sowers. 2010. Longitudinal, epidemiologic studies of female reproductive aging. *Ann. N.Y. Acad. Sci.* **1204:** 188–197.
36. Gorney, C. April 12, 2010. "Estrogen Dilemma." *New York Times*. p. MM52.
37. Goldman, N. 2010. New evidence rekindles the hormone therapy debate. *J. Fam. Plann. Reprod. Health Care* **36:** 61–64.

38. Whiteman, M.K. *et al.* 2008. Inpatient hysterectomy surveillance in the United States, 2000–2004. *Am. J. Obstet. Gynecol.* **198:** 34.e1–e7.
39. Friedman-Koss, D., C.J. Crespo, M.F. Bellantoni & R.E. Andersen. 2002. The relationship of race/ethnicity and social class to hormone replacement therapy: results from the Third National Health and Nutrition Examination Survey 1988–1994. *Menopause* **9:** 264–272.
40. Harlow, S.D. 2010. Health consequences of reproductive aging: a commentary. *Ann. N.Y. Acad. Sci.* **1204:** 163–168.
41. Ellison, P.T. 2010. Life historical perspectives on human reproductive aging. *Ann. N.Y. Acad. Sci.* **1204:** 11–20.
42. Penn, D.J. & K.R. Smith. 2007. Differential fitness costs of reproduction between the sexes. *Proc. Natl. Acad. Sci. USA* **104:** 553–558.
43. Austad, S.N. 2010. Animal models of reproductive aging: what can they tell us? *Ann. N.Y. Acad. Sci.* **1204:** 123–126.
44. Ottinger, M.A. 2010. Mechanisms of reproductive aging: conserved mechanisms and environmental factors. *Ann. N.Y. Acad. Sci.* **1204:** 73–81.
45. Wood, J.W. 1994. *Dynamics of Human Reproduction: Biology, Biometry, Demography*. Aldine de Gruyter. New York.
46. Kaplan, H. & M. Gurven. 2008. Top-down and bottom-up research in biodemography. *Demogr. Res.* **19:** 1587–1602.

Life historical perspectives on human reproductive aging

Peter T. Ellison

Department of Human Evolutionary Anthropology, Harvard University, Cambridge, Massachusetts

Address for correspondence: Peter T. Ellison, Department of Human Evolutionary Biology, Peabody Museum, Harvard University, 11 Divinity Avenue, Cambridge, Massachusetts 02138. pellison@fas.harvard.edu

A commentary is offered on the chapters that comprise the section on Theoretical Foundations, emphasizing novel contributions of each. Three additional points are then made. First, while the biology of reproductive aging may be common to all human populations, its actual course can be expected to vary between individuals and between populations depending on ecological conditions and developmental histories. Second, increasing fertility (such as that typical of humans compared with hominoid relatives and imputed ancestral species) decreases the opportunity and impact of contributions from ascendant relatives and increases the opportunity and impact of contributions from collateral and descendent relatives in promoting the fitness of a focal individual. Finally, an argument is made that the major change in human life history physiology in the Pleistocene has been the extension of adult lifespan, not any change in ovarian physiology or rate of reproductive senescence, and that extended lifespan created a selection pressure for the emergence of indirect reproductive effort among postreproductive individuals, not the reverse.

Keywords: evolution; life history; senescence; reproduction; menopause; ecology

All of the papers in this section are framed in terms of life history theory, the branch of evolutionary biology that seeks to understand the evolution of mortality and fertility patterns. The phenomenon of reproductive aging clearly falls within this domain, but the facts of human reproductive aging have been difficult to square with the dominant life history models. A long postreproductive life, in particular, has been viewed as an anomaly in need of explanation. The papers in this section range from theoretical treatments of general and reproductive senescence to more specific attempts to grapple with the evolution of human reproductive aging in particular. In this commentary I will make a few observations on some of the key contributions of these papers, and then offer a few related observations of my own.

Kirkwood and Shanley[1] provide an excellent overview of the evolutionary theory of senescence. One of their principal points is to underscore that senescence—decline in somatic function—is a consequence of the inevitable accumulation of mortality, rather than mortality being a result of the inevitable accumulation of senescence. Conventional wisdom would suggest that elderly organisms often "die of old age," that is, that deteriorating systems lead to increasing vulnerability and eventually to death. While this may be the direction of causation for a particular organism, it is not the direction of evolutionary causation. Organisms do not evolve longer or shorter lifespans because they senesce more rapidly or more slowly; rather, they evolve a pace of senescence that reflects how quickly they die. Because of necessarily diminishing probability of survivorship with increasing age the fitness value of late-age investment in somatic maintenance and repair necessarily declines. Organisms that devote energy and other limiting resources to preserving somatic integrity for an old age that rarely comes are clearly at a selective disadvantage.

Kirkwood and Shanley[1] suggest that reproductive aging should be viewed as a variant of senescence, a decline in the functional capacity of the germ cells, their supporting tissues, and/or other somatic components of the reproductive system. If reproductive tissues themselves require maintenance and repair, including the germ line itself, then the general evolutionary theory of senescence would apply. Once again the expectation, almost universally met, is that organisms in the wild will ordinarily die before reproductive senescence causes an extreme deterioration of fecundity. Maintaining fecundity for an old age that rarely comes is also energetically wasteful, and therefore deleterious for fitness.

The problem, then, is to understand the few cases in which the expectation is not met, particularly humans, where a long postreproductive life is common. Kirkwood and Shanley[1] make two observations that should be incorporated by any evolutionary theory of postreproductive life:

(1) There is little or no *value* to postreproductive life unless it is not truly "postreproductive" but rather includes a capacity for continued reproductive investment.
(2) There is little or no *cost* to postreproductive life unless postreproductives are somehow "parasitic," lowering the fitness of their offspring and relatives.

They point out that while there is considerable evidence for continued reproductive investment by postreproductive individuals, it is unlikely to have greater fitness consequences than continued direct reproduction would have had, and therefore can not provide a "causal" explanation for the emergence of postreproductive life.

Tuljapurkar and Steiner[2] take on the question of phenotypic heterogeneity in fitness traits among individuals. Most life history models, like most demographic and population genetics models, incorporate assumptions of "fixed heterogeneity," for example, a fixed distribution of fecundity among couples or of frailty among individuals. Tuljapurkar and Steiner note that, in fact, the phenotypes of individuals change over time in trajectories that incorporate large stochastic components. They turn their attention to the development of alternative models that take such "dynamic heterogeneity" into account.

Dynamic heterogeneity models must surely represent a promising direction, very much in keeping with contemporary efforts to better understand phenotypic variation as something over and above genetic variation. The model developed so far by Tuljapurkar and Steiner[2] might be termed a "fully stochastic" model in which fitness traits of interest are allowed to vary stochastically and independently at the level of the individual. They find that data from a population of English swans fits a dynamic heterogeneity model better than a fixed heterogeneity model. That, of course, does not mean that their model is the best possible. I hope and expect that, as this family of models matures, Tuljapurkar and Steiner and others will investigate other possible structures for dynamic heterogeneity. One important variant would be a model in which positive correlations between fitness traits are possible. This would reflect the phenomenon of "phenotypic correlation" in which traits that one might expect to be negatively correlated—survival and fertility, for example—are in fact positively correlated. This often occurs, for example, when both traits are affected by common conditions of energy abundance or scarcity. Other sources of phenotypic variation, including epigenetic developmental plasticity, might also be incorporated into future versions of the dynamic heterogeneity models.

Bribiescas[3] raises the issue of phenotypic plasticity in his paper on human male reproductive aging, noting that in some ways the reproductive system of the aging male appears to display a diminishing range of phenotypic responsiveness. Circadian variation appears to declines with age, for example, as does the anabolic response to exercise. Bribiescas also makes the important point, also noted in passing by Kirkwood and Shanley[1] and reiterated by Kaplan *et al.*[4] that human males do, in fact, display significant reproductive aging with minimal fertility in the sixth decade and beyond. Although male germ cells continue to be produced late in life, declining function of Sertoli cells, prostate and accessory glands, declining potency and libido all combine to dramatically curtail the fecundity of most males. Rather than thinking of males and females as being on very different trajectories of reproductive aging, their trajectories are in fact quite parallel, as predicted by Kirkwood's disposable soma theory[5] extended to reproductive tissues. The key sex difference is not in the fact of reproductive aging, but in the strategy of gamete production in the first place, contrasting a male strategy of persistent germ cell mitosis (which I term "iterogametogenesis" in parallel with the term iteroparous referring to repeated parity) with a female pattern of limited germ cell mitosis (which I correspondingly term "semelgametogenesis") resulting in a finite, and eventually depletable, oocyte supply. Virtually all mammals display this sex difference in gamete production, with the result that virtually any female mammal, if kept alive long enough, will undergo true menopause.

Hawkes and Smith[6] underscore the fact that a limited oocyte supply together with a constant rate of atresia will necessarily result in menopause in any long-lived female mammal. They go further by reviewing evidence to indicate that the rate of atresia in chimpanzees is virtually identical to that in humans and the predicted average age of oocyte depletion virtually the same. Emery Thompson et al.[7] report that late-age fecundity and late-age survivorship may be correlated in chimpanzees, based on a relatively small sample of wild animals, and from this observation speculate that reproductive aging may be quantitatively different in chimpanzees compared with humans. But Hawkes and Smith[6] point to similar observations among Utah Mormons. Indeed, as noted earlier, the observation of phenotypic correlation between survivorship and fertility at the individual level should not surprise us, whether in chimps or humans. Hawkes and Smith[6] in contrast argue convincingly that the fixed oocyte supply and rate of atresia observed in humans should be considered an ancestral trait, not a derived one. Rather the extended adult life expectancy of humans is the derived trait. If anything about human menopause needs an evolutionary explanation it is not its appearance, but rather its persistence under conditions of a dramatically extended adult life expectancy.

Kaplan et al.[4] directly address the issue of postreproductive life and the persistence of menopause in humans, approaching it from the perspective of a life-course energetic model wherein the energetic costs of fertility must be supported by caloric productivity in excess of consumption at the family level. Kaplan et al. have been developing models of this type for some time with considerable success in both empirical and theoretical terms.[8] They use demographic and behavioral data from a number of well-studied forager populations to inform their models. One general observation supported by these models is that young individuals consume more than they produce while older individuals produce the surplus necessary to support the young. This surplus production is enough to maintain a relatively high rate of fertility for humans. Here they use these models to argue that, compared with the current reality of female menopause around age 50, any extension of female fecundity into the sixth decade would increase total caloric demand by increasing the number of offspring dependent on old parents while diminishing productivity to the point of pushing families into net caloric deficit, an unsupportable scenario. They also note that male productivity is sustained at much higher levels than female productivity late in life even as male fertility declines to insignificance. The interpretation they advance is that the diversion of energy flows from late age individuals to current offspring and grandoffspring is favored by natural selection whereas the extension of female fecundity is not. They also argue that the diversion of male productivity in this way is quantitatively much more important that the diversion of female productivity, though the principle holds for both sexes.

In terms of Kirkwood and Shanley's[1] two observations, Kaplan et al.[4] find positive benefit in the continued reproductive investment of postreproductive individuals, but primarily in postreproductive males rather than females. Thus they would attribute positive selective pressure for postreproductive life primarily to a "grandfather effect." They also find negative fitness consequences of extended female fecundity, both in terms of increased numbers of dependent children and decreases in late-age female productivity.

Phenotypic variation in patterns of reproductive aging

The papers in this section make a very coherent set with a number of overlapping themes and specific arguments that dovetail nicely. I would like to make a few additional points that I hope complement those already made.

My first observation pertains to the notion of phenotypic variation in reproductive aging. One of the most important contributions of reproductive ecology, and physiological ecology in general, to our understanding of human biology has been to emphasize the broad range of observable phenotypic variation associated with different environmental and developmental conditions. Perhaps the best example of this is variation in female ovarian function associated with variation in energy availability. Far from being a homeostatically canalized trait, human ovarian function, characterized by hormonal profiles across the menstrual cycle, is facultatively variable within individuals. A continuum of ovarian function has been described

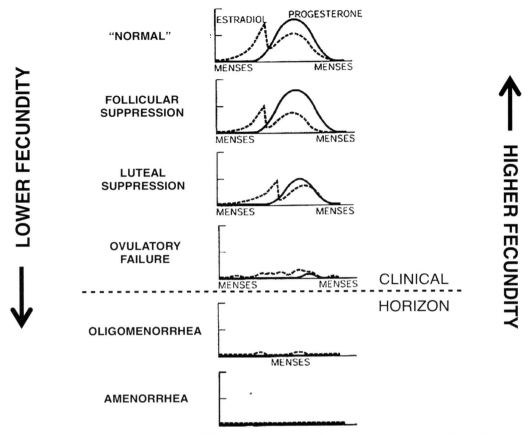

Figure 1. The continuum of ovarian function.[9,14] Ovarian function, as characterized by ovarian steroid profiles across the menstrual cycle, varies in a progressive fashion through grades of follicular suppression, luteal suppression, ovulatory failure, oligomenorrhea (infrequent menstruation), and amenorrhea (absent menstruation). These grades represent different levels of fecundity. Individual women move across this trajectory with age, energetic status, and breastfeeding status. Ordinarily the clinical horizon for recognizing suppressed ovarian function is at the level of disrupted menstrual regularity, but fecundity may vary considerably before this clinical horizon is reached.

ranging from an idealized, "textbook" menstrual cycle through stages described as follicular suppression, luteal suppression, ovulatory failure, oligomenorrhea, and amenorrhea[9] (Fig. 1). This continuum can be rendered quantitatively by measures of ovarian steroid hormones, particularly estradiol and progesterone, made across the entire cycle. Variation along the continuum represents variation in fecundity, or the probability of successful conception given exposure to intercourse, a probability that drops to zero if ovulation does not occur, but which is also quantitatively lower when hormonal profiles are suppressed.[10–12]

Individual variation along the continuum of ovarian function is highly correlated with variation in metabolic energy availability. Even moderate decreases in energy balance or increases in energy expenditure are regularly observed to result in decreases in ovarian function.[13–15] This variation has been documented under a wide variety of conditions in a wide variety of populations.[16–20] Within any given population individual women probably form a distribution along the continuum of ovarian function and the distribution as a whole may shift in response to prevailing conditions that affect most people (Fig. 2). For example, seasonal changes in energy balance associated with seasonal changes in food availability in subsistence populations have been linked to seasonal variation in the frequency of ovulation and seasonality in conceptions and births (Fig. 3).[21,22] Developmental effects have also been documented.[23] Smaller size at birth has been

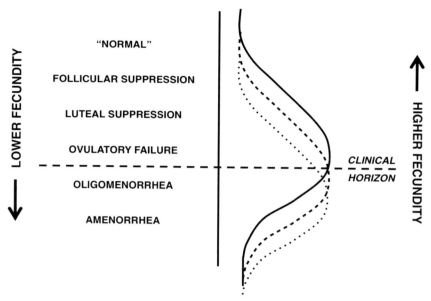

Figure 2. Population variation along the continuum of ovarian function.[14] The distribution of women along the continuum of ovarian function can vary between populations or within a population over time. Changes in the frequency of women at any given grade may indicate shifts in the overall population distribution that affect population level fecundity.

association with lower levels of ovarian function in adulthood in Polish women.[24,25] Bangladeshi migrants to London have higher levels of ovarian function as adults if they migrate as children compared with those who migrate as adolescents or adults.[26]

As Bribiescas notes,[3] variation in male testicular function appears to be less acute in response to short-term, moderate variations in energy availability. Variation between populations in average testosterone levels, however, does appear to reflect shifting distributions of male testicular activity in response to chronic energetic conditions.[27,28] In both females and males variation in gonadal function associated with energetics can be interpreted as facultative modulation of reproductive effort.[29–31] In females, variation in ovarian function modulates the probability of conception and thus the rate of reproduction and its high metabolic demands. In males, variation in testicular function modulates somatic investment in sexually dimorphic muscle mass as well as behavioral investment in mating effort.

In addition to variation correlated with energetics, both sexes show consistent patterns of variation in gonadal function associated with age.[27,32–34] In females, ovarian function follows a parabolic trajectory with age, peaking in the early to mid 20s and declining noticeably by the mid to late 30s. In males, testosterone levels peak in the early 20s and decline steadily from there.

Importantly, however, the trajectory of decline in gonadal function in both sexes depends on the height of the peak attained in early adulthood. By the sixth decade ovarian function ceases in women of all populations, and testosterone levels in men of all populations become similarly low. But populations differ in the average rate of decline. In males, this appears to have significant consequences for changes in body composition.[35] In populations where testosterone levels decline substantially with age the tendency toward increasing adiposity is greater. The consequences of differential rates of decline in ovarian steroids in women have not been explored but could be equally important. Population variation in the experience of menopausal symptoms, for example, might well be influenced by the quantitative nature of changes in circulating ovarian steroids. Women in westernized, developed societies may experience these symptoms more acutely in part because of an experience of chronically higher exposure during their reproductive years.

It is also possible that changes in bone mineral metabolism will show variation associated with the

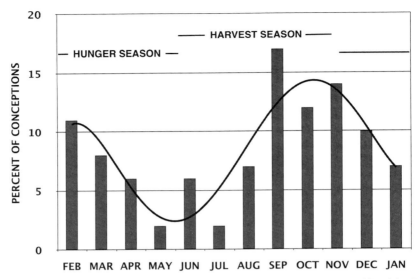

Figure 3. Seasonal variation in conceptions among the Lese of the Ituri Forest, Democratic Republic of the Congo.[21] Variation in conception frequency is associated with changes in energetic status among Lese women driven by seasonal changes in food availability. These changes in energetic status are in turn associated with shifts in the population distribution along the continuum of ovarian function.

rate of decline in circulating steroid levels. High exposure to ovarian steroids during the reproductive years may lead to lower receptor densities or other[36] changes in tissue sensitivity, leaving individuals at greater risk of the consequences of steroid withdrawal.

The take-home message is that although the biology of reproductive aging may be common to all human populations, the reality of reproductive aging may differ between individuals and populations depending on the ecological conditions under which they live. Phenotypic heterogeneity in gonadal function is associated with ecological conditions for adaptive reasons, and because environments differ, so will patterns of reproductive aging.

Energy flows and human demography

Most contemporary models of human evolution incorporate the observation that human reproductive rates are high and birth intervals short compared with other hominids. The energetic burden of this reproductive rate is higher than an individual female can sustain based on her own metabolic productivity. Thus most evolutionary models consider energetic inputs in addition to the mother's to be crucial to her reproductive success and the emergence of modern human life history patterns. Different scholars have championed different candidates for the source of this additional energetic input. Lovejoy[37] hypothesized the importance of contributions from monogamously pair-bonded males. Hawkes and colleagues[38–40] have emphasized contributions from maternal grandmothers. Hrdy[41] has included contributions from female peers, especially female relatives. Kramer[42] has emphasized the contributions of children in underwriting their own costs in the family. In their chapter of this volume, Kaplan et al.[4] emphasize the contributions of males, highlighting those of postreproductive males in particular. Hill and Hurtado[43] argue that energetic contributions from outside a mated pair are necessary to support human reproductive rates while also arguing that grandmothers can not quantitatively account for all of this input. Reiches et al.[44] propose that humans are characterized by a broad network of shared productive activities that essentially result in a *pooled energy budget* that all members of a cooperating productive/reproductive unit share in. At different stages of life all individuals may contribute energy surpluses to this pooled budget, either through their own efforts or through liberating more productive members of the group from less productive tasks. Indirect reproductive effort thus becomes an energy allocation category available to juveniles and adults as well as postreproductive individuals, with consequences that resolve many of the anomalies of human life history patterns.

Discriminating among these competing hypotheses is not easy. Demographic models and assumptions are often invoked and challenged by those engaged in debate over these issues. For example, Hill and Hurtado[45,46] argue that the probability of living, co-resident grandmothers is too low for their contributions to have a major effect, while Hrdy[36,41] argues against assumptions of patrilocality among proto-humans. The highly specific models of Kaplan *et al.*[4] and Hill and Hurtado[43] use observed fertility and mortality rates among specific extant foragers to model energy consumption and production across the lifespan. The gains in precision from this style of modeling have costs in making the resulting models less generalizable.[47]

An additional demographic argument can be made, relevant to this debate, that does not depend on any specific assumptions about levels of mortality or fertility or patterns of residence, based simply on the probabilities of living relatives of various kinds and the way those probabilities change with changing conditions of mortality and fertility. The formal mathematics of these probabilities has been developed by Keyfitz[48] but the key points are intuitive. (See also Howell's[49] use of these probabilities in analyzing kinship networks among the Dobe !Kung.) One can think of relatives in three categories: ascendant lineal relatives (parents, grandparents, etc.), descendant lineal relatives (offspring, grandoffspring, etc.), and collateral relatives (uncles, aunts, brothers, sisters, cousins, nephews, nieces, etc.). One can then ask how, for a proband of a given age, the probability of having a living relative of a given type varies with changes in mortality rates and fertility rates.

The most important difference is this regard is between ascendant relatives on the one hand and collateral and descendant relatives on the other. The probability of living ascendant relatives depends only on mortality rates and is independent of fertility rates. Any individual has at most two living parents, four living grandparents, etc., and the expected number alive depends only on adult mortality rates. The number of current mates is usually limited in the same way (given the predominant pattern of serial monogamy among humans). The numbers of collateral and descendant relatives are not limited, however, and are sensitive to both fertility and mortality rates, particularly pre-adult mortality rates.

This simple demographic reality has consequences for any model of supplemental energetic inputs to human reproductive success. As fertility rates increase and mortality rates (particularly pre-adult mortality rates) decrease from chimpanzee-like levels to modern human levels, the numbers of expected collateral and descendant relatives increase markedly. The number of expected living ascendant relatives may increase as well, but can never exceed two parents, four grandparents, etc. On the other hand, the number of descendant relatives who are potential recipients of parental and grandparental inputs increases rapidly, diluting the potential contributions from ascendant relatives to any given proband. One can conclude, then, that as fertility levels increased in the human lineage, energetic contributions from collateral and descendant relatives likely became increasingly important for reproductive success and the contributions of ascendant relatives of diminishing importance.

Climbing Mount Postreproductive life

I believe Hawkes and Smith[6] are correct in arguing that the ovarian physiology of female reproductive aging is an ancestral, and not a derived, trait in humans. It is not the timing of female reproductive aging that has changed since the last common ancestor of chimpanzees and humans (the "LCA"), but rather the rate of adult mortality. The evolutionary cause of our extended lifespan is another puzzle that would lead beyond the focus of this chapter, but it is likely that it did not occur until millions of years after the split from the chimpanzee lineage. It is useful, however, to contemplate the situation that arose when lifespan first began to extend beyond the end of reproduction. What selection pressures can we infer as a consequence when significant numbers of postreproductive individuals begin to appear?

I can think of three possibilities:

(1) To the extent that postreproductive individuals are parasitic on younger relatives, decreasing their fitness, there would be selection to increase postreproductive mortality to remove them.
(2) There would be selection to extend reproductive life.

(3) There would be selection on postreproductive individuals to make a net positive contribution to the fitness of their younger relatives, over and above any costs that their presence imposes.

Often we are tempted, when confronted with alternatives like these to try to calculate the relative fitnesses of each option in some absolute fashion, like comparing the heights of different adaptive peaks on a Sewall Wright-style fitness landscape. That is, however, the wrong way to approach the question, as Wright himself argued. Natural selection does not act to move a population toward the highest peak; it moves it in the direction of the steepest immediate path of ascent, a path that may well lead to a peak that is not the highest in the neighborhood. So in this case we should not ask ourselves, "Which is highest, Mount Parasitic Elimination, Mount Extended Fecundity, or Mount Indirect Reproductive Effort?" Rather we should ask which provides the steepest initial path of ascent.

The slope of the path up Mount Parasitic Elimination depends on how serious the parasitism is, and can only result in restoring fitness to what it was before the parasitic elders appeared. The path up Mount Extended Fecundity may be very shallow initially, or may even lead through an impassable valley. For females, it would involve changing either the initial oocyte supply or the rate of follicular atresia. Because the rate of atresia is exponential, increasing follicular supply is very ineffective in producing changes in the ultimate age at follicular exhaustion. Doubling the follicular supply, the equivalent of two additional ovaries, would only delay the age of menopause by 3 or 4 years at most. Changing the rate of atresia would be more effective, but this trait seems to be highly conserved.

The path up Mount Indirect Reproductive Effort seems likely to be the steepest of the three, assuming that by the time extended lifespan appears humans are already living in groups with pooled energy budgets represented by cooperative foraging and food sharing. Selection would favor those postreproductive elders who engaged in indirect reproductive effort by making positive net contributions to their relatives' fitness. By this logic, extended lifespan creates the selective pressure for grandparental contributions to their offspring, not the other way around, as is often argued.

Two additional observations are worth making about this scenario. While Mount Extended Fecundity may not provide a viable path for females, it may be much more tenable for males given their very different gametogenic physiology. Some mixed strategy of direct and indirect reproductive effort may provide the steepest path for males. And the path up Mount Indirect Reproductive Effort may be made shallower by the presence of freeloaders, postreproductive individuals who do not contribute much if anything to the pooled energy budget but whose inclusive fitness may benefit from the contributions of other postreproductive individuals.

In summary, armchair reasoning can only advance our understanding so far. In thinking about the evolution of patterns of human reproductive aging, however, it does help to assemble all the contributions of papers like the ones in this section and to consider their combined consequences. When we do, I think we come to appreciate the three observations made here: (1) phenotypic variation exists in patterns of reproductive aging that is expressed at the level of populations and that may be associated with chronic conditions of energy availability; (2) humans share in a pooled energy budget that helps to support an accelerated rate of reproduction and that increases the potential contributions of collateral and descendant relatives over ascendant relatives as it becomes more successful; and (3) the major change in human life history physiology since the LCA has been the extension of the adult lifespan, not any change in ovarian physiology, with the consequence that extended lifespan has produced selective pressure for the emergence of indirect reproductive effort among postreproductive individuals, not the reverse.

Conflicts of interest

The author declares no conflicts of interest.

References

1. Kirkwood, T.B.L. & D.P. Shanley. 2010. The connections between general and reproductive senescence and the evolutionary basis of menopause. *Ann. N.Y. Acad. Sci.* **1204:** 21–29.
2. Tuljapurkar, S. & U.K. Steiner. 2010. Dynamic heterogeneity and life histories. *Ann. N.Y. Acad. Sci.* **1204:** 65–72.
3. Bribiescas, R.G. 2010. An evolutionary and life history perspective on human male reproductive senescence. *Ann. N.Y. Acad. Sci.* **1204:** 54–64.

4. Kaplan, H. *et al.* 2010. Learning, menopause, and the human adaptive complex. *Ann. N.Y. Acad. Sci.* **1204:** 30–42.
5. Kirkwood, T.B.L. 1977. Evolution of ageing. *Nature* **270:** 301–304.
6. Hawkes, K. & K.R. Smith. 2010. Do women stop early? Similarities in fertility decline in humans and chimpanzees. *Ann. N.Y. Acad. Sci.* **1204:** 43–53.
7. Emery Thompson, M. *et al.* 2007. Aging and fertility patterns in wild chimpanzees provide new insights into the evolution of menopause. *Curr. Biol.* **17:** 2150–2156.
8. Kaplan, H. & A. Robson. 2002. The co-evolution of intelligence and longevity and the emergence of humans. *PNAS* **99:** 10221–10226.
9. Ellison, P.T. 1990. Human ovarian function and reproductive ecology: new hypotheses. *Am. Anthropol.* **92:** 933–952.
10. Lipson, S.F. & P.T. Ellison. 1996. Comparison of salivary steroid profiles in naturally occurring conception and non-conception cycles. *Hum. Reprod.* **11:** 2090–2096.
11. Venners, S.A. *et al.* 2006. Urinary estrogen and progesterone metabolite concentrations in menstrual cycles of fertile women with non-conception, early pregnancy loss or clinical pregnancy. *Hum. Reprod.* **21:** 2272–2280.
12. Baird, D.D. *et al.* 1997. Preimplantation hormonal differences between the conception and non-conception menstrual cycles of 32 normal women. *Hum. Reprod.* **12:** 2607–2613.
13. Ellison, P.T. 1994. Salivary steroids and natural variation in human ovarian function. *Ann. N.Y. Acad. Sci.* **709:** 287–298.
14. Ellison, P.T. 2008. Energetics, reproductive ecology, and human evolution. *Paleoanthropology* **2008:** 172–200.
15. Jasienska, G. 2003. Energy metabolism and the evolution of reproductive suppression in the human female. *Acta Biotheor.* **51:** 1–18.
16. Jasienska, G. & P.T. Ellison. 1998. Physical work causes suppression of ovarian function in women. *Proc. Biol. Sci.* **265:** 1847–1851.
17. Ellison, P.T. *et al.* 1993. Population variation in ovarian function. *Lancet* **342:** 433–434.
18. Ellison, P.T., N.R. Peacock & C. Lager. 1986. Salivary progesterone and luteal function in two low-fertility populations of Northeast Zaire. *Hum. Biol.* **58:** 473–483.
19. Panter-Brick, C., D.S. Lotstein & P.T. Ellison. 1993. Seasonality of reproductive function and weight loss in rural Nepali women. *Hum. Reprod.* **8:** 684–690.
20. Vitzthum, V.J. *et al.* 2002. Salivary progesterone levels and rate of ovulation are significantly lower in poorer than in better-off urban-dwelling Bolivian women. *Hum. Reprod.* **17:** 1906–1913.
21. Bailey, R.C. *et al.* 1992. The ecology of birth seasonality among agriculturalists in central Africa. *J. Biosoc. Sci.* **24:** 393–412.
22. Ellison, P.T., N.R. Peacock & C. Lager. 1989. Ecology and ovarian function among Lese women of the Ituri Forest, Zaire. *Am. J. Phys. Anthropol.* **78:** 519–526.
23. Ellison, P.T. 1996. Age and developmental effects on human ovarian function. In *Variability in Human Fertility.* L. Rosetta & C.G.N. Mascie-Taylor, Eds.: 69–90. Cambridge University Press. Cambridge, UK.
24. Jasienska, G., I. Thune & P.T. Ellison. 2006. Fatness at birth predicts adult susceptibility to ovarian suppression: an empirical test of the Predictive Adaptive Response hypothesis. *Proc. Natl. Acad. Sci. USA* **103:** 12759–12762.
25. Jasienska, G. *et al.* 2006. High ponderal index at birth predicts high estradiol levels in adult women. *Am. J. Hum. Biol.* **18:** 133–140.
26. Núñez-de la Mora, A. *et al.* 2007. Childhood conditions influence adult progesterone levels. *PLoS Med.* **4:** e167.
27. Ellison, P.T. *et al.* 2002. Population variation in age-related decline in male salivary testosterone. *Hum. Reprod.* **17:** 3251–3253.
28. Bribiescas, R.G. 1996. Salivary testosterone levels among Aché hunter/gatherer men and a functional interpretation of population variation in testosterone among adult males. *Hum. Nat.* **7:** 163–188.
29. Ellison, P.T. 2001. *On Fertile Ground.* Harvard University Press. Cambridge, MA.
30. Ellison, P.T. 2003. Energetics and reproductive effort. *Am. J. Hum. Biol.* **15:** 342–351.
31. Bribiescas, R.G. 2006. *Men: Evolutionary and Life History.* Harvard University Press. Cambridge, MA.
32. Ellison, P.T. 1994. Advances in human reproductive ecology. *Annu. Rev. Anthropol.* **23:** 255–275.
33. Ellison, P.T. 1996. Developmental influences on adult ovarian function. *Am. J. Hum. Biol.* **8:** 725–734.
34. Lipson, S.F. & P.T. Ellison. 1992. Normative study of age variation in salivary progesterone profiles. *J. Biosoc. Sci.* **24:** 233–244.
35. Campbell, B.C., P.B. Gray & P.T. Ellison. 2006. Age-related patterns of body composition and salivary testosterone among Ariaal men of Northern Kenya. *Aging Clin. Exp. Res.* **18:** 470–476.
36. Hrdy, S.B. 1999. *Mother Nature: A History of Mothers, Infants, and Natural Selection.* Pantheon. New York.
37. Lovejoy, C.O. 1981. The origin of man. *Science* **211:** 341–350.
38. Hawkes, K. 2003. Grandmothers and the evolution of human longevity. *Am. J. Hum. Biol.* **15:** 380–400.
39. Hawkes, K. 2004. Human longevity: the grandmother effect. *Nature* **428:** 128–129.
40. Hawkes, K. *et al.* 1998. Grandmothering, menopause, and the evolution of human life histories. *Proc. Natl. Acad. Sci. USA* **95:** 1336–1339.
41. Hrdy, S.B. 2009. *Mothers and Others: The Evolutionary Origins of Mutual Understanding.* Harvard University Press. Cambridge, MA.
42. Kramer, K. 2005. *Maya Children: Helpers at the Farm.* Harvard University Press. Cambridge, MA.
43. Hill, K. & A.M. Hurtado. 2009. Cooperative breeding in South American hunter-gatherers. *Proc. Biol. Sci.* **276:** 3863–3870.
44. Reiches, M.W. *et al.* 2009. Pooled energy budget and human life history. *Am. J. Hum. Biol.* **21:** 421–429.

45. Hill, K. & A.M. Hurtado. 1996. *Aché Life History: The Ecology and Demography of a Foraging People*. Aldine De Gruyter. Hawthorne, NY.
46. Hill, K. & A.M. Hurtado. 1991. The evolution of reproductive senescence and menopause in human females. *Hum. Nat.* **2:** 315–350.
47. Levins, R. 1968. *Evolution in Changing Environments*. Princeton University Press. Princeton, NJ.
48. Keyfitz, N. 1977. *Applied Mathematical Demography*. John Wiley & Sons. New York.
49. Howell, N. 1979. *Demography of the Dobe !Kung*. Academic Press. New York.

ANNALS OF THE NEW YORK ACADEMY OF SCIENCES
Issue: *Reproductive Aging*

The connections between general and reproductive senescence and the evolutionary basis of menopause

Thomas B. L. Kirkwood and Daryl P. Shanley

Institute for Ageing and Health, Newcastle University, Campus for Ageing and Vitality, Newcastle upon Tyne, United Kingdom

Address for correspondence: Thomas B. L. Kirkwood, Institute for Ageing and Health, Newcastle University, Campus for Ageing and Vitality, Newcastle upon Tyne NE4 5PL, United Kingdom. tom.kirkwood@ncl.ac.uk

We consider the relationship between the factors responsible for the general biology of aging and those that specifically influence the aging of the reproductive system. To understand this relationship it is necessary to be clear about the evolutionary forces acting on both sets of factors. Only in this way can the correct causal connections be established. Of particular significance is the existence in some species of a distinct period of postreproductive life. This is most striking in the case of the human menopause, for which a particular combination of biological and sociobiological factors appear to be responsible.

Keywords: aging; evolution; menopause; life history; reproduction; senescence

Introduction

The aging process is commonly defined as a progressive decline within adult organisms of the functional capacity of most, if not all, organ systems, resulting in an age-specific increase in mortality rate and a decline in fertility.[1] An increase in mortality and decline in fertility are detrimental to Darwinian fitness. Thus, aging should be selected against and its widespread occurrence has long been regarded as a key puzzle in evolutionary theory. The fact that longevity is clearly under genetic control raises the interest in understanding the evolutionary roots of aging.[2] A clear contrast between this control of organismal *survival* and an aging program, which involves regulation of mechanisms for organismal *death*, is essential for this understanding. Linked with this is the need to understand the connections between the longevity of reproductive viability and the length of life itself.

At a superficial level, the forces governing general and reproductive senescence are commonly seen to be causally interconnected according to a rather loose logic that has characterized some of the popular thinking around these questions. This has caused enduring confusion. The answer to why aging occurs is often answered by suggesting that once an organism runs out of reproductive viability, or even when it has fulfilled some necessary "quota" of reproduction, it is surplus to requirement and may die. This extends even to the suggestion that postreproductive survival is a drain on a species' resources and therefore that there is likely to be active programming of the aging process in order to get rid of superfluous consumers. These ideas are founded, for the most part, on an inadequate understanding of how natural selection operates and they deserve mention here only on account of their perennial resurfacing, like weeds in a garden. There are, however, deep interconnections between the biology of general and reproductive senescence that merit careful attention. In addressing these intriguing questions it is of paramount importance to organize the logic into the correct causal sequence. In this review, we summarize first why aging is thought to occur at all. We then consider the implications of the evolutionary theory of aging for the mechanisms affecting reproductive senescence. Finally, we address the significance within this framework of postreproductive survival and in particular of the menopause.

Why general senescence occurs

There is general acceptance today that underlying the evolution of aging is the inescapability of

doi: 10.1111/j.1749-6632.2010.05520.x

death from extrinsic hazards, such as predators, challenging environmental conditions, and infectious disease.[3,4] As a result, the cumulative probability of surviving to older and older ages grows smaller and smaller.[5] As Medawar[5] pointed out, and others have elaborated more formally,[6,7] it follows from this empirical observation that, if traits affecting evolutionary fitness are expressed in an age-specific manner, the power of natural selection to affect the evolutionary fate of these traits will gradually diminish with increasing age.

From this recognition of the waning power of natural selection, *even in the absence of original senescence*, it follows that the door is opened to the evolution of specific factors that might cause senescence to arise. There are two "classical" formulations of how this might occur. In the first, the "mutation accumulation" theory, deleterious alleles that affect survival or reproduction only very late in life, when selection is weak, could accumulate in the genome over evolutionary time by mutation pressure checked only weakly by mutation-selection balance.[5] In the second, alleles with "antagonistically pleiotropic" effects, such that they enhance fitness early in life when selection is strong, but depress it late in life when selection is weak, can be favored by natural selection. The reasoning is that the early beneficial effects will, as a direct result of the differential weighting caused by the action of extrinsic mortality, count for more than the later deleterious effects, even when the early effects are smaller in absolute magnitude than the later effects.[8]

These classical theories for the evolution of senescence are essentially "mechanism-free," in the sense that they postulate only the age-specific character of the hypothetical alleles. This is both a strength of the theories, in that they are neutral with respect to the specific nature of the molecular and cellular mechanisms, but it is also a limitation, in that their predictions are of only a general kind. This limitation was overcome in a subsequent theory that was based on recognizing the physiological costs of an organism's long-term maintenance. The "disposable soma" theory[9,10] recognized that such maintenance is costly and from this derived an integrated hypothesis explaining both why and how aging is caused. Taking account of the attrition in survival caused by extrinsic mortality, somatic maintenance needs only to be good enough to keep the organism in sound physiological condition for as long as it has a reasonable chance of remaining alive in the wild environment. For example, since more than 90% of wild mice die in their first year,[11] a mouse that invests in mechanisms for survival beyond this age has only a 10% chance of receiving any benefit—clearly not a worthwhile return. Nearly all of the mechanisms required for somatic maintenance and repair (DNA repair, antioxidant systems, protein turnover, etc.) require metabolic resources. Resources are scarce, and organisms must tradeoff investment in maintenance and repair with other physiological demands, such as reproduction[12] and immunity.[13]

An abundance of empirical evidence has accumulated in support of these theories. A prediction of classical theories is that interference with the schedule of either reproduction or mortality may impact upon the actions of natural selection on the determinants of longevity. This has been verified through a series of artificial selection experiments, most notably in the fruit fly *Drosophila melanogaster* where selection for late reproduction increased longevity[14–16] and imposing different mortality regimes resulted in the expected effects on longevity.[17] Additional support is provided by "natural experiments," such as the longer lifespans observed in island than mainland populations of opossums that had higher rates of mortality presumably as the result of greater predation pressure.[18] Furthermore, these selection experiments and a variety of comparative studies[19,20] have confirmed repeatedly that, in line with the predictions of the disposable soma theory, increased longevity is associated with an increased investment in the mechanisms underpinning somatic durability and maintenance.

The status of an "aging program"

Although the logical and empirical underpinnings of the evolutionary theory of senescence are extremely strong, there has continued to be a tendency to seek explanation of aging in terms of some kind of adaptive genetic program that specifically limits the individual's lifespan. This has led to recurring misunderstandings about the genetic basis of aging and longevity.[21] It is beyond doubt that genetic factors are important influences on the length of life. This is indicated by the interspecific differences in species' lifespans, the discovery of mutations affecting lifespan, and by the clear heritability of human longevity.[22,23] However there is an essential distinction between this well-established genetics of aging and longevity, which recognizes that genes influence the mechanisms that underpin longevity, and the genetics that is inferred for an aging program, which would involve genetically specified mechanisms that actually result in the destruction of living systems. This distinction was articulated as long ago as 1982:

> The issue that distinguishes programmed from nonprogrammed ageing is not *whether* the factors that determine longevity are specified within the genome, but rather, *how* this is arranged. An organism which undergoes programmed ageing is regarded as having a specific mechanism to limit its duration of life, whereas an organism which is not programmed in this way does not. In the latter type of organism, duration of life may be determined, for example, simply by the efficiency of somatic repair (p. 114).[24]

The attractions of the program concept are easily understood. First, aging is phylogenetically a very widely distributed trait and in species where senescence occurs, it affects every individual that lives long enough to experience its adverse impacts on fertility and vitality. To many, it therefore seems to make sense that aging exists "for a purpose." Second, there are, as already observed, clear genetic effects on longevity and this leads naturally to supposing that the relevant genes specify some kind of "aging clock." Third, in a postgenome era, when new evidence of genetic causality is being uncovered in many realms of biology, the default assumption that aging is *caused by* gene action preexists in the minds of most of those who come afresh to considering why aging occurs, although this argument is undercut by recent observations that regulation of gene expression deteriorates with age.[25] Finally, and despite the evidence that the details of the aging process are intrinsically variable from one individual to another, there is sufficient broad reproducibility about the manifestations of senescence that it naturally lends itself to the intuition that it has somehow to be programmed. It may reflect the fact that the idea of programmed aging is so apparently intuitive that few attempts have been made to develop a formal logic to support this idea. The commonest suggestions are that possession of a fixed limit to lifespan (i) is beneficial, or even necessary, to prevent the species from overcrowding its environment,[26] or (ii) promotes long-term evolutionary fitness by securing the necessary turnover of generations that allows novel adaptations to be selected.[27]

For either of these suggestions to work, it is a necessary prerequisite that intrinsic aging should make a sufficient contribution to natural mortality that the hypothesized selection process is feasible. If an individual dies before senescent effects are apparent, it makes no difference whether or not that individual is endowed with genes that program aging. Such a program can only be fashioned by selection acting to realize the hypothesized benefits of a program for aging *in those individuals who survive to an age when the program takes effect*. It is therefore a problem for program theories of aging that although some degree of senescence (age-related functional decline and increase in mortality rates) has been reported in many natural populations of species that show evident aging in a captive setting, relatively few individuals survive long enough to be affected by it.[28,29] The exception occurs in semelparous species, such as Pacific salmon, that have evolved a life history plan in which there is only a single bout of reproduction. In such species, death of the parent usually occurs rather quickly after reproduction. Indeed, an important source of misunderstanding of the evolutionary theory of aging has been to regard postreproductive death in semelparous species as an instance of programmed aging, when in fact its evolutionary explanation appears likely to be very different.[30]

How reproductive senescence relates to general senescence

As Weismann[31] recognized, there is often an important distinction in multicellular animals between the cells that constitute the reproductive lineage, or "germline," and those that make up the rest of the body, or "soma." It is an essential requirement of the germline that it can propagate itself indefinitely, or the branch of life that it represents would quickly die out. The essence of the disposable soma theory is the recognition that Weismann's soma/germline distinction has deep implications for investments in the long-term maintenance of the soma. These investments are predicted to be limited, leading to somatic senescence. In effect, the primary function of the soma is to support the germline in its all-important reproductive role.

The necessary immortality of the germline raises interesting questions about its relationship to reproductive senescence. Although the germline must be protected from the long-term accumulation of faults that would lead, over generations, to its eventual failure, it does not follow that in an individual organism reproductive function needs to be sustained indefinitely. The gonads, which include somatic as well as germ cells, are vulnerable to broadly the same kinds of molecular damage that affect other organs. There can be little adaptive value in securing a nonaging reproductive system while every other system of the body is falling apart, except where reproductive success is unaffected by somatic failure, which is generally unlikely. The important quality of the germline is simply that while reproductive viability is sustained, germ cells should as far as possible be free from molecular defects that might compromise the viability of offspring. Babies need to be born young, not old.

How the germline secures its immortality involves a combination of factors and is not completely understood. The overproduction of gametes provides a mechanism to select only the most viable cells and it is possible that such selection acts to screen out significant numbers of potentially defective germ cells from contributing to reproduction.[32] Selection also acts at all stages of pregnancy from implantation of the fertilized egg through to *in utero* and neonatal mortality, which further helps to reduce the threat of accumulating faults in the germline.[33] There also appear to be elevated levels of maintenance and repair in germ cells, as compared with somatic cells.[34] This is evidently the case with the enzyme telomerase, which acts to maintain telomeres in germ cells but which is commonly switched off or downregulated in somatic cells.[35] A specific prediction of the disposable soma theory was that for reasons of energy efficiency there should be a switching off of the mechanisms responsible for high fidelity maintenance "at or around the time of differentiation of somatic cells from the germ-line" (p. 303).[9] It is therefore striking that exactly such a process has been reported when mouse or human embryonic stems cells undergo early differentiation, which is accompanied by a general reduction in the levels of key cellular maintenance systems, such as DNA repair and antioxidant defenses.[34,36]

Although the germline maintains its potential for immortality throughout the fertile period, it is clear that the germ cell population does indeed undergo significant aging from a statistical point of view, even while reproductive viability is maintained. In the case of the human ovary the rate of follicular loss accelerates from around age 35,[37] and male fertility begins to decline, at a more gradual rate, from around age 40.[38] There is also an increase in the frequency of chromosomal abnormalities in newborn children as a function of maternal and, to a lesser extent, paternal age.[39] Nevertheless, healthy children born to older parents are not prematurely aged, although there is some suggestion that daughters' (but not sons') longevity is adversely affected by advanced paternal age.[40] The specific molecular mechanisms underlying age-related deterioration in the germ cell population need to be better understood. For example, oocytes from aged humans show a decline in the ability to segregate chromosomes synchronously,[41] and studies in mice are investigating the underlying reasons for this (M. Herbert *et al.*, unpublished observation).

In view of the centrality of reproduction within the organism's life history, it is natural that there should be intercommunication between the gonads and the rest of the body that may have important consequences for senescence. This is seen dramatically in the case of semelparous species, where the entire life cycle is geared toward maximizing success during the one and only bout of reproduction. The rapid deterioration of Pacific salmon after mating is a byproduct of a life history that has been geared by natural selection to stake everything on the success or failure of a single bout of reproduction. The first phase of a semelparous life history is devoted to growth and to acquiring the resources necessary for reproduction. As soon as the signal to reproduce is triggered, a massive effort is made to mobilize all available resources to maximize reproductive success, even if this leaves the adult so severely depleted or damaged that death ensues. Once a species has evolved down the pathway that results in semelparity (this is most likely to occur where ecological circumstances decree that the chance of surviving to breed again would in any case be small), there is no reason to hold back resources for postreproductive adult survival. Although instances may conceivably occur where the death of the adult directly benefits its young,[30] there is little evidence that semelparous organisms are *actively* destroyed once reproduction is complete. They are, in effect, extreme examples of the "disposable soma." It is striking that in Pacific salmon, removal of the gonads before reproduction can occur results in significant extension of lifespan.[42] In iteroparous species, castration has less dramatic effects on senescence although there is some evidence that male castration reduces or removes the longevity disadvantage experienced by males in humans and some domesticated species.[43] Elegant studies using germ cell ablation by direct or genetic methods have shown that in the nematode *Caenorhabditis elegans*, germ cells exert significant effects on somatic aging.[44,45] As with the studies in semelparous species, these experiments reveal not that the reproductive system actually programs aging per se but that the allocation of resources between reproduction and maintenance may be tuned to signals that take account of the organism's status with respect to its physiological and maturational status. A further interesting example of interplay between the reproductive and general somatic systems that well illustrates the connections between resource allocation and senescence is seen in the case of rodent dietary restriction. During periods of food shortage, mice (and to a slightly lesser extent rats) switch off fertility and appear to divert any resources thereby saved into increased somatic maintenance. There is some evidence that such a switch in resources is the result of a selective adaptation,[46] which delays general and reproductive senescence while deferring fertility until the environmental is more favorable.

Natural selection and postreproductive life

A period of postreproductive life is seen when either there is a specific shut-down of fertility well ahead of general biological senescence or when reproductive system-specific senescence leads to a decline in reproductive function faster than general senescence leads to mortality. Even if reproductive system-specific senescence runs, on the average, no faster than general senescence, variation in the relative timing of senescent effects in different organ systems will mean that postreproductive individuals may be found who are observed to be alive but no longer fertile. Since reproduction is physiologically demanding, it is unlikely that organisms will remain fertile up to death. These considerations mean that the biological significance of, and reasons for, postreproductive survival need to be reviewed with care (e.g., Reznick *et al.*).[47] This is particularly relevant for female postreproductive survival, since in many species the store of potential oocytes is fixed early in development and postreproductive life begins once this store has been exhausted. Artifactual postreproductive survival may occur when disease accelerates ovarian depletion, or, as seems likely to have happened in laboratory rodent strains, intensive breeding has selected for increased early fecundity leading to more rapid ovarian exhaustion than would occur in natural populations.[48]

If a significant period of postreproductive survival is seen in a majority of surviving individuals within a population, this raises intriguing challenges to explain why such a state should exist. Two general observations help to focus the analysis of such examples:

Observation 1. There is little or no advantage to be gained from an organism's survival after its reproduction is complete unless the survival of the adult contributes to the success of its offspring, in which case survival is not strictly postreproductive since parental care is an integral part of the package. "Offspring" may, in social organisms, be generalized to include genetic kin and their progeny as well as direct descendants.

Observation 2. There is little to be gained by causing the death of a postreproductive adult except where such survival adversely affects the success of its offspring, as defined in 1.

Both observations may be further qualified by the fact that any potential advantage of either postreproductive survival or death will be strongly modulated by the strength of selection acting at the relevant ages. This is particularly relevant in iteroparous species where the force of selection declines throughout reproductive life. In the case of semelparous life histories, the force of selection is maximal until reproduction commences. This is because even though many individuals will die before this can happen, all reproduction is still in the future until this point and therefore any differences between alternative genotypes will feel the full force of natural selection. Immediately after semelparous reproduction is completed, the force of natural selection reduces to zero, except where parental care is operative (although, as noted above, in such cases it can not yet be said that reproduction is actually completed). Thus, there is a spectrum of possibilities that is seen across the range of semelparous organisms. In some instances, death and reproduction are intimately linked, for example if the adult body is consumed as a food source by the young. In the mite *Adactylidium* the young hatch inside the body of the mother and eat their way out.[24] In other cases, the parent indulges in short-term parental care before dying, as in the female *Octopus hummelinckii* that propel water over their eggs to ventilate them.[24] Often, however, as in Pacific salmon, the semelparous adult simply dies although not necessarily at once, generally from side-effects of extreme reproductive effort, with no obvious benefit being generated either by death or temporary survival. Thus, the first point to note in considering the significance of postreproductive survival across the species range is that the underlying life history pattern, in particular whether it is iteroparous or semelparous, is an essential consideration.

Observation 2 above is relevant to considering whether cessation of reproduction should result in an abrupt increase in mortality. Such a prediction was considered by Hamilton[6] who referred to the possibility as a "wall of death." If reproduction ends, the effect on the force of selection is such that within the terms of the mutation-accumulation theory for evolution of senescence, there should be a steep increase in the number of late-acting deleterious mutations beyond this age. This formal concept is, however, questionable. First, despite significant effort to demonstrate that mutation accumulation contributes to senescence, the great majority of studies have proved negative.[49] Second, as was pointed out by Kirkwood,[9] the concept begs the physiological question of what might be the timing mechanism to control the action of late-acting mutations if we disallow, as previous considerations tell us we must, the notion of a program for aging.

The really interesting question concerning postreproductive life is whether the exceptions noted with respect to Observation 1 are sufficient to make an extended period of survival positively advantageous. This is of greatest interest in social animals where the concept of "inclusive fitness"[50]—namely, the idea that an individual can contribute to the success of its genes not only through its own reproduction but by aiding the reproduction of its kin—comes into its own. Studies in social animals, such as lions, baboons, and killer whales,[51–53] have demonstrated significant survival of postreproductive females in the natural environment. Interestingly, chimpanzees do not seem to have a significant postreproductive survival in the wild and this may also be true in captivity.[54,55] The clearest evidence of postreproductive life follows fertility loss in women (menopause), which occurs at a remarkably similar age—around 50 years—in all human populations,[56] and which is preceded by a period of 10–15 years of declining fertility. When compared with other species, the decline in human fertility and ultimately the menopause happens unusually early in the lifespan.[56,57] Its proximate cause (as in other mammalian females) is the exhaustion of ovarian oocytes, accompanied by degenerative changes in reproduction-associated elements of the neuroendocrine system.[58]

Two broad hypotheses have been advanced to explain menopause in terms of active selection for a period of postreproductive survival. These are founded on the extreme altriciality of human infants and the extensive opportunities for intergenerational cooperation within kin groups.[8,59–61] The altriciality of human offspring appears to be the result of a compromise driven by the evolution of an increasingly large brain in the hominid ancestral lineage and the pelvic constraint on the birth canal. On the one hand, the human neonatal brain size is near to the limit that is compatible with safe delivery, and even so presents considerable mortality risk to the mother in cases of birth complications. On the other hand, the newborn human infant still requires its brain to grow and develop for a considerable period before it is capable of any kind of independent existence, which renders it highly dependent on adult (usually maternal) attention for its survival. Given that maternal mortality increases with age and that maternal death will seriously compromise the survival of any existing dependent offspring, it appears to make sense to cease having more children when the risks outweigh the benefits. Nevertheless, *Homo sapiens* is unique in the extent to which kin assist in care and provisioning of young.[61,62] Thus, an alternative theory is that menopause enhances fitness by producing postreproductive grandmothers who can assist their adult offspring by sharing in the burden of provisioning and protecting their grandchildren. A further contribution to inclusive fitness may also be made within kin groups if postreproductive women contribute similar support to the survival and reproduction of other relatives.

At the core of any plausible evolutionary hypothesis must be, in addition to a verbal statement of the potential adaptive benefit, a quantitative demonstration that there is indeed an associated increase in fitness under natural fertility and mortality conditions representative of our evolutionary past. This validation is often lacking but without it, the hypothesis remains a matter of speculation. Even to demonstrate, for example, that postreproductive women result in a reduction in grandchild mortality *does not establish that menopause is adaptive unless it can be demonstrated that overall fitness is actually enhanced*. It is therefore highly significant that attempts mathematically to model the fitness benefits resulting from menopause in terms either of the maternal survival or grandmother hypotheses have shown that the magnitude of the contributions from individual sources might have to be unrealistically high to make the necessary difference.[59,63,64] Only when the effects of menopause on maternal mortality and the grandmother contribution were combined, was an increase in fitness observed.[64] Subsequent analysis combining life history modeling with data from a West African population highlighted the importance to fitness of the observed grandmaternal contribution in reducing of grandchild mortality but it also revealed that this contribution was only just sufficient to offset the fitness benefits that might otherwise accrue from continued reproduction of the grandmother herself.[65] Using a different approach, Lahdenperä et al.[66] analyzed multigeneration records from Finland and Canada to show that women with a prolonged reproductive lifespan had more grandchildren. There is, however, an important caveat in all such studies that individual variations in material circumstances and health, including exposure to infectious diseases, will tend to generate positive associations between longevity and reproductive success.

The demonstration that menopause can, in quantitative terms, result in enhanced fitness lends support to the idea that there may be something special about postreproductive life in humans. This needs further study. In particular, theory needs to take account of how individuals move between kin groups, since this has a bearing on the degree of relatedness. Cant and Johnstone[67] have examined the situation where intergroup transfer is chiefly via younger women joining the kin group of their male partners. In this situation new arrivals will have little biological incentive initially to contribute effort to the group's fitness but the incentive will increase over time as a result of interbreeding, since, as the female accumulates her own offspring, these will share genes with the group. Lee[68] has proposed that analyses should take account of intergenerational resource transfers in modeling the benefits of postreproductive life. There is less novelty here than at first there seems to be, since what Lee deals with via transfers has already been represented in state-dependent life history models as costs and benefits for the relevant individuals. Nevertheless, the idea of transfers is congruent with data gathered by anthropologists and this may have advantages. More problematic in the specific model developed by Lee[69] is the reliance on mutation accumulation as the process through which effects on senescence are assumed to occur, since there is poor support for such a mechanism.

Finally, an area where much greater attention needs to be focused is on the connections between mechanisms of general biological senescence and reproductive decline. In relation specifically to menopause, Pavard et al.[70] point out that the increased failure rate in reproduction resulting from senescence (stillbirths, birth defects, etc.) may result in an age-related decline in offspring quality that undermines the fitness contribution of later born children. Shanley et al.[65] have also noted, in line with the disposable soma theory, that the metabolic costs of the extra maintenance that would be required to support reproductive function for longer may be an additional factor contributing to the evolutionary advantages of menopause. It is likely to be through better understanding the mechanisms responsible for reproductive senescence, and the selection forces acting upon them, that further advances will be made.

Conflicts of interest

The authors declare no conflicts of interest.

References

1. Maynard Smith, J. 1962. Review lectures on senescence: I. The causes of aging. *Proc. R. Soc. Lond. B Biol. Sci.* **157:** 115–127.
2. Kuningas, M. *et al.* 2008. Genes encoding longevity: from model organisms to humans. *Aging Cell* **7:** 270–280.
3. Promislow, D.E.L. 1991. Senescence in natural populations of mammals: a comparative study. *Evolution* **45:** 1869–1887.

4. Ricklefs, R.E. & A. Scheuerlein. 2001. Comparison of aging-related mortality among birds and mammals. *Exp. Geront.* **36:** 845–857.
5. Medawar, P.B. 1952. *An Unsolved Problem of Biology*. H.K. Lewis. London.
6. Hamilton, W.D. 1966. The moulding of senescence by natural selection. *J. Theor. Biol.* **12:** 12–14.
7. Charlesworth, B. 1980. *Evolution in Age-Structured Populations*. Cambridge University Press. Cambridge.
8. Williams, G.C. 1957. Pleiotropy, natural selection, and the evolution of senescence. *Evolution* **11:** 398–411.
9. Kirkwood, T.B.L. 1977. Evolution of ageing. *Nature* **270:** 301–304.
10. Kirkwood, T.B.L. & R. Holliday. 1979. The evolution of aging and longevity. *Proc. R. Soc. Lond. B Biol. Sci.* **205:** 531–546.
11. Phelan, J.P. & S.N. Austad. 1989. Natural selection, dietary restriction and extended longevity. *Growth Dev. Aging* **53:** 4–6.
12. Speakman, J.R. 2008. The physiological costs of reproduction in small mammals. *Philos. Trans. R. Soc. Lond. B Biol. Sci.* **363:** 375–398.
13. Ricklefs, R.E. & M. Wikelski. 2002. The physiology/life-history nexus. *Trends Ecol. Evol.* **17:** 462–468.
14. Partidge, L. & K. Fowler. 1992. Direct and correlated responses to selection on age at reproduction in Drosophila melanogaster. *Evolution* **46:** 76–91.
15. Zwaan, B.J., R. Bijlsma & R.F. Hoekstra. 1995. Direct selection on life span in Drosophila melanogaster. *Evolution* **49:** 649–659.
16. Rose, M.R. 1984. Laboratory evolution of postponed senescence in Drosophila melanogaster. *Evolution* **38:** 1004–1010.
17. Stearns, S.C., M. Ackermann, M. Doebeli & M. Kaiser. 2000. Experimental evolution of aging, growth, and reproduction in fruitflies. *Evolution* **97:** 3309–3313.
18. Austad, S.N. 1993. Retarded senescence in an insular population of Virginia opossums. *J. Exp. Zool.* **229:** 695–708.
19. Kapahi, P., M.E. Boulton & T.B.L. Kirkwood. 1999. Positive correlation between mammalian life span and cellular resistance to stress. *Free Radic. Biol. Med.* **26:** 495–500.
20. Murakami S., A. Salmon & R.A. Miller. 2003. Multiplex stress resistance in cells from long-lived dwarf mice. *FASEB J.* **17:** 1565–1566.
21. Longo, V.D., J. Mitteldorf & V.P. Skulachev. 2005. Programmed and altruistic ageing. *Nat. Rev. Genet.* **6:** 866–872.
22. Finch, C.E. & R.E. Tanzi. 1997. Genetics of aging. *Science* **278:** 407–411.
23. Christensen, K., T.E. Johnson & J.W. Vaupel. 2006. The quest for genetic determinants of human longevity: challenges and insights. *Nat. Rev. Genet.* **7:** 436–448.
24. Kirkwood, T.B.L. & T. Cremer. 1982. Cytogerontology since 1881: a reappraisal of August Weismann and a review of modern progress. *Hum. Genet.* **60:** 101–121.
25. Bahar, R. et al. 2006. Increased cell-to-cell variation in gene expression in ageing mouse heart. *Nature* **441:** 1011–1014.
26. Wynne-Edwards, V.C. 1962. *Animal Dispersion in Relation to Social Behaviour*. Oliver & Boyd. Edinburgh.
27. Libertini, G. 1988. An adaptive theory of the increasing mortality with increasing chronological age in populations in the wild. *J. Theor. Biol.* **132:** 145–162.
28. Finch, C.E. 1990. *Longevity, Senescence and the Genome*. Chicago University Press. Chicago.
29. Brunet-Rossinni, A.K. & S.N. Austad. 2006. Senescence in wild populations of mammals and birds. In *Handbook of the Biology of Ageing*. E.J. Masoro & S.N. Austad, Eds.: 243–266. Academic Press. San Diego.
30. Kirkwood, T.B.L. 1985. Comparative and evolutionary aspects of longevity. In *Handbook of the Biology of Aging*. C.E. Finch & E.L. Schneider, Eds.: 27–44. Van Nostrand Reinhold. New York.
31. Weismann, A. 1889. *Essays Upon Heredity and Kindred Biological Problems*. Clarendon Press. Oxford.
32. Hartshorne, G.M. et al. 2009. Oogenesis and cell death in human prenatal ovaries: what are the criteria for oocyte selection? *Mol. Hum. Reprod.* **15:** 805–819.
33. Forbes, L.S. 1997. The evolutionary biology of spontaneous abortion in humans. *Trends Ecol. Evol.* **12:** 446–450.
34. Saretzki, G. et al. 2008. Downregulation of multiple stress defense mechanisms during differentiation of human embryonic stem cells. *Stem Cells* **26:** 455–464.
35. Woodring, E.W. et al. 1996. Telomerase activity in human germline and embryonic tissues and cells. *Dev. Genetics* **18:** 173–179.
36. Saretzki, G. et al. 2004. Stress defense in murine embryonic stem cells is superior to that of various differentiated murine cells. *Stem Cells* **22:** 962–971.
37. Faddy, M.J. et al. 1992. Accelerated disappearance of ovarian follicles in mid-life: implications for forecasting menopause. *Hum. Reprod.* **7:** 1342–1346.
38. Hassan, M.A.M. & S.R. Killick. 2003. Effect of male age on fertility: evidence for the decline in male fertility with increasing age. *Fertil. Steril.* **79:** 1520–1527.
39. Risch, N., E.W. Reich, M.M. Wishnick & J.G. McCarthy. 1987. Spontaneous mutation and parental age in humans. *Am. J. Hum. Genet.* **41:** 218–248.
40. Gavrilov, L.A. & N.S. Gavrilova. 1997. Parental age at conception and offspring longevity. *Rev. Clin. Gerontol.* **7:** 5–12.
41. Battaglia, D.E., P. Goodwin, N.A. Klein & M.R. Soules. 1996. Fertilization and early embryology: influence of maternal age on meiotic spindle assembly oocytes from naturally cycling women. *Hum. Reprod.* **11:** 2217–2222.
42. Robertson, O.H. 1961. Prolongation of the life span of kokanee salmon (Oncorhynchus nerka kennerlyi) by castration before beginning of gonad development. *Proc. Natl. Acad. Sci. USA* **47:** 609–621.
43. Brown-Borg, H.M. 2007. Hormonal regulation of longevity in mammals. *Ageing Res. Rev.* **6:** 28–45.
44. Arantes-Oliveira, N., J. Apfeld, A. Dillin & C. Kenyon. 2002. Regulation of life-span by germ-line stem cells in Caenorhabditis elegans. *Science* **295:** 502–505.
45. Yamawaki, T.M. et al. 2008. Distinct activities of the germline and somatic reproductive tissues in the regulation of Caenorhabditis elegans' longevity. *Genetics* **178:** 513–526.
46. Shanley, D.P. & T.B.L. Kirkwood. 2000. Calorie restriction and aging: a life history analysis. *Evolution* **54:** 740–750.
47. Reznick, D., M. Bryant & D. Holmes. 2006. The evolution of senescence and post-reproductive lifespan in guppies (*Poecilia reticulata*). *PLoS Biol.* **4:** 136–143.

48. Flurkey, K. *et al.* 2007. PohnB6F1: a cross of wild and domestic mice that is a new model of extended female reproductive life span. *J. Gerontol. A Biol. Sci. Med. Sci.* **62:** 1187–1198.
49. Kirkwood, T.B.L. & S.N. Austad. 2000. Why do we age? *Nature* **408:** 233–238.
50. Hamilton, W.D. 1964. The genetical evolution of social behaviour. *J. Theor. Biol.* **7:** 1–16.
51. Packer, C., M. Tatar & A. Collins. 1998. Reproductive cessation in female mammals. *Nature* **392:** 807–811.
52. Cohen, A.A. 2004. Female post-reproductive lifespan: a general mammalian trait. *Biol. Rev.* **79:** 733–750.
53. Ward, E.J. *et al.* 2009. The role of menopause and reproductive senescence in a long-lived social mammal. *Front. Zool.* **6:** 4.
54. Atsalis, S. & E. Videan. 2009. Reproductive aging in captive and wild common chimpanzees: factors influencing the rate of follicular depletion. *Am. J. Primat.* **71:** 271–282.
55. Herndon, J.G. & A. Lacreuse. 2009. "Reproductive aging in captive and wild common chimpanzees: factors influencing the rate of follicular depletion" by S. Atsalis and E. Videan. *Am. J. Primat.* **71:** 891–892.
56. Pavelka, M.S.M. & L.M. Fedigan. 1991. Menopause: a comparative life history perspective. *Yearb. Phys. Anthropol.* **34:** 13–38.
57. Caro, T.M. *et al.* 1995. Termination of reproduction in non-human and human female primates. *Int. J. Primatol.* **16:** 205–220.
58. Wise, P.M., K.M. Krajnak & M.L. Kashon. 1996. Menopause: the aging of multiple pacemakers. *Science* **273:** 67–70.
59. Hill, K. & A.M. Hurtado. 1991. The evolution of premature reproductive senescence and menopause in human females: an evolution of the "grandmother" hypothesis. *Hum. Nature* **2:** 313–350.
60. Peccei, J.S. 1995. The origin and evolution of menopause: the altriciality- lifespan hypothesis. *Ethol. Sociobiol.* **16:** 425–449.
61. Hawkes, K. *et al.* 1998. Grandmothering, menopause, and the evolution of human life histories. *Proc. Natl. Acad. Sci. USA* **95:** 1336–1339.
62. Hawkes, K., J.F. O'Connell & N.G. Blurton Jones. 1997. Hazda women's time allocation, offspring provisioning, and the evolution of long postmenopausal life spans. *Curr. Anthropol.* **38:** 551–577.
63. Rogers, A.R. 1993. Why menopause? *Evol. Ecol.* **7:** 406–420.
64. Shanley, D.P. & T.B.L. Kirkwood. 2001. Evolution of the human menopause. *BioEssays* **23:** 282–287.
65. Shanley, D.P., R. Sear, R. Mace & T.B.L. Kirkwood. 2007. Testing evolutionary theories of menopause. *Proc. R. Soc. Lond. B Biol. Sci.* **274:** 2943–2949.
66. Lahdenperä, M. *et al.* 2004. Fitness benefits of prolonged post-reproductive lifespan in women. *Nature* **428:** 178–181.
67. Cant, M.A. & R.A. Johnstone. 2008. Reproductive conflict and the separation of reproductive generations in humans. *Proc. Natl. Acad. Sci. USA* **105:** 5332–5336.
68. Lee, R. 2003. Rethinking the evolutionary theory of aging: transfers, not births, shape senescence in social species. *Proc. Natl. Acad. Sci. USA* **100:** 9637–9642.
69. Lee, R. 2008. Sociality, selection, and survival: simulated evolution of mortality with intergenerational transfers and food sharing. *Proc. Natl. Acad. Sci. USA* **105:** 7124–7128.
70. Pavard, S., C.J.E. Metcalf & E. Heyer. 2008. Senescence of reproduction may explain adaptive menopause in humans: a test of the "mother" hypothesis. *Am. J. Phys. Anthropol.* **136:** 194–203.

Learning, menopause, and the human adaptive complex

Hillard Kaplan,[1] Michael Gurven,[2] Jeffrey Winking,[3] Paul L. Hooper,[1] and Jonathan Stieglitz[1]

[1]Department of Anthropology, University of New Mexico, Albuquerque, New Mexico. [2]Department of Anthropology, UC Santa Barbara, Santa Barbara, California. [3]Department of Anthropology, Texas A&M, College Station, Texas

Address for correspondence: Hillard Kaplan, MSC01-1040 Department of Anthropology, University of New Mexico, Albuquerque, New Mexico 87131. hkaplan@unm.edu

This paper presents a new two-sex learning- and skills-based theory for the evolution of human menopause. The theory proposes that the role of knowledge, skill acquisition, and transfers in determining economic productivity and resource distribution is the distinctive feature of the traditional human ecology that is responsible for the evolution of menopause. The theory also proposes that *male* reproductive cessation and post-reproductive investment in descendants is a fundamental characteristic of humans living in traditional foraging and simple horticultural economies. We present evidence relevant to the theory. The data show that whereas reproductive decline is linked to increasing risks of mortality in chimpanzees, human reproductive senescence precedes somatic senescence. Moreover under traditional conditions, most human males undergo reproductive cessation at the same time as their wives. We then present evidence that after ceasing to reproduce, both men and women provide net economic transfers to children and grandchildren. Given this pattern of economic productivity, delays in menopause would produce net economic deficits within families.

Keywords: menopause; fertility; senescence; intergenerational transfers; human life history

Introduction

Available demographic evidence from hunter-gatherers and forager-horticulturalists without access to modern medicine shows that men and women can expect to live an additional two decades upon reaching age 45.[1,2] This adult age-specific mortality profile is rather uniform across extant traditional societies, and there is paleodemographic evidence suggesting the existence of older adults throughout the upper Paleolithic.[3] This implies that survival into old age is a fundamental feature of human biology. Behavioral data also show that older postreproductive adults of both sexes are quite productive,[1,4–6] and tend to produce more energy than they consume until about age 70. Reproductive senescence, however, occurs at much earlier ages in women and is largely complete by age 45. This pattern is also rather uniform across human populations, and there is surprisingly little variation in age of menopause cross-culturally.[7] The existence of two nonreproductive decades of adult life raises the fundamental evolutionary question: under what conditions will organisms evolve for whom general somatic senescence proceeds much more slowly than reproductive senescence?

Evolutionary theories of menopause that propose an adaptive function for reproductive cessation must show that the acceleration in reproductive senescence relative to mortality-related senescence results in higher fitness than the standard simultaneous decline in survival-related and reproductive functions (see Refs. 8–10 for reviews). Such theories need to provide a reason why direct reproduction will yield lower fitness than investing in alternatives, such as existing children and grandchildren. Special conditions must come into play for the following reason. In a diploid sexually reproducing organism, a female will be related to her offspring with Wright's coefficient of genetic relationship, r, of 0.5, whereas her grandchildren will only be half as related to her ($r = 0.25$). Therefore, according to inclusive fitness theory,[11] her investments in grandchildren will have to produce twice the fitness effect as in children for selection to favor investment in grandchildren.

It is clear that the high dependence of human offspring alone is not sufficient to explain menopause. Women undergo menopause about the time that they have reproducing daughters. If offspring need were the sole driver, selection would more likely favor "helping at the nest" by adult daughters and sons (a very common pattern among nonhumans) rather than reproductive cessation by the older female. After all, an individual is related to its sibling by an r of 0.5 if the two siblings share the same mother and father. Holding all else constant, an individual should be indifferent between direct reproduction and helping her mother produce a sibling. Thus, if children need additional investment, why is it that young females (and males) do not defer reproduction to help their mothers reproduce rather than vice versa? Another way to frame the question about menopause is to ask, "Why should women cease to reproduce and help descendants, instead of continuing to reproduce with the help of descendants?" An adaptive theory of menopause must specify the conditions that provide an answer to that question.

This paper presents a new learning- and skills-based theory for the joint evolution of human menopause and extended postreproductive life. *The theory proposes that the role of knowledge, skill acquisition, and transfers in determining economic productivity and resource distribution is the distinctive feature of the traditional human ecology that is responsible for the evolution of menopause.* Moreover, we argue that the traditional hunter-gatherer pattern of production, reproduction, and parental investment depends fundamentally on a cooperative division of labor between men and women. The theory therefore proposes that in addition to female menopause, *male* reproductive cessation and postreproductive investment in descendants is a fundamental characteristic of humans living in traditional foraging and simple forager-horticultural economies. The theory builds on existing ideas—specifically the Grandmother and Mother hypotheses[12,13]—in proposing that menopause and the decrease in fertility with age that precedes it are evolved human traits that have been maintained by selection because women will leave more descendants by ceasing to reproduce and investing in existing descendants. However, the specific causal hypotheses that the theory integrates are new.

Table 1. GEE logistic model of older women's likelihood of giving birth by BMI tercile ($N = 537$ person-years across 224 women aged 35–54 years)

Variable	B	SE	Wald χ^2	P
Intercept	4.723	0.9990	22.348	<0.001
Age	−0.159	0.0243	43.026	<0.001
High BMI	0.642	0.2622	5.990	0.014
Middle BMI	0.236	0.2692	0.766	0.381
Low BMI (baseline)	0	–	–	–

The paper begins with a brief presentation of the theory, followed by a discussion of the evidence upon which the theory is built. We begin with a comparative analysis of chimpanzee and human female reproductive senescence. We then examine the age-specific fertility of men and the likelihood of reproducing following menopause of wives. This is followed by behavioral evidence concerning food production and resource transfers across generations by women and men. The next section examines the total expected net caloric consumption of families as it varies over the life cycle, then simulates the caloric effects of adjusting the age schedule of women's fertility, delaying the onset of menopause. The paper concludes by linking these observations to the theory, and discussing directions for future theoretical and empirical research.

A "learning" theory of human reproductive decline and cessation

Although human foragers have lived in virtually all the world's terrestrial habitats, they always occupy one extreme feeding niche, eating the highest quality, most nutrient dense, and difficult to acquire plant and animal foods in their environment.[1,14] More than any other species, humans rely on brain-based skills and knowledge to acquire food from the environment. Those mental abilities combine with physical abilities—such as strength, coordination, and balance—to determine the rate of energy acquisition per unit time. In a series of papers, we have shown that peak physical condition in humans occurs in the early to mid-20s, but that peak economic productivity does not occur until after age 40. This is due to the fact that skill acquisition and learning continue to increase after peak physical condition is reached. Thus, peak economic productivity

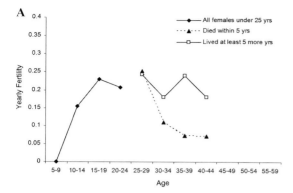

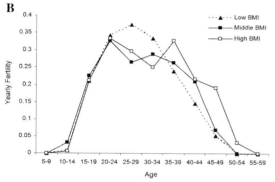

Figure 1. Impact of physical condition on chimpanzee and human fertility rates. (A) Probability of giving birth among female chimpanzees, stratified by those who died within 5 years of giving birth, and those robust enough to live at least 5 additional years following a birth. Adapted from Ref. 31. (B) Probability that a Tsimane woman gives birth using prospective data collected from 2002 to 2008, stratifying women into three groups based on their baseline BMI. (Sample includes 1,267 females between the ages of 5 and 59 and represents a total of 3,121 observation years. Because of the rapid change in BMI across adolescence, females under age 20 were separated into BMI terciles within 1-year age intervals, whereas older women were separated into BMI terciles within 5-year age intervals.)

bsetween 40 and 50 years of age can be more than four times as high as at age 20. After age 50, however, declines in physical condition begin to outpace gains from learning, and people cease to be net producers by around age 70 (see Refs. 1 and 5 and the "Evidence" section below for supporting data for these claims). Our theory proposes that the nature of the high-skill human foraging niche has a series of implications, which, taken together, disfavor old-age reproduction and favor old-age production and kin investment for both women and men, and thus drive the evolution of human reproductive decline and menopause.

Most theories of menopause and the empirical tests they stimulate estimate the age-specific cost of reproduction by the probability of dying in childbirth.[15–18] We propose that the cost of reproducing at advanced ages also includes increased risks of future mortality and reduced expected future productivity due to maternal depletion.[19] For example, maternal immune responses are lowered during pregnancy[20]; as women age and experience immunosenescence, the costs of immunosuppression are likely to increase. The energetic costs of lactation also probably occupy a greater proportion of a woman's physiological reserves as she ages. For these reasons, the cost of reproduction, both in terms of future mortality and future economic production, is likely to be higher for a 45-year-old woman than for her 20-year-old daughter.

Although most species are likely to evidence increasing costs of reproduction with age, late-age reproduction may be particularly costly for humans. Because human productivity is determined by both physical condition and long-developing skills, it is more important to survive long enough and maintain good enough condition to reap the rewards of earlier investments in skill development. This is accomplished by favoring somatic maintenance (and thus future production) over reproduction as the body begins to age.

The payoff to late-age reproduction in humans is also reduced by declining oocyte quality with age. There is significant evidence that oocyte quality declines with age in most mammals.[21,22] Because human offspring require extraordinary levels of investment to reach independence, the cost of continuing to reproduce from a deteriorating stock of oocytes should weigh more heavily in the human case than for most other species. An older mother producing highly dependent offspring may either: (a) risk investment in particularly low-quality offspring; or (b) ensure that she produces only sufficiently high-quality offspring, either by investing more energy in maintaining the quality of her oocytes, or by being more selective in allowing oocytes to implant or come to term. All of these options entail energetic costs, lost investments, or reduced fertility for older females that should be greater in species with heavier parental investment. Evidence presented by Ellison[23] and Haig[24] suggests that much of the burden in maintaining pregnancy prior to implantation depends on chemical signals produced by the embryo to maintain the corpus luteum and spur progesterone production. They suggest that maternal physiology utilizes these signals to detect quality differences in embryos and terminate low-quality pregnancies. We propose that human reproductive physiology may be particularly sensitive to embryo quality, and employ a more stringent selective sieve to prevent inferior embryos from implanting. Particularly long-lived animals may additionally face greater relative declines in oocyte quality over the lifespan (see Refs. 21 and 22 for reviews), which would also lower the returns to direct reproduction at advanced ages in humans.

While the returns to late-age reproduction are reduced, the returns to old-age kin investment are increased for humans relative to other animals. Because the skills required for efficient food production take time to learn, children in foraging societies do not produce as much food as they consume until they are 18–20 years of age.[1] This means that they must rely on subsidies from other individuals. As the number of overlapping dependents in a young mother's household grows with each birth, total caloric need is expected to outpace a single couple's combined productivity, creating a demand for calories from sources outside the immediate household (see Ref. 25 and "Evidence" section). Older kin enjoying high levels of learning-based productivity and facing increasing costs of direct reproduction are in a prime position to meet this demand. This is true to a greater extent for humans than for most other mammals as a result of the life history characteristics—high productivity late in life, high offspring need, and the simultaneous dependency of multiple offspring—which coevolved with the skills-based human foraging strategy.

Finally, the skills-based foraging niche also provides the conditions that lead most men in foraging societies to undergo reproductive cessation at the same time as their wives. The skills-based economy of humans is associated with unusually high male energetic investment in offspring. In fact, men provide the majority of energetic support for reproduction in most hunting and gathering groups.[26,27] Despite the high need for protein and lipids to support brain growth during development, the mobility, danger, and long-term skill investments involved in human hunting make it largely incompatible with the primate female's evolved commitment to carrying (rather than caching) infants and lactation-on-demand. This generates a complementarity between male and female inputs into offspring success, a sex-specific specialization in hunting by men, and high returns to male parental investment. Woman, in turn, specialize in a mix of childcare and foraging for plant resources.

The returns to male parental investment and the long overlapping dependence of children interact in producing a dominant pattern of long-term pair-bonding and male reproductive cessation in traditional foraging societies. Given that children remain dependent after their younger siblings are born, men and women in foraging societies face higher costs from switching mates than in many other species. A mother who begins a new union often suffers reduced paternal investment from the father of her previous children. Conversely, for a woman who is about to initiate reproduction, a man who has children from a previous union is less attractive because he already has vessels in which to invest. Consider a 20-year-old woman who is about to begin reproducing. For her, a 50-year-old man is less attractive than a 25-year-old, even though the older man may currently be more economically productive. The 50-year-old has two disadvantages: first, he already has his peak dependency load of existing children; second, his food production will decrease in the future and his mortality risk will increase. If the 20-year-old prefers to have all her children with one man, the younger man is preferable, because of his expected future contributions. This also implies that older men, who also face a tradeoff between investing in existing children and grandchildren and seeking a new mate, most often "choose" to remain married and cease reproducing when their wives reach menopause.[28]

Our theory is that these altered age-specific benefits and costs of fertility, production, and kin-investment—which derive from the specialized skills-based foraging niche and its attendant shift in economic productivity toward older ages—combine to favor "early" reproductive cessation in both men and women. This pattern has only evolved once. Even in those toothed whales that evidence female menopause, there is no such equivalent in males. In those species, males typically have much shorter lifespans than females and do not invest in offspring.[29] Humans are an outlier species in many senses, from brain size to lifespan to menopause to male parental investment. Special conditions are necessary to produce such an outlier. The combination of a brain-based, knowledge-, and skill-intensive foraging niche with a primate heritage selected for this complex of traits (large brains, long lifespan, long offspring dependence, high selectivity of oocyte quality, high male parental investment, and bisexual reproductive cessation).

In a recent paper, Kaplan and Robson[30] present a formal bioeconomic model for the evolution of aging. They show that reproductive cessation can be optimal prior to the optimal time to cease investing in mortality reduction and future longevity. The Kaplan–Robson model does not include all the considerations elaborated earlier, but provides an analytical result demonstrating the conditions under which menopause can evolve by natural selection. It shows that if (a) the energetic costs of reproduction increase with declining physical condition due to senescence, (b) economic transfers can allow surplus productivity at one point in the life course to be utilized at another point in the life course, and (c) individuals remain economically productive after reproductive cessation, then there is an age at which fitness—measured in terms of the instantaneous growth rate, r, of the lineage—can be maximized by reproductive cessation and the allocation of remaining resources to mortality reduction, physical maintenance, and intergenerational transfers. The present theory is based partially on the insights derived from that formal model.

The remaining sections of the paper will focus on the empirical evidence related to the theory.

Evidence

Chimpanzee and human reproductive decline and its link to somatic senescence

This section provides a comparative analysis of chimpanzee and Tsimane fertility. A recent analysis by Emery Thompson et al. showed that while mean chimpanzee fertility rates decline toward the end of life, females in good physical condition show no significant fertility decline with age (Fig. 1A, adapted from Ref. 31). Among females aged 25 and older, healthy individuals have significantly higher fertility than females who died within 5 years of the birth- or risk-year considered. Their findings suggest that chimpanzee reproductive senescence is tightly linked to somatic senescence and vulnerability to mortality. Using a similar approach for

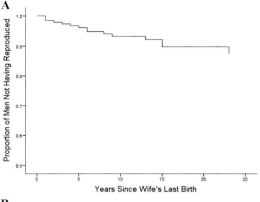
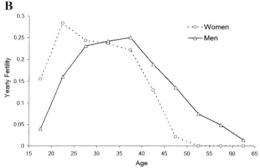
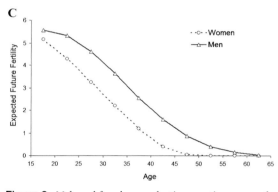

Figure 2. Male and female reproductive cessation among the Tsimane. (A) Probability that a Tsimane man did not reproduce after his wife had her last birth (see text for details). (Sample based on retrospective reproductive histories including 188 final female births and 182 husbands; six were married polygynously to two wives.) (B) Age-specific fertility rates for Tsimane men and women, given in 5-year intervals. (Sample based on retrospective reproductive histories of 431 women and 391 men covering the period 1950–2002; this includes 12,394 risk-years and 2,238 births for women, and 12,514 risk-years and 1,943 births for men aged 15–64.) (C) Expected future fertility by age considers the cumulative sum of remaining future reproduction discounted by the probability of surviving to those ages. Survivorship data are from Ref. 46.

traditional humans, we expect to see a decoupling of somatic and reproductive senescence.

To compare a traditional human case with Emery Thompson et al.'s[31] results, we performed a prospective analysis of the effect of physical condition, represented by body mass index (BMI), on age-specific fertility among Tsimane women. We examined the probability of a live birth occurring in each full calendar year following a woman's first nonpregnant BMI measure based on census data collected between 2002 and 2008. Figure 1B shows the mean fertility of Tsimane women by age divided into low, middle, and high BMI terciles. Although women with high BMI have slightly higher fertility at the end of their reproductive careers, all three condition levels show a characteristic decline in fertility that reaches zero for all women by the late 40s or early 50s, regardless of condition, unlike the chimpanzee case.

To further examine decline in fertility, the generalized estimating equations (GEE) method in SPSS 16 was used to test for the effects of age and BMI on the likelihood of giving birth in each year from ages 35 to 54. The woman's identity was included as a random repeated subject variable. Women in the highest BMI tercile have almost twice the likelihood of giving birth than women in the lowest BMI tercile (Table 1). Inclusion of a BMI-by-age interaction term does not yield a significant parameter estimate nor improve the model's goodness of fit, indicating that the slope of the decline in fertility by age is not significantly affected by BMI. This finding again contrasts with the chimpanzee case, in which the fertility of low-condition females declines with age significantly faster than that of high-condition females, who show no significant decline in fertility with age.

Male reproductive cessation

Consistent with the high levels of male parental investment, the majority of marriages among foragers and forager-horticulturalists are monogamous.[32] Although in some societies, such as the Ache and Hadza, there are frequently a series of short-term unions in early adulthood, this is generally followed by a single long-term reproductive union. In a sample of 145 hunter-gather societies, the modal percentage of polygynous marriages in a society is 0–4%, and in most societies fewer than 10% of marriages are polygynous.[33] One of the consequences of

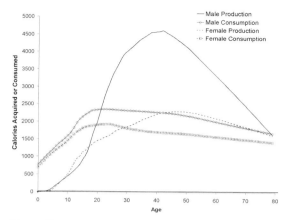

Figure 3. Age-specific caloric production and consumption profiles for the Tsimane. Daily production was estimated for non-rice foods from interviews covering the previous 2 days of food production. These data covered 43,656 sample days over 749 individuals. Rice production was estimated from interviews concerning the amount of rice harvested in the previous year. This data covered 589 individuals from the nonrice sample. Credit for rice production was based on the proportional time spent in field labor from the 2-day production interviews. Loess curves were fit over the daily nonrice and rice production rates by age and sex. The loess prediction curves were then summed to produce the final curves. Consumption was estimated by first calculating the total energy expenditure (TEE) based on the age, sex, and weight of individuals.[47] These were plotted by age and the maximal consumption level was estimated to be 2,770 calories per day for the Tsimane. The TEE of each individual was divided by this to determine the proportion of consumer (POC). The number of production days sampled was multiplied by each individual's POC and these were then summed to determine the total number of consumer days. The total production during the sampling period was then divided by this sum to determine the true caloric intake of the maximal consumer, which equaled 2,661 calories per day. Each individual's POC was multiplied by 2,661 to determine their consumption level. We then fit a loess curve to the consumption levels by age and sex.

monogamy is reproductive cessation among men after their wives reach menopause. For example, Ache foragers have high initial divorce rates when they are young; nevertheless, 90% of men who had more than one child with a woman did not reproduce after their wives reached menopause.[34]

Tsimane demographic data show that 90% of Tsimane men whose wives reached menopause did not reproduce again after their wife's last birth. Of the 10% who did reproduce, half (5.2%) were polygynously married and had a child with a younger co-wife, still within the bonds of marriage. The remainder had affairs outside of marriage (3.1%) or reproduced after the wife's death (1.5%). Given that some men at risk of reproducing after their partner reached menopause are still alive and may reproduce in the future, we conducted a survival analysis of male reproduction following menopause. From the survival curve (Fig. 2A) it is evident that the greatest chance of reproduction is in the first 5 years after the wife's last child, consistent with the pattern of polygynous men reproducing with the younger co-wife. Because the younger co-wives were often reaching middle age as well, most of these men only reproduced once after their first wife reached menopause.

The linkage of men's reproductive schedules with women's can also be seen from the age-specific fertilities of the two sexes. The male curve is shifted to the right of the female curve by about 5 years, consistent with the age differences among spouses (Fig. 2B). The tail of the male curve stretches out a bit from the female curve due to some men being more than 5 years older than their spouse and the few men who reproduce after their wife's menopause. The male and female curves for expected future fertility (i.e., reproductive value) are strikingly similar, after age differences in marriage are taken into account (Fig. 2C).

Physical condition, age profiles of productivity, and intergenerational transfers

Age profiles of productivity and intergenerational transfers among human hunter-gatherers and forager-horticulturalists have been documented in a series of publications.[1,5,6,35,36] Those data show that children remain dependent on their parents until 18–20 years of age, with a peak dependency in early adolescence (from birth, caloric requirements grow faster than productivity until about age 12 or 13).

Peak productivity in adulthood for both men and women occurs well after strength and physical condition peak. For example, among Ache foragers of Paraguay, men's strength peaks at around 25 years of age but both meat acquired and hunting return rates (amount acquired per hour spent hunting) peak between 40 and 50 years of age.[37] Data on strength and hunting ability among Tsimane men show the same pattern[5]; moreover, skill in successfully pursuing prey is the most important determinant of hunting success.

Figure 3 shows the age-profiles of food production and consumption among Tsimane men and women. Both male and female production peaks after age 40. Males produce as much as they consume

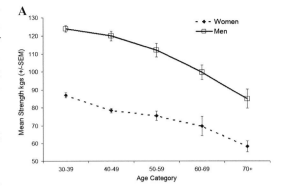

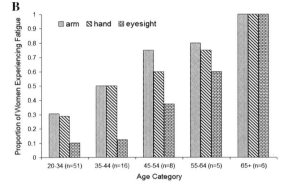

Figure 4. Physical decline with age among Tsimane adults. (A) Strength is the sum of chest, shoulder, thigh, leg, and hand grip strength, measured using the Lafayette Manual Muscle Tester and Smedley III Analog Grip Strength Tester. (Sample includes 416 women and 428 men.) (B) Proportion of Tsimane women that report experiencing physical problems during rice pounding. Problems include arm and hand pain and poor eyesight. (Sample includes 104 women.)

by about age 20, and females by age 28. The caloric deficit in childhood is compensated for by a caloric surplus in adulthood. The increase in total food production is driven by two effects. First, there is an increase in efficiency (production per unit time) until the mid-40s. Second, there is a corresponding increase in work effort, probably reflecting the increase in dependency load. The decrease in production with age is driven primarily by declines in efficiency.

Figure 4 examines physical decline with age. Panel A shows the decline in strength with age for both men and women, and Panel B shows pain-related fatigue among women while they pound rice. Both figures show considerable declines before peak productivity is reached.

Food sharing data allow for a more direct understanding of inter-generational wealth flows. Figure 5

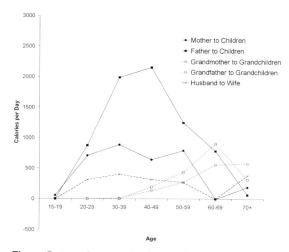

Figure 5. Net caloric transfers between kin groups across three generations. Transfers were calculated using data from 3,850 consumption events by 674 individuals during instantaneous scan observations. The number of events in which individual A was the acquirer of food consumed by individual B divided by the total number of times individual A was named an acquirer was interpreted as the proportion of individual A's production that went to individual B. (For foods with multiple acquirers, each acquirer was assigned a proportion of credit, and these credits were the values actually tallied.) The proportional distribution to each kin member was then calculated for each aggregated age-sex group, as the number of observations per individual was low. To capture the observed population age structure, each individual alive in a 25-community census was assigned their age- and sex-specific daily production and proportional distribution levels. Daily production (represented in Fig. 3) was multiplied by proportional distribution to determine gross transfers. These were summed in both directions for each kin dyad to determine net transfers. Averages were then calculated for each age-sex group.

plots the *net* transfers between pairs of related individuals. Net transfers are calculated by taking the total amount of food given from individual A to individual B and then subtracting the total amount given from B to A. Those amounts are derived from data on the consumers of food acquired by all family members. In the figure we present those nets from fathers to children, mothers to children, grandfathers to grandchildren, and grandmothers to grandchildren. Even though food is transferred in both directions between these pairs of individuals, the figure shows that *net transfers flow downward across generations*. The downward flow from both mothers and fathers to their children continues into adulthood, even when their children become adults and have children of their own. During the postre-productive period of life (after age 45), transfers to existing children dominate during middle age, with an increasing proportion of resources being transferred to grandchildren with age; the absolute volume of transfers to grandchildren peaks in the 60s. Net transfers approach zero after age 70.

From Figure 5, it can also be seen that men transfer more calories to descendants than do women. However, women's work in childcare, food processing, and household maintenance exceeds that of men, and both sexes spend similar amounts of total time in work.[38] This division of labor appears to be universal in foraging societies, although the relative energetic contributions of the two sexes vary according to local ecology. In the sample of 10 forager societies for which quantitative data exist, men, on average, acquire 68% of calories and almost 88% of protein, while women acquire the remaining 32% of calories and 12% of protein.[26]

Transfers, calories, and menopause

Just as food production increases with age during the reproductive period, so too do the caloric demands of dependents. In fact, the caloric demands on parents increase faster than does their productivity.[25] Figure 6A shows data from the Tsimane on the net productivity of parents, the net caloric demands of children and the net surplus or deficit of families as a function of a woman's age. This figure shows that as families grow, their net deficit increases, even though parental productivity is increasing as well. Most importantly, it can be seen that the deficit of growing families is compensated for by the net surplus of postreproductive individuals, who provision descendent kin (see also Fig. 5).

Figure 6B simulates the caloric effects of a delay in reproductive decline and menopause. The average net caloric demand of children in families headed by mothers in their 30s was extended throughout the 40s; the net caloric demand of children beyond the 40s then continued 10 years behind schedule (so that a 60 year old was experiencing the typical progeny dependency of a 50 year old). In this case, the surplus provided by older people continues to be consumed by their dependent children. Figure 7 shows the cumulative net caloric balance, given the Tsimane sample and the delayed menopause simulation. The "contrary to fact" delayed menopause simulation shows genetic lineages

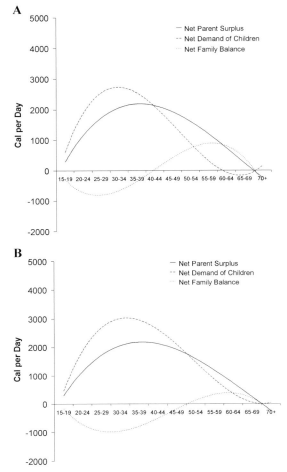

Figure 6. Parental production, children's demands, and net family production. (A) Observed Tsimane pattern. These calculations are based on rice and non rice production data from 106 families, including 561 individuals who were sampled for an average of 67 days. (B) Simulation based on delayed menopause (see text). Daily caloric production and estimated consumption levels were summed for parents and children within families. Parental production and child demand levels were then aggregated over 5-year age intervals to calculate overall family balances. Because cumulative mortality risk leads to a larger number of families headed by younger parents than families headed by older parents, those families that do survive must produce surpluses that more than make up for their previous deficits. To take into account the effects of mortality, summed net balances for all age intervals were divided by the number of families in the 15–19 age interval in an attempt to include in the denominator those families that were lost to mortality. Third-order polynomial curves were fit to the mean values of each age interval.

with a fertility and economic transfer regime that would be in net economic deficit, and therefore could not support itself.

Discussion and conclusions

The theory presented in this paper builds on existing adaptive hypotheses for the evolution of menopause. Most adaptive explanations have focused on women's roles as mothers and grandmothers.[4,12,13,15,16,39–41] The mother version emphasizes the long period of juvenile dependence in humans, and its possible links to brain development.[13,39] According to this view, women stop reproducing at the expected age at which they will be able to raise their last child to maturity before dying. If children require 20 years of parental investment, then ceasing to reproduce at age 45 would make sense with an expected age of death of 65, given survival to that age. The grandmother version proposes that women cease reproducing in order to invest in grandchildren and help their daughters reproduce.[12,40,41] According to Hawkes et al.,[41] the strength-intensive nature of human foraging means that grandmothers can acquire more than children and help provision them.

The present theory extends and modifies those ideas in three important ways. First, it specifies the unique ecological conditions responsible for the evolution of the temporal separation of reproductive and somatic senescence in humans. Second, it identifies the important role that men play in the human life history strategy, and highlights that premature reproductive cessation occurs in men as well as women. Third, it addresses the question of why reproductive cessation and downward kin-investment by elders should be favored over the alternative of continued reproduction supported by younger, nonreproductive "helpers at the nest."

The fundamental premise of our theory is that the role of brain-based skills and learning in economic production for both men and women during our evolutionary past is at the far end of the evolutionary continuum. Skills and knowledge are accumulated throughout life, but physical condition, from strength to immune function, declines throughout adulthood. As a result, human economic productivity—which is a function of both cumulatively learned abilities and physical strength and endurance—continues to increase even after physical condition begins to decline. We

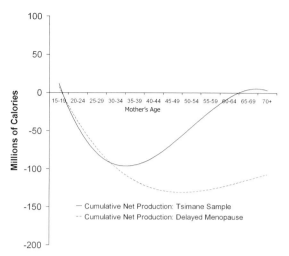

Figure 7. Cumulative net caloric balance of families given the Tsimane sample and the delayed menopause simulation from Figure 6.

propose that this disjunction between economic and physiological aging is the ecological key to human menopause. It simultaneously generates two conditions: (1) the physiological cost of later reproduction is rendered high for women, but their economic productivity and that of their husbands remains high; and (2) infants, juveniles, and adolescents produce less food than their growing bodies require.

A key feature of our theory is that it incorporates declining oocyte quality and increasing physiological costs of reproduction. We argue that declining oocyte quality with age has a larger impact on the tradeoff between reproducing and investing in descendants for humans than it does for chimpanzees and most other mammals. Here we base our argument on evidence showing that human reproductive physiology is replete with mechanisms designed to ensure that investment in low-quality oocytes and embryos is curtailed, from the follicular development phase through implantation and placental development.[23,42] We further propose that it is the length and volume of human parental investment that selected for more stringent mechanisms of quality control, to direct investment toward high-quality offspring rather than "waste" it on lower quality offspring. The logical extension of this argument is that across species, optimal levels of selectivity with respect to oocyte quality will increase as parental investment increases.

To this logic, we add the observation that the physiological cost of reproduction increases with maternal age, not only due to increased risks of death in childbirth, but also due to maternal depletion that should affect survival and productivity at future ages. Given the disjunction between physiological aging and economic aging in humans and given the low productivity of children who have yet to learn, those physiological costs of reproduction should weigh more heavily on women than on females of other species. Human females have more to give by living longer, and thus should be less willing to risk death than other species.

Finally, our theory is two-sex, in that it proposes that reproductive cessation occurs regularly among human males as well as females. We argue that human males also face a similar tradeoff between investment in existing descendants and continued reproduction. However, instead of facing increased physiological costs of reproduction with age, males become less attractive as mates as they age. This is due to two reasons. First, the importance of male investment in offspring and the long-term dependence of young in humans have resulted in long-term monogamous pair-bonds between men and women. As a result, marriage to an older man is less attractive to a young woman, because he is likely to die before she completes her reproductive career. Second, given that older men are likely to have existing dependent young, their investment in children produced by a new marriage will likely be lower.

We compared chimpanzee decline in fertility with age to that of Tsimane females. The data compiled by Emery Thompson *et al.* show that aging chimpanzees in relatively good condition do not reduce fertility with age (or reduce fertility at later ages).[31] In contrast, the Tsimane data show that while women with higher BMI, one measure of condition, do have higher fertility late in life, the decline in fertility with age in healthy women is more dramatic than among healthy chimpanzees. This is even more striking given that chimpanzee females in relatively good condition in their 30s are still in worse condition than most women at that age, showing much more advanced signs of aging. This suggests that in response to declining oocyte quality, chimpanzee female reproductive physiology is less selective than that of human females. We used data from the Ache and Tsimane to show that men have a low probability of reproducing after their wives reach menopause, and that their age-related fertility decline is very similar to that of their wives.

We then presented evidence for both men and women that caloric production in traditional economies does not exceed consumption until adulthood, and that middle- and old-age adults produce a caloric surplus. The data also show that intergenerational wealth flows are downward within families, and that both men and women invest in existing children and then grandchildren after they cease reproducing. We then examined the joint economic and reproductive life histories of families by plotting the expected caloric demands of children, the net productivity of parents, and the resulting household net caloric surplus (or deficit) as functions of a woman's age. Those results showed that energetic burden of reproducing families produces a caloric deficit, which is compensated for by the caloric surpluses of postreproductive individuals.

The impacts of reproductive cessation on calorie balance were then illustrated by simulating the continued reproduction of women at their 30-year-old rate until age 50. That simulation revealed that all of the caloric surplus of older people would be consumed by the extension of the reproductive period, and the whole family lineage would remain in caloric deficit.

The evidence presented in this paper can not be considered a test of the theory, because the theory was developed in response to the evidence. In addition, most of the evidence is "circumstantial" in that it is consistent with the theory, but does not demonstrate that the relative importance of foraging for high-quality resources using learning-intensive acquisition strategies is the primary ecological driver of menopause. Given that menopause has evolved so infrequently and its particular two-sex form in humans is unique, ecological tests may prove elusive.

Nevertheless, individual components of the theory may be testable with comparative data. For example, there is a growing corpus of data on whales that should allow for comparative tests. Some toothed-whales show clear evidence of menopause and a long postmenopausal lifespan in females.[29,43] It is interesting to note that this branch of the cetacean line shows some broad similarities in its foraging niche to humans. Killer whales, for example, demonstrate ecologically diverse foraging strategies, strongly based on cultural traditions passed through matrilineal kin from old to young (see Ref. 44 for a detailed review of learned cultural traditions in cetaceans). Their foraging strategies and brains also reflect complex cognitive processes.[45] Similarly, comparative research on complementarity, male parental investment and the linkage between male and female reproductive strategies could test other components of the theory.

Future research should focus on investigating the costs of reproduction, selectivity with respect to oocyte quality, and economic transfers. We still know very little about maternal depletion in traditional natural fertility societies, and how aging affects the costs of reproduction in terms of future longevity, health, and productivity. Another area for investigation is species differences in oocyte quality control. Do humans and chimpanzees differ in the selectivity of oocytes prior to ovulation, during fertilization and the completion of meiosis, or during embryogenesis? Food sharing in traditional societies is also very complex. There are both within- and between-family transfers, and the mix of kinship, reciprocity and other factors determining those transfers is still poorly understood. A clearer understanding of those phenomena will help evaluate the present theory and provide insight into the evolution of human reproductive cessation.

Acknowledgments

This research supported by the National Science Foundation (BCS-0422690) and the National Institute on Aging (R01AG024119–01). The authors thank Sam Bowles, Ken Wachter, Melissa Emery Thompson, and Benjamin Hanowell for their very helpful comments, as well as collaborators on previous papers, Jane Lancaster, Arthur Robson, Kim Hill, and Magdalena Hurtado, for their help in developing the ideas and data presented here.

Conflict of Interest

The authors declare no conflicts of interest.

References

1. Kaplan, H., K. Hill, J. Lancaster & A.M. Hurtado. 2000. A theory of human life history evolution: diet, intelligence, and longevity. *Evol. Anthropol.* **9:** 156–185.

2. Gurven, M. & H. Kaplan. 2007. Longevity among hunter-gatherers: a cross-cultural comparison. *Popul. Dev. Rev.* **33:** 321–365.
3. Konigsberg, L.W. & N.P. Herrmann. 2006. The osteological evidence for human longevity in the recent past. In *The Evolution of Human Life History*. K. Hawkes & R. Paine, Eds.: 267–306. School of American Research Press. Santa Fe.
4. Hawkes, K. 2003. Grandmothers and the evolution of human longevity. *Am. J. Hum. Biol.* **15:** 380–400.
5. Gurven, M., H. Kaplan & M. Gutierrez. 2006. How long does it take to become a proficient hunter? Implications on the evolution of delayed growth. *J. Hum. Evol.* **51:** 454–470.
6. Kaplan, H. 1994. Evolutionary and wealth flows theories of fertility: empirical tests and new models. *Popul. Dev. Rev.* **20:** 753–791.
7. Thomas, F. *et al.* 2001. International variability of ages at menarche and menopause: patterns and main determinants. *Hum. Biol.* **73:** 271–290.
8. Austad, S. N. 1997. *Why We Age*. Wiley. New York.
9. Armstrong, E. & D. Falk. 1982. *Primate Brain Evolution: Methods and Concepts*. Plenum Press. New York.
10. Sherman, P.W. 1998. The evolution of menopause. *Nature* **392:** 759–761.
11. Hamilton, W.D. 1964. The genetical evolution of social behavior. *J. Theor. Biol.* **7:** 1–52.
12. Hawkes, K. *et al.* 1998. Grandmothering, menopause, and the evolution of human life histories. *Proc. Natl. Acad. Sci. USA* **95:** 1336–1339.
13. Peccei, J.S. 2001. Menopause: adaptation or epiphenomenon? *Evol. Anthropol.* **10:** 43–57.
14. Kaplan, H.S. 1997. The evolution of the human life course. In *Between Zeus and the Salmon: The Biodemography of Longevity*. K. Wachter & C. Finch, Eds.: 175–211. National Academy of Sciences. Washington, DC.
15. Hill, K. & A.M. Hurtado. 1991. The evolution of reproductive senescence and menopause in human females. *Hum. Nat.* **2:** 315–350.
16. Rogers, A. 1993. Why menopause? *Evol. Ecol.* **7:** 406–420.
17. Sear, R., F. Steele, I. McGregor & R. Mace. 2002. The effects of kin on child mortality in rural Gambia. *Demography*. **39:** 43–63.
18. Sear, R., R. Mace & I.A. McGregor. 2000. Maternal grandmothers improve the nutritional status and survival of children in rural Gambia. *Proc. Biol. Sci.* **267:** 1641–1647.
19. Tracer, D.P. 1991. Fertility-related changes in maternal body composition among the Au of Papua New Guinea. *Am. J. Phys. Anthropol.* **85:** 393–405.
20. Heyborne, K. & R.M. Silver. 1996. Immunology of postimplantation pregnancy. In *Reproductive Immunology*. R.A. Bronson, N.J. Alexander & D. Anderson, Eds.: 383–417. Blackwell Science. Oxford.
21. Ottolenghi, C. *et al.* 2004. Aging of oocyte, ovary, and human reproduction. *Ann. N.Y. Acad. Sci.* **1034:** 117–131.
22. vom Saal, F.S., C.E. Finch & J.F. Nelson. 1994. Natural history and mechanisms of aging in humans, laboratory rodents and other selected vertebrates. In *Physiology of Reproduction*, Vol. 2. E. Knobil, J. Neill & D. Pfaff, Eds.: 1213–1314. Raven Press. New York.
23. Ellison, P.T. 2001. *On Fertile Ground: A Natural History of Human Reproduction*. Harvard University Press. Cambridge, MA.
24. Haig, D. 1993. Genetic conflicts in human pregnancy. *Q. Rev. Biol.* **68:** 495–532.
25. Gurven, M. & R. Walker. 2006. Energetic demand of multiple dependents and the evolution of slow human growth. *Proc. Biol. Sci.* **273:** 835–841.
26. Kaplan, H.S., K. Hill, A.M. Hurtado & J. Lancaster. 2001. The embodied capital theory of human evolution. In *Reproductive Ecology and Human Evolution*. P.T. Ellison, Ed.: 293–317. Aldine de Gruyter. Hawthorne, NY.
27. Marlowe, F. 2001. Male contribution to diet and female reproductive success among foragers. *Curr. Anthropol.* **42:** 755–760.
28. Winking, J., H. Kaplan, M. Gurven & S. Rucas. 2007. Why do men marry and why do they stray? *Proc. Biol. Sci.* **274:** 1643–1649.
29. Foote, A. 2008. Mortality rate acceleration and post-reproductive lifespan in matrilineal whale species. *Biol. Lett.* **4:** 189–191.
30. Kaplan, H. & A. Robson. 2009. We age because we grow. *Proc. Biol. Sci.* **276:** 1837–1844.
31. Emery Thompson, M. *et al.* 2007. Aging and fertility patterns in wild chimpanzees provide insights into the evolution of menopause. *Curr. Biol.* **17:** 2150–2156.
32. Marlowe, F.W. 2004. The mating system of foragers in the standard cross-cultural sample. *Cross-Cult. Res.* **37:** 282–306.
33. Binford, L.R. 2001. *Constructing Frames of Reference*. University of California Press. Berkeley, CA.
34. Hill, K. & A.M. Hurtado. 1996. *Ache Life History: The Ecology and Demography of a Foraging People*. Aldine. Hawthorne, NY.
35. Gurven, M.D. & H.S. Kaplan. 2006. Determinants of time allocation to production across the lifespan among the Machiguenga and Piro Indians of Peru. *Hum. Nat.* **17:** 1–49.
36. Kaplan, H. & A. Robson. 2002. The emergence of humans: the coevolution of intelligence and longevity with intergenerational transfers. *Proc. Natl. Acad. Sci. USA* **99:** 10221–10226.
37. Walker, R., K. Hill, H. Kaplan & G. McMillan. 2002. Age-dependency in skill, strength and hunting ability among the Ache of eastern Paraguay. *J. Hum. Evol.* **42:** 639–657.
38. Gurven, M. *et al.* 2009. A bioeconomic approach to marriage and the sexual division of labor. *Hum. Nat.* **20:** 151–183.
39. Lancaster, J.B. & B.J. King. 1985. An evolutionary perspective on menopause. In *In Her Prime: A View of Middle Aged Women*. V. Kerns & J.K. Brown, Eds.: 13–20. Bergen and Garvey. Garden City, NJ.
40. Williams, G.C. 1957. Pleiotropy, natural selection and the evolution of senescence. *Evolution*. **11:** 398–411.
41. Hawkes, K., J.F. O'Connell & N. Blurton Jones. 1989. Hardworking Hadza grandmothers. In *Comparative Socioecology of Humans and Other Mammals*. V. Standen & R.A. Foley, Eds.: 341–366. Basil Blackwell. London.

42. Hunt, P. & T. Hassold. 2008. Human female meiosis: What makes a good egg go bad? *Trends Genet.* **24:** 86–93.
43. Ward, E.J. *et al.* 2009. The role of menopause and reproductive senescence in a long-lived social mammal. *Front. Zool.* **6:** 1–10.
44. Rendell, L. & H. Whitehead. 2001. Culture in whales and dolphins. *Behav. Brain Sci.* **24:** 309–382.
45. Marino, L. *et al.* 2007. Cetaceans have complex brains for complex cognition. *PLoS Biol.* **5:** 0966–0972.
46. Gurven, M., H. Kaplan & A. Zelada Supa. 2007. Mortality experience of Tsimane Amerindians: regional variation and temporal trends. *Am. J. Hum. Biol.* **19:** 376–398.
47. FAO. 2001. Human energy requirements. Food and Nutrition Technical Report Series.

ANNALS OF THE NEW YORK ACADEMY OF SCIENCES
Issue: *Reproductive Aging*

Do women stop early? Similarities in fertility decline in humans and chimpanzees

Kristen Hawkes[1] and Ken R. Smith[2,3]

[1]Department of Anthropology, University of Utah, Salt Lake City, Utah. [2]Department of Family and Consumer Studies, University of Utah, Salt Lake City, Utah. [3]Huntsman Cancer Institute, University of Utah, Salt Lake City, Utah

Address for correspondence: Kristen Hawkes, Department of Anthropology, University of Utah, 270 South 1400, East, Stewart 102, Salt Lake City, Utah 84112-0060. hawkes@anthro.utah.edu

Two kinds of evidence suggest that female fertility may end at an earlier age in modern people than in ancestral populations or in our closest living relatives, chimpanzees. We investigate both to see whether fertility schedules or ovarian follicle counts falsify the alternative hypothesis that the age of terminal fertility changed little in the human lineage while greater longevity evolved due to grandmother effects. We use 19th century Utah women to represent non-contracepting humans, and compare their fertility by age with published records for wild chimpanzees. Then we revisit published counts of ovarian follicular stocks in both species. Results show wide individual variation in age at last birth and oocyte stocks in both humans and chimpanzees. This heterogeneity, combined with interspecific differences in adult mortality, has large and opposing effects on fertility schedules. Neither realized fertility nor rates of follicular atresia stand as evidence against the hypothesis that ages at last birth changed little while greater longevity evolved in our lineage.

Keywords: grandmother hypothesis; fertility decline; menopause; heterogeneity; follicular depletion; chimpanzee comparisons

Introduction

Women usually outlive their fertility by decades, a feature often described as the distinctively human mid-life menopause. Williams[1] famously proposed that earlier fertility termination evolved in humans as a consequence of other evolutionary changes that made late births increasingly risky. Now we know that ages at last birth are similar in humans and the other living great apes.[2] Childbearing years extend into the 40s, and end there—medical interventions notwithstanding—in our own species and our closest living relatives. We also know that long adult lifespans distinguish us from other apes.[2] This is regularly obscured by inferences from the global increases in human life expectancies since the mid-19th century. Many assume—erroneously—that when average lifespan is less than 50 there must be few old people. Instead, the global increases in life expectancies to the mid-20th century were largely driven by declines in infant and juvenile mortality.[3] Hunter-gatherers provide a key line of evidence about the mortality experience humans faced before the origins of agriculture about 10,000 years ago. Among the best-studied hunter-gatherer populations life expectancies at birth are less than 40 years, yet for the girls who survive to adulthood, most—63–77%—outlive the childbearing years.[4–7]

Consistent with those findings, an alternative hypothesis about the evolution of human life history proposes that grandmother effects increased longevity in our lineage without changes in the age of female fertility decline.[8,9] A key stimulus to this grandmother hypothesis was the economic productivity of older women among Hadza hunter-gatherers.[10] The subsidies these elders provided for young children whose mothers were nursing a new infant suggested a similar role for ancestral grandmothers when ecological changes in the Plio-Pleistocene reduced the availability of foods youngsters could handle for themselves.[11] Those ecological changes opened a novel fitness window for females whose own fertility was declining. By subsidizing their grandchildren they also enhanced the fertility of their daughters. Through those grandmother effects, more robust elders, able to help more, left more descendants.

However, some observations seem inconsistent with that grandmother hypothesis. Here we consider two that appear initially to support Williams' "stopping early" hypothesis instead. One is the recent report that free ranging chimpanzees maintain high fertility through age classes that are associated with declining fertility in women.[12] Comparisons between humans and chimpanzees are of special relevance for reconstructing human life history evolution[2] for at least two reasons. They are our closest living relatives, and, because of similarities in body and brain size, they are the favored model for estimating maturation and aging rates in australopithecines, the fossil genus ancestral to our own.[13,14] The evidence that age-specific fertility remains high in chimpanzees long after it starts to decline in women might indicate that humans stop childbearing earlier than chimpanzees do. If so, and if chimpanzees are more similar to our common ancestor, it would contradict the proposition of the grandmother hypothesis that women's fertility does not end earlier now than in our presapiens past. The other contradiction is an apparently sharp acceleration in the depletion rate of ovarian follicle stocks around the age of 38 in women, a pattern that has been interpreted as evidence of a programmed shift to menopause "20 years early."[15–18]

To compare human and chimpanzee fertility schedules we use 19th century records from the Utah Population Database (UPDB)[19] to represent natural fertility in humans; and for wild chimpanzees we rely on reports from Boesch and Boesch-Achermann[20] from one study site and Emery Thompson et al.[12] for a compilation of records from six more. We use a single population to represent the human pattern because the decline in age-specific fertility is very similar among natural fertility human populations, including women in the UPDB.[21] The Utah database has the great advantage for our inquiry of both large size necessary for assessing demographic parameters and of individual records that allow us to investigate associations between one's fertility rate and her age at last birth. Both humans and chimpanzees display wide individual variation in these features and in both species this heterogeneity in fertility is associated with variation in survival rates.[12,22–27] However, adult mortality is much lower in human populations than in chimpanzees.[28] We show that because of this difference in mortality (women usually outlive the childbearing ages and chimpanzees do not), heterogeneity has opposing effects on age-specific fertility rates (ASFRs) through the fourth and fifth decade—the 30s and 40s—in these two species.[29]

Then we turn to the available data on oocyte counts with age. All mammalian females develop a fixed stock of oocytes near the time of birth that is subsequently depleted almost entirely by atresia throughout juvenile and adult life.[16] The issue here is the biphasic model proposed to characterize depletion rates in women. In the most widely cited model, an initial exponential rate of loss persists to the late 30s and then accelerates to reduce stocks to menopause levels around the age of 50.[15,30,31] The sharp rate change in the late 30s is interpreted as possible evidence of a shift from later ages of menopause in the past.[15–18] More than 10 years ago, Leidy et al.[32] pointed out that a biphasic model that best fits all the human data has an inflexion point not at 38 but 10 years later, much closer to the average age at menopause. Their analysis justified comparison of chimpanzee and human follicular depletion rates with simple exponential models across the age range from birth to 47 years.[33] Both the similarity in rates of decline in humans and chimpanzees and the questionable inflexion point at 38 in humans are relevant here. Leidy et al.[32] and others, including Faddy and Gosden[17] who were themselves architects of the most widely cited biphasic model, pointed out that a single inflexion point at any age is biologically unrealistic. More recently Hansen et al.[34] have published new follicle counts and shown that a power function is a better fit for the decline with age in both the new cases and those previously analyzed. However as they note, the wide variation in counts for women of the same age makes their power model "inadequate for predicting the reproductive lifespan for an individual" (p. 706).[34] The individual variation in fertility rates and age at last delivery highlighted here underscore that caveat.

Our analyses can not demonstrate that fertility decline in modern women remains at ancestral ages. We nonetheless conclude that available data are best interpreted as evidence *against* the stopping

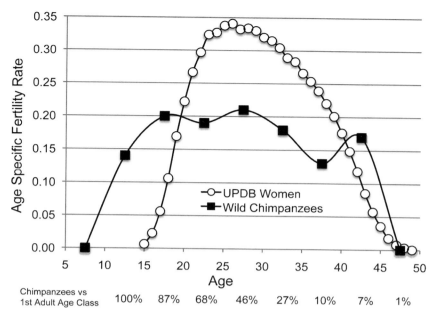

Figure 1. Natural fertility humans and wild chimpanzees compared. The humans are 42,493 UPDB parous women born 1845–1890 who were monogamously married, neither divorced nor widowed before 50, and lived to at least that age. Chimpanzees are 165 females from four study sites compiled by Emery Thompson et al.[12] There are 627.3 risk years in the 10–14 year chimpanzee interval, that dwindle to 7.8 in the 45–49 year interval (Ref. 12 and supplemental data Table S2). The relative number of risk years in each 5-year interval is represented by the percentage relative to the initial adult interval (10–14) below the horizontal axis.

early hypothesis. Instead they are consistent with a parsimonious inference of the grandmother hypothesis that greater longevity evolved in the human lineage while fertility declines with age changed little from our last common ancestor with other great apes.

Age-specific fertility

Records of ASFRs are far richer for humans than for any other primate. While fertility levels in our species vary widely, the changes in rate by age take characteristic shapes depending on whether fertility is natural or controlled.[21] Controlled fertility reflects practices of family limitation, and results in a concave decline in age-specific fertility as more women reach their desired family sizes. In natural fertility populations, family size limitation is absent and the decline is convex. These two age patterns of decline are strikingly constant across populations,[21] with the potential for continued child bearing reflected by actual births in natural fertility populations.

The mid-19th century settlers in Utah (UPDB[19]), comprising both Mormons and non-Mormons alike, had high fertility as is often found in colonizing populations[35,36] and the convex decline of natural fertility. Figure 1 shows the ASFRs of 42,493 parous, monogamously married UPDB women born between 1845 and 1890. To remove the very small influence of divorce and the complications of remarriage for this population, only women who married once are included. To observe completed fertility that is not curtailed by early death and to remove the effects of husband's death, this analysis also excludes women who died or were widowed at or before the age of 50.

Available samples are always much smaller for chimpanzees. ASFRs for wild chimpanzees from six study sites, based on 165 females[12] are also shown in Figure 1. The comparison illustrates some of the differences between humans and other great apes highlighted by the grandmother hypothesis: Humans have later ages at first birth, shorter birth intervals (higher fertility peaks), but similar ages at last birth compared to chimpanzees. In both species fertility approaches zero around the age of 45.

A closer look at the comparison also suggests a difference in the decline with age. Human ASFRs

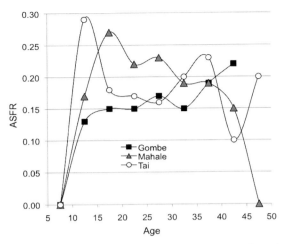

Figure 2. Chimpanzee ASFRs at three study sites. Gombe and Mahale data from Emery Thompson et al.[12]; Tai data from Boesch and Boesch-Achermann.[20]

have a peaked shape, beginning to decline around the age of 30 to reach less than a third of maximum in the early 40s. Chimpanzee ASFRs are flat instead, persisting at about the rate reached before the age of 20 for two more decades. The chimpanzee samples are very small and become miniscule at the older ages. The survival restriction imposed on the UPDB women in Figure 1—all survived at least to 50—can not be imposed on the chimpanzees. Along the horizontal axis of Figure 1 we indicate the relative sample size for each 5-year interval in Emery Thompson et al.'s[12] chimpanzee compilation as a percentage of the number of risk years in the first adult category. The 627.3 risk years for 10–14 year olds declines to 7.8 risk years in the 45–49 year interval, an index of the slim chance that a young adult female survives the childbearing years (Ref. 12 supplemental information table S2).

The small sample of chimpanzees, especially at older ages, raises questions about whether the plateau shape might be just an artifact of sample size. To explore that possibility we plot chimpanzee populations separately. Figure 2 shows ASFRs for three chimpanzee study sites. Gombe and Mahale (the two longest running East African sites) are the only populations included in Emery Thompson et al.[12] compilation that have more than one individual known to be over 40 years old. The Tai population in the Ivory Coast is not included in that synthesis, but "age-specific fertility does not diminish in Tai chimpanzees" (p. 62).[20]

The recurrence of the same trend in all these cases underscores the contrast with humans. As Coale and Demeny[37] reported in their classic compendium, "In all [human] populations where reliable records have been kept, fertility is zero until about age 15, rises smoothly to a single peak, and falls smoothly to zero by age 45–50" (p. 35).

Elsewhere[29] we elaborated a suggestion about the flat age-specific fertility in wild chimpanzees from Emery Thompson et al.[12] They cited correlations between later fertility and longer survival in humans and suggested that similar linked heterogeneity in chimpanzees would mean that especially fertile females are more likely to survive into their late 30s and 40s. If low fertility females die at younger ages, their selective removal increases average fertility rates at older ages—even if the fertilities of the survivors themselves are usually declining from their own rates when younger. We noted a similar bias recognized in assessments of variation in birth intervals with parity in humans.[35,36] When women of all parities are pooled, birth intervals can appear to be constant, or even decreasing across parities. This is because women that reach higher parities have shorter intervals, and high parity intervals can only come from those women. Following Emery Thompson et al.,[12] we hypothesized that if high fertility chimpanzee females are more likely to survive to older ages, the fertility rates at older intervals come disproportionately from those females because low fertility females have died.

Here we further explore the effects of the difference in adult mortality rates between humans and chimpanzees on ASFRs. Women vary widely in the ages at which they reach physiological thresholds associated with declining fertility.[38] Half the women in natural fertility populations have had their last delivery around 40, and reach menopause about 10 years later.[38–40] Consider the distribution around those thresholds. For the UPBD women whose age-specific fertility is reported above, Figure 3 plots the proportion of women who have already lived past their last parturition at each age across the childbearing years (filled circles). More than 10% have no more births after the age of 30, about 27% have no more after 35. ASFRs plotted on the same figure (open circles) show that the downward slope of the ASFR is driven by the increasing proportion of

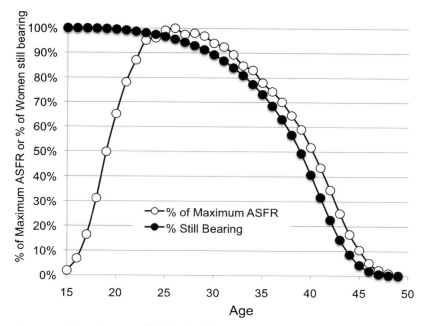

Figure 3. Decreasing proportions of women still giving birth drive down human ASFR. Percentage of women not yet past their last delivery and percentage of maximum age-specific fertility by age for 42,493 UPDB parous women born 1845–1890 who were monogamously married, neither divorced nor widowed before 50, and lived to at least that age.

women at each succeeding age who have no more births. This association has been noted in other human populations as well.[21,35] While the sloping decline in ASFRs is readily interpreted as similar declining fertility among all women, that inference is incorrect. Instead it is strongly dependent on the expanding fraction of women in these age classes who are no longer bearing offspring.

In chimpanzees by contrast, females rarely outlive their own fertility, and those with higher fertility rates live longer. Emery Thompson et al.[12] looked for associations between fertility rates and survival in females over the age of 25 by dividing their female observation years from the six-site sample into healthy and unhealthy years. An observation year for a given chimpanzee was considered healthy if she survived an additional 5 years or more, unhealthy if she did not. Their Figure 2 (ref. 12, p. 2152) shows that fertility was about twice as high in females who would survive at least 5 more years than in those who would not. Similar heterogeneity in which those with higher fertility also have later ages at last birth is suggested in the human pattern in Figure 3. At age 39, less than half the women are still delivering (49%), and yet age-specific fertility

is 59% of maximum. Women who continue to give birth are doing so almost 20% faster than the overall average at the age of maximum fertility.

To more directly parallel Emery Thompson's probe for an association between higher fertility rates and later parturitions in chimpanzees, we divided the Utah women according to their ages at last birth. As Figure 4 shows, the women who bore their last baby before 35 also had lower age-specific fertility in the preceding years than those who continued bearing past 45.

In Figure 5 we compare age-specific fertility at each age to the ASFR of just the women who would still bear offspring beyond that age (filled circles). Women who had last births at a given age (and their last babies) are excluded from this subset to avoid ascertainment bias. Including them would overestimate the fertility of the women continuing to give birth because those women, identified by parturitions at exactly that age, must have an ASFR of 1. The fertility schedule for the subset of women continuing to bear in future is flatter than the usual schedule that includes all the women in each age class (open circles). At the age of 30 the ASFR of those who will continue to have babies past that age is 6% higher

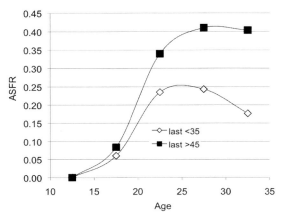

Figure 4. Women's fertility rates vary with their ages at last birth. ASFR for two subsets of the 42,493 UPDB parous women born 1845–1890 who were monogamously married, neither divorced nor widowed before 50, and lived to at least that age. Fertility rates before 35 of the women whose fertility ended by 35 ($n = 10,440$) are compared to the rates at the same ages of the women who were fertile past 45 ($n = 2659$).

than the standard schedule. By 35 the rate of those that will continue is 17% higher; by 40 it is 32% higher. Over the last 3 years for which this subset can be distinguished (ages 45–47), the rate averages 96% higher than the ASFR as usually calculated. Women that continue to breed have a flatter fertility schedule than the standard human schedule for the same reason the chimpanzee schedule is flat. In both species, females that continue bearing to older ages are the ones that produce at higher rates. In chimpanzees, individuals near the end of their fertility are excluded by mortality. Here we have excluded women past their own fertility by manipulation, representing only those with future fertility in the black circles in Figure 5. According to the grandmother hypothesis, those excluded survivors are the evolutionary legacy of ancestral grandmother effects.

Because chimpanzees, even under the most benign conditions, rarely survive beyond their own fertility, a manipulation to make chimpanzee ASFRs look more human can only be partial. However, captivity provides an opportunity to see what happens to chimpanzee ASFR when mortality is reduced. Higher survival than found in free-ranging populations[41] leaves more frail individuals alive at older ages. If, as Emery Thompson et al. have shown,[12] frailty affects both fertility and mortality, then more females with lower fertility levels and earlier fertility termination should survive to older ages among

captives than survive in the wild. Observations at Taronga Park Zoo accumulated since the mid-1960s are suggestive.[42] This population of chimpanzees experienced "conditions of near-optimal nutrition . . . [and] natural breeding conditions" (p. 282).[42] While infant mortality was similar to the wild, "the major contrast is the greater life expectancy for female adults at the zoo" (p. 294).[42] The Taronga Park ASFR begins to slope down before the age of 30, more like the usual human ASFR than like the flat schedule for chimpanzees in the wild. This peaked shape is consistent with the expectation of greater heterogeneity when more females survive into their late 30s and 40s. Also consistent with that expectation, pregnancy outcomes in captives show increasing failures with age. Roof et al.[43] examined 1255 pregnancies in 272 females from three Primate Research Centers and found a clear rise in spontaneous abortions and stillbirths with increasing maternal age, a result that parallels evidence of increasing fetal loss with increasing age in women.[35,44]

Follicle stocks

Fertility in all female mammals declines with age as oocyte stocks are depleted.[16] Most of the initial stock is lost to atresia. As stocks decline in women, numbers reach thresholds associated first with reduced fecundability, then secondary sterility, and finally menopause at different ages in different individuals.[17,45,46] The classic human records of ovarian follicle stocks show that among females of the same age, remaining primordial follicle pools can vary by two orders of magnitude.[34,47–50]

Recognizing the variation, analysts nevertheless initially characterized the general pattern of follicle depletion in women as "biphasic," with a sharp acceleration in the rate of loss before menopause. This invited speculation that the acceleration might indicate an evolutionary shift to earlier menopause.[15,16,49,50] Recently Cant and Johnstone[18] used that inference to support their hypothesis that mid-life menopause evolved in humans when reproductive competition between the generations pushed older ancestral females to terminate their fertility early.

The initial oocyte stock and rate of follicular attrition in human females are commensurate with a longer reproductive life span: specifically, an age at menopause of ∼70 years. . . . The onset of the accelerated phase of reproductive

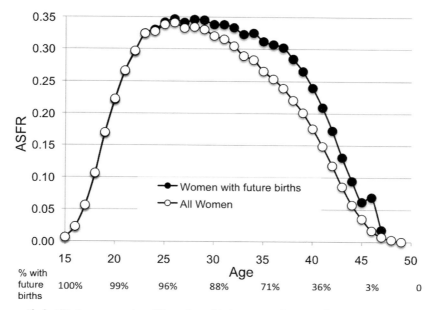

Figure 5. Age-specific fertility for women who will have future births compared to ASFR for all. ASFR for all 42,493 UPDB parous women born 1845–1890 who were monogamously married, neither divorced nor widowed before 50, and lived to at least that age (open circles), and ASFR for just the subset who will still bear offspring beyond each age (filled circles). Percentage of the 42,493 sample who will still deliver at a later age is indicated below the x-axis.

senescence that leads to menopause coincides with the age at which, in natural-fertility populations, human females can first expect to encounter reproductive competition from the next generation.... [This] early reproductive cessation reflects 'the ghost of reproductive competition past' (p. 5333–5334).[18]

The biphasic model[15] used by Cant and Johnstone[18] found the inflexion point at 37.5 years. As Leidy et al.[32] noted, when all the data are included in the sample, the inflexion point in a best-fit biphasic model moves from 37.5 to 48 years. They concluded that,

> The data emphatically do not support an abrupt change in the exponential rate of decay at age 37.5.... [But] defending a biphasic model with a critical age of 48 seems equally absurd.... The idea that a process of cellular degeneration that begins at birth, when there are approximately three quarters of a million follicles in the ovaries, and ... accelerates in women with 3,000 follicles just as these women approach menopause, seems clinically uninteresting if not biologically implausible (p. 857).[32]

Faddy and Gosden[17] themselves noted the biological improbability of a biphasic shift and tailored a subsequent model to the data in which "the step-change in the rate of follicle attrition was replaced by a model which assumed that this rate changes more gradually with the size of the follicle store" (p. 1484). Hansen et al.[34] have now shown the fit of a simpler power model. Figure 6 plots the classic follicle counts on original measurement scales, Panel A includes all the cases in the classic data sets, Panel B restricts attention to subjects age 30 and older (excluding one conspicuous outlier from Block's[47] counts visible in Panel A: 224,500 for a 37 year old). This display underscores the wide scatter in the counts across all ages to emphasize the improbability of a widely shared acceleration in the rate of loss at age 37.5.

Consider that each count represents one snapshot, of one infant, girl, or woman, along her own interrupted trajectory from a maximum initial stock of oocytes to exhaustion of that stock sometime after the age of menopause. The cross-sectional counts can not reveal individual fertility trajectories. However, they can still be useful for comparing humans to other species. Follicle counts from archived ovarian sections taken at necropsy from captive chimpanzees provide an index of the rate of decline in primordial follicles with age in that species.[33] The

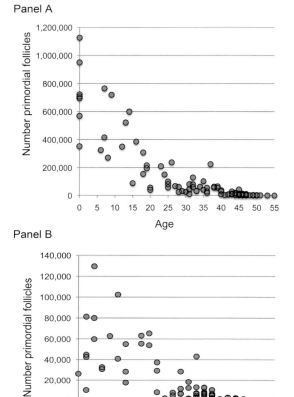

Figure 6. Follicle counts in humans by age. Counts are from Block,[47,48] Richardson et al.,[49] Gougeon et al.[50] Panel A includes all cases, Panel B includes all cases for subjects over the age of 30 except one conspicuous outlier from Block's[47] counts evident in Panel A: 224,500 for a 37 year old.

available sections span the ages of 0–47, a range that falls within the first phase of the Leidy et al.[32] observation that a single exponential rate to the age of 48 was a better fit to the human data than a biphasic shift at 37.5. On those grounds we fit single exponential regressions to the follicle counts across the 0–47 age range in both species. Measured this way, across this age range, the decrease in primordial follicle counts with age in the chimpanzee sample is indistinguishable from the rate of depletion with age in classic human data sets (Fig. 7).[33]

The intercepts—the heights—of the two regression lines are necessarily different because the chimpanzee counts are for single sections while the human counts are for whole ovaries. Variation in ovary size, section thickness, and estimation protocols make it impossible to specify precisely the fraction of an ovary represented in a section,[52] but the order of magnitude of the difference can be estimated. An average section represents about 1/2000 of an average human ovary.[47,49] If the single sections were from human ovaries they should differ from whole ovaries by about that much. The difference in heights of the two lines in Figure 7 thus indicates at least an order of magnitude similarity in human and chimpanzee stock sizes. That general similarity and the similar slopes of declining stocks with age are consistent with a wider body of findings, including hormone and cycling data from captivity[51,53–55] but see Ref. 56 and the fertility data from the wild discussed above. All suggest that chimpanzees would reach menopause at about the same ages humans do—if they lived long enough.

Discussion

Women's decline in fertility with age is conventionally represented by fertility rates for each age class, ratios of the number of births to women of that age divided by the number of women in that age class. As we have illustrated with the UPDB data, declines in ASFRs at the population-level are not the same as declines in the fertility of individual women. Human ASFRs fall smoothly from a peak near 30 to reach zero near 45 because increasing proportions of women in these age classes are no longer bearing offspring. Considering only the women who will still give birth in future intervals, the peaked pattern flattens, staying higher at subsequent ages than in the standard fertility schedule. While the birth intervals of the women still breeding increase with parity, their average fertility rate remains higher because lower fertility women leave the risk pool. This clarifies the difference in the shape of the fertility schedules of human and chimpanzee populations. The sloping decline after an earlier fertility peak in humans seems to suggest earlier termination of fertility in humans than in free-ranging chimpanzees who have much flatter ASFRs. However, when the women who have passed their last parturition are removed from subsequent age classes, human ASFRs start to approximate those of chimpanzees. The fact that women usually outlive their fertility makes the shape of the fertility schedule different, even if the fertility declines with age experienced by individuals in both species closely overlap.

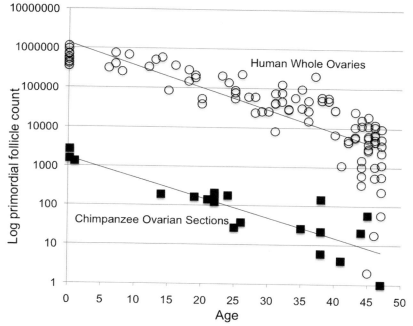

Figure 7. Follicle stock depletion from birth to age 47 in humans compared to chimpanzees. The human counts are for whole ovaries from the classic sources (open circles),[46–49] and the chimpanzee counts are for single ovarian sections taken at chimpanzee necropsies (closed squares),[33] displayed here on logarithmic scale. Heights of the lines must differ because a section is only a thin slice of an ovary. The human ovaries contained about 2000 sections per ovary[46,48] (see text). Rates of follicular depletion with age are indicated by the slopes of the lines. The slope and 95% confidence interval for whole human ovaries is −0.05594 (−0.06421053, −0.04767339). For the chimpanzee sample of ovarian sections, the slope and 95% confidence interval are −0.05079 (−0.06494935, −0.03662765). These slopes are statistically indistinguishable.[33]

Assumptions that ovarian follicular depletion rates show a distinctively sharp acceleration in women in their late 30s provide another line of evidence used to infer that fertility declines in humans are now earlier than they were in ancestral populations. Leidy et al.[32] noted that models identifying an inflexion point at 37.5 years have been well received in the medical literature as the timing corresponds with age-related increases detected in chromosomal abnormalities, fetal loss, and rising levels of pituitary FSH. In addition to the evolutionary questions of direct interest here, the influence of the biphasic model on those giving medical advice[30,31] is additional reason to highlight the lack of support for it.

Although we discussed, and even engaged in, fitting age-specific curves to the follicle counts ourselves, we nevertheless emphasize their limits. Just as the ratio of births per woman gives the fertility rate of an age interval, but not the fertility rate of the individual women in it, so the central tendency of follicle counts across subjects of the same age is not the same as the counts for individuals. Central tendencies can, however, provide an index for cross-species comparisons. Expanding the sample of counts for humans and especially chimpanzees whenever possible can provide direct evidence for (and against) hypotheses about a derived early fertility decline in humans. Chimpanzee samples currently available indicate similar distributions of follicle stock sizes and similar rates of follicular depletion with age in both species.[33]

We conclude that neither age-specific fertility trends nor rates of follicular depletion are consistent with the hypothesis that menopause is earlier in women than in our ancestors or our closest living relatives. Of course neither can prove that humans do not stop earlier than our ancestors did. However, for now, the burden of evidence indicates that declines in fertility with age do not differ much between humans and chimpanzees. Similar individual variation in declining fertility that is pruned out by mortality in chimpanzees is preserved by greater longevity in humans. This individual variation, especially important to women who are delaying pregnancy, is obscured by emphasis on average age at last birth or average age at menopause.[30,31,38,39] Perhaps paradoxically, a grandmother hypothesis to explain the evolution of human longevity contributes to exploring this heterogeneity in fertility, confirming the continuing utility of this working hypothesis.

Acknowledgments

This study was supported in part by National Science Foundation grant 0850951 (Chimpanzee Reproductive and Physiological Aging) and by National Institute of Aging Grant AG022095 (The Utah Study of Fertility, Longevity, and Aging). The content is solely our responsibility and does not necessarily represent the official views of the supporting agencies. We thank the Pedigree and Population Resource (funded by the Huntsman Cancer Foundation) for its role in the ongoing collection, maintenance and support of the Utah Population Database (UPDB). We also acknowledge Dr. Geraldine P. Mineau and Alison Fraser, MSPH, for their careful management of and assistance with the data used for this study. We are grateful to James Coxworth, Richard Paine, and Alan Rogers for analytical assistance and advice, and thank Jim Herndon and Sarah Hrdy for useful comments. This study in funded by NIH.

Conflicts of interest

The authors declare no conflicts of interest.

References

1. Williams, G.C. 1957. Pleiotropy, natural selection, and the evolution of senescence. *Evolution* **11:** 398–411.
2. Robson, S.L., C.P. van Schaik & K. Hawkes. 2006. The derived features of human life history. In *The Evolution of Human Life History*. K. Hawkes & R.R. Paine, Eds: 17–45. School of American Research Press. Santa Fe.
3. Oeppen, J. & J. Vaupel. 2002. Broken limits to life expectancy? *Science* **296:** 1029–1031.
4. Howell, N. 1979. *Demography of the Dobe !Kung*. Academic Press. New York.
5. Hill, K. & A.M. Hurtado. 1996. *Ache Life History: the Ecology and Demography of a Foraging People*. Aldine de Gruyter. New York.
6. Blurton Jones, N.G., K. Hawkes & J.F. O'Connell. 2002. Antiquity of postreproductive life: are there modern impacts on hunter-gatherer postreproductive life spans? *Am. J. Hum. Biol.* **14:** 184–205.
7. Hawkes, K. & N.G. Blurton Jones. 2005. Human age structures, paleodemography, and the grandmother hypothesis. In *Grandmotherhood: the Evolutionary Significance of the Second Half of Life*. E. Voland, A. Chasiotis & W. Schiefenhovel, Eds.:118–140. Rutgers University Press. New Brunswick.
8. Hawkes, K. et al. 1998. Grandmothering, menopause, and the evolution of human life histories. *Proc. Natl. Acad. Sci. USA* **95:** 1336–1339.
9. Hawkes, K. 2003. Grandmothers and the evolution of human longevity. *Am. J. Hum. Biol.* **15:** 380–400.
10. Hawkes, K., J.F. O'Connell & N.G. Blurton Jones. 1997. Hadza women's time allocation, offspring provisioning and the evolution of post-menopausal lifespans. *Curr. Anthropol.* **38:** 551–577.
11. O'Connell, J.F., K. Hawkes & N.G. Blurton Jones. 1999. Grandmothering and the evolution of Homo erectus. *J. Hum. Evol.* **36:** 461–485.
12. Emery Thompson, M. et al. 2007. Aging and fertility in wild chimpanzees provide insights into the evolution of menopause. *Curr. Biol.* **17:** 2150–2156.
13. Smith, B.H. & R.L. Tompkins. 1995. Toward a life history of the Hominidae. *Ann. Rev. Anthropol.* **24:** 257–279.
14. Robson, S.L. & B. Wood. 2008. Hominin life history: reconstruction and evolution. *J. Anat.* **212:** 394–425.
15. Faddy, M.J. et al. 1992. Accelerated disappearance of ovarian follicles in mid-life: implications for forecasting menopause. *Hum. Reprod.* **7:** 1342–1346.
16. vom Saal, F.S., C.E. Finch & J.F. Nelson. 1994. Natural history and mechanisms of reproductive aging in humans, laboratory rodents, and other selected vertebrates. In *The Physiology of Reproduction*, 2nd ed. E. Knobil & J.D. Neill, Eds.: 1213–1314. Raven Press. New York.
17. Faddy, M.J. & R.G. Gosden. 1996. A model conforming the decline in follicle numbers to the age of menopause in women. *Hum. Reprod.* **11:** 1484–1486.
18. Cant, M.A. & R.A. Johnstone. 2008. Reproductive conflict and the separation of reproductive generations in humans. *Proc. Natl. Acad. Sci. USA* **105:** 5332–5336.
19. Bean, L.L., G.P. Mineau & D.L. Anderton. 1990. *Fertility Change on the American Frontier, Adaptation and Innovation*. University of California Press. Berkeley.
20. Boesch, C. & H. Boesch-Achermann. 2000. *The Chimpanzees of the Tai Forest: Behavioural Ecology and Evolution*. Oxford University Press. Oxford.
21. Wood, J.W. 1989. Fecundity and natural fertility in humans. *Oxf. Rev. Reprod. Biol.* **11:** 61–109.
22. Perls, T.T., L. Alpert & R.C. Fretts. 1997. Middle aged mothers live longer. *Nature (London)* **389:** 133.
23. Müller, H.-G., J.-M. Chiou, J.R. Carey & J.-L. Wang. 2002. Fertility and life span: late children enhance female longevity. *J. Gerontol. A Biol. Sci. Med. Sci.* **57:** B202–B206.
24. Smith, K.R., G.P. Mineau & L.L. Bean. 2002. Fertility and post-reproductive longevity. *Soc. Biol.* **49:** 185–205.
25. Jacobsen, B.K., I. Heuch & G. Kvale. 2003. Age at natural menopause and all cause mortality: a 37-year follow-up of 19,731 Norwegian women. *Am. J. Epidemiol.* **157:** 923–929.

26. Gagnon, A. et al. 2009. Is there a trade-off between fertility and longevity? A comparative study of women from three large historical databases accounting for mortality selection. *Am. J. Hum. Biol.* **21:** 533–540.
27. Smith, K.R. et al. 2009. Familial aggregation of survival and late female reproduction. *J. Gerontol. Biol. Sci. Med. Sci.* **64:** 740–744.
28. Hill, K. et al. 2001. Mortality rates among wild chimpanzees. *J. Hum. Evol.* **39:** 1–14.
29. Hawkes, K., K.R. Smith & S.L. Robson. 2009. Mortality and fertility rates in humans and chimpanzees: how within-species variation complicates cross-species comparisons. *Am. J. Hum. Biol.* **21:** 578–586.
30. Al-Azzawi, F. 2001. The menopause and its treatment in perspective. *Postgrad. Med. J.* **77:** 292–304.
31. Lobo, R. 2005. Potential options for preservation of fertility in women. *N. Engl. J. Med.* **353:** 64–73.
32. Leidy, L.E., L.R. Godfrey & M.R. Sutherland. 1998. Is follicular atresia biphasic? *Fertil. Steril.* **70:** 851–859.
33. Jones, K.P. et al. 2007. Depletion of ovarian follicles with age in chimpanzees: similarities to humans. *Biol. Reprod.* **77:** 247–251.
34. Hansen, K.R. et al. 2008. A new model of reproductive aging: the decline in ovarian non-growing follicle number from birth to menopause. *Hum. Reprod.* **23:** 699–708.
35. Wood, J.W. 1994. *Dynamics of Human Reproduction: Biology, Biometry, Demography.* Aldine de Gruyter. New York.
36. Mineau, G.P., L.L. Bean & M. Skolnick. 1979. Mormon demographic history II: the family life cycle and natural fertility. *Pop. Studies* **33:** 429–446.
37. Coale, A.J. & P. Demeny. 1983. *Regional Model Life Tables and Stable Populations*, 2nd ed. Princeton University Press. Princeton.
38. Te Velde, E.R. & P.L. Pearson. 2002. The variability of female reproductive ageing. *Hum. Reprod. Update* **8:** 141–154.
39. Broekmans, F.J. et al. 2007. Female reproductive ageing: current knowledge and future trends. *Trends Endocrinol. Metab.* **18:** 58–65.
40. Bongaarts, J. & R.G. Potter. 1983. *Fertility, Biology, and Behavior: An Analysis of the Proximate Determinants.* Academic Press. New York.
41. Dyke, B. et al. 1995. Model life table for captive chimpanzees. *Am. J. Primatol.* **37:** 25–37.
42. Littleton, J. 2005. Fifty years of chimpanzee demography at Taronga Park Zoo. *Am. J. Primatol.* **67:** 281–298.
43. Roof, K.A. et al. 2005. Maternal age, parity, and reproductive outcome in captive chimpanzees (Pan troglodytes). *Am. J. Primatol.* **67:** 199–207.
44. Holman, D.J. & J.W. Wood. 2001. Pregnancy loss and fecundability in women. In *Reproductive Ecology and Human Evolution.* P.T. Ellison, Ed.: 15–38. Aldine de Gruyter. New York.
45. O'Connor, K.A., D.J. Holman & J.W. Wood. 2001. Menstrual cycle variability and the perimenopause. *Am. J. Hum. Biol.* **13:** 465–478.
46. Sievert, L.L. 2006. *Menopause: a Biocultural Perspective.* Rutgers University Press. New Brunswick.
47. Block, E. 1952. Quantitative morphological investigations of the follicular system in women: variations at different ages. *Acta Anat.* **14:** 108–123.
48. Block, E. 1953. A quantitative morphological investigation of the follicular system in newborn female infants. *Acta Anat.* **17:** 201–206.
49. Richardson, S.J., V. Senikas & J.F. Nelson. 1987. Follicular depletion during the menopausal transition: evidence for accelerated loss and ultimate exhaustion. *J. Clin. Endocrinol. Metab.* **65:** 1231–1237.
50. Gougeon, A., R. Ecochard & J. Thalabard. 1994. Age-related changes of the population of human ovarian follicles: increase in the disappearance rate of non-growing and early-growing follicles in aging women. *Biol. Reprod.* **50:** 653–663.
51. Graham, C.E. 1979. Reproductive function in aged female chimpanzees. *Am. J. Phys. Anthropol.* **50:** 291–300.
52. Tilly, J.L. 2003. Ovarian follicle counts—not as simple as 1,2,3. *Reprod. Biol. Endocrinol.* **1:** 11.
53. Gould, K.G., M. Flint & C.E. Graham. 1981. Chimpanzee reproductive senescence: a possible model for evolution of the menopause. *Maturitas* **3:** 157–166.
54. Lacreuse, A. et al. 2008. Menstrual cycles continue into advanced old age in the common chimpanzee (Pan troglodytes). *Biol. Reprod.* **79:** 407–412.
55. Walker, M.L. & J.G. Herndon. 2008. Menopause in nonhuman primates? *Biol. Reprod.* **79:** 398–406.
56. Videan, E.N., J. Fritz, C.B. Heward & J. Murphy. 2006. The effects of aging on hormone and reproductive cycles in female chimpanzees (Pan troglodytes). *Comp. Med.* **56:** 275–283.

ANNALS OF THE NEW YORK ACADEMY OF SCIENCES
Issue: *Reproductive Aging*

An evolutionary and life history perspective on human male reproductive senescence

Richard G. Bribiescas

Department of Anthropology, Yale University, New Haven, Connecticut

Address for correspondence: Richard G. Bribiescas, Yale University, Department of Anthropology, New Haven, Connecticut 06511. richard.bribiescas@yale.edu

Unlike menopause, male reproductive senescence does not involve an acute drop in fertility. Men do, however, manifest distinct changes in somatic and gonadal function with age. Moreover, population variation in male reproductive senescence reveals phenotypic plasticity resulting from environmental, lifestyle, and genetic factors. An evolutionary and life history perspective is vital for understanding male reproductive senescence because aging involves biological constraint as well as adjustments to reproductive strategies and the allocation of somatic resources. An awareness of life history–related tradeoffs between energetic and time constraints is especially useful because biological aspects of male senescence are products of environmental challenges and natural selection. This article reviews the adaptive significance of the evolutionary biology of human male senescence with particular attention to population variation. An evolutionary perspective cannot only shed light on the origins and biology of human male senescence but also provide insights into contemporary issues of male aging and health.

Keywords: men; evolution; testosterone; hormones; health; aging

Introduction

Male reproductive senescence is largely understood through the lens of clinical medicine, yielding valuable insights into the basic biology of male reproductive function and aging. But researchers have not traditionally viewed male reproductive senescence as a product of evolution by natural selection nor has this area of research, with few exceptions, received attention from evolutionary human biologists.[1] Consequently, our comprehension of the biology of male reproductive senescence in regards to within and between species variation or how this variation potentially affects fitness and health has been limited.

A more thorough grasp of human diversity and the interaction between human male biology and ecological contexts is needed. First, a brief introduction into the basic aspects of evolutionary and life history theory is presented. Subsequent discussions of the biology of male reproductive senescence are outlined including our contemporary understanding of human and comparative variation of key biological traits. Finally, health issues common to the aging human male are discussed.

Evolution by natural selection requires a trait to exhibit variation, heritability, and differential reproductive success between individuals. These tenets are behind the evolution of virtually all biological processes found within human males. Therefore, traits that are central to male reproductive senescence, such as testicular and hormonal function, are products of natural selection. The biology of male senescence has not and does not occur in a vacuum. It is the result of millions of years of change and stasis in response to environmental challenges. Moreover, traits central to male reproductive senescence exhibit a range of phenotypic plasticity that responds to environmental stochasticity.

The concept of "life history traits" will serve as important touchstones for understanding male senescence from the perspective of an evolutionary biologist. Briefly, life history theory is a perspective that aims to understand evolution through the organization and timing of key events, such as age of reproductive maturation, age of cessation of growth, and age of death, just to name a few.[2] Important life history traits include size at birth, rate of growth until maturation, adult size, as well as number, size, and sex of offspring. The organization and scheduling of these life history traits and events are affected by the harvesting, usage, and internal allocation of energetic resources.[3] Organisms that allocate resources in an efficient manner should have a selective advantage resulting in the proliferation of traits that efficiently manage energetic resources and time allocation.[4]

Under circumstances of resource constraint, tradeoffs are expected between physiological needs. For example, a common male tradeoff is investment in resources that benefit survivorship versus reproductive effort. In many vertebrates, this is often observed through the positive effects of testosterone on mating behaviors while compromising survivorship.[5] As males age, senescence causes phenotypic and genotypic degradation, as well as reducing plasticity, thereby constraining males' responses to environmental challenges.

Life history evolution of the human male

The evolution of male life histories has covaried with the evolution of general human life history traits. Humans are distinguished from most other mammals and primates by having long lives, short interbirth intervals compared to other great apes, large altricial offspring, an extended period of juvenile growth and development, and high degrees of parental care. The evolution of these life history traits is contingent on several aspects of human evolution including a significant amount of paternal care, provisioning by males, and lower extrinsic mortality. Otherwise, sustaining a high level of fertility, offspring size, and slow growth would be unlikely.[6]

The costs of human male reproductive function are fundamentally different compared to females. Males do not expend significant amounts of metabolic energy toward reproduction. Unlike females who invest considerable metabolic energy toward menstruation, gestation, and lactation, males only invest in the marginal metabolic needs of spermatogenesis to conceive an offspring. Mammalian male reproductive success is also largely constrained by access to females whereas female fitness is more readily limited by access to energetic resources. This does not imply that female fitness can not benefit from multiple mating partners in order to decrease the risk of infanticide or procure parental care or provisioning. However, females can not increase their reproductive output by garnering more mates.

Consequently, the much broader range of fitness variation in males compared to females has resulted in natural selection acting strongly on male traits that allow them to more effectively compete with other males and attract females. Human males therefore invest a significant amount of energy in sexually dimorphic somatic tissue that is comparable to energy expended by female reproductive function over the course of a lifetime. Key and Ross have shown that in nonhuman primates, when body size differences between males and females exceeds 60%, the difference in metabolic costs associated to sexual dimorphism over the lifetime of a male is roughly equivalent to the reproductive costs of females.[7] Given the relatively low amount of metabolic investment in direct reproduction (i.e., spermatogenesis), selection has operated on physiological mechanisms that monitor and regulate the metabolic costs of somatic investment that affects male fitness (Fig. 1).[3,8,9] In other words, somatic commitment in sexually dimorphic tissue in human males is investment in reproductive effort and diminishment of that somatic allocation with age is an integral part of male reproductive senescence, perhaps even more so than age-associated changes in spermatogenesis.[1]

Male reproductive senescence can be understood from four perspectives that incorporate proximate biological factors and demographic outcomes. First, age-related changes in direct reproductive function, such as spermatogenesis and the ability to fertilize an egg and produce viable offspring, are obviously important. Second, age-related changes in somatic investment provide insights into how aging constrains male somatic phenotypic plasticity in response to energetic challenges and reproductive possibilities. Third, age-related changes in fertility as indicated by demographic data illustrates the outcomes of physiological constraints and adjustments with age.[10] Male fertility variation between populations also aids in our understanding of how male reproduction can be maintained at older ages, perhaps providing insights into the evolution of general human senescence and postreproductive life in

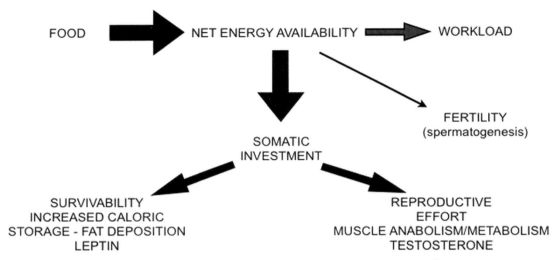

Figure 1. A model of central life history tradeoffs and their hormonal constituents in human males.[8]

females.[10] Finally, male reproductive behavior should adjust in response to the constraints and challenges of aging. This paper will focus on the first three aspects of male aging and biodemography. For representative discussion of male behavioral strategies and aging see Ref. 11.

Sperm quality, conception, and aging

Fathering offspring relies on producing viable and motile sperm, procuring a potential mate, and completing intercourse to ejaculation. Older men do exhibit signs of compromised fertility. In couples undergoing *in vitro* fertilization in which female fertility was established, male age was positively associated with risk of conception failure.[12] Older fathers also have a greater risk of producing genetically damaged sperm, spontaneous abortions, preterm births, low birth weight, and children with genetic defects.[13–15] Risk of Down's syndrome also increases when both parents are over 35 compared to when only the mother is older.[16] The ability to complete sexual intercourse is often reduced in older men, as is evident with greater incidences of erectile dysfunction (ED) with age.[17] Although ED is widely recognized in industrialized societies, it is common across societies.[18]

Female choice is an integral component of mate access. In a cross-cultural study of preferred somatotypes that varied on an endomorphic–ectomorphic scale, university females in Great Britain and Sri Lanka ranked somewhat muscular and hirsute somatotypes to be most attractive. Less muscular and more endomorphic physiques were ranked the least desirable.[19] Decreases in testosterone and increases in estradiol (E2) promote greater adiposity and muscle mass loss, cues that are likely to diminish a male's attractiveness. Age-associated increases in adiposity may also serve as a cue for compromised fertility.[20]

Sex hormone variation

Hormones act as a liaison between the environment, genetic expression, and phenotypic plasticity.[21] Variation in energy availability, resulting from changes in food availability or activity, results in changes in hormone levels, such as insulin, luteinizing hormone (LH), testosterone, E2, cortisol, leptin, and growth factors, all of which incur complementary or competing effects on growth, immune function, and reproduction. Hormones are key biological components that modulate tradeoffs within organisms, including human males.[22,23]

Testosterone is a central hormone governing male reproductive function and somatic investment in sexually dimorphic tissue. Testosterone in well-fed, sedentary men, including all age classes is higher compared to nonwestern males.[8,24–26] Given the high costs of skeletal muscle tissue (approximate 20% of male basal metabolic rate [BMR]) and the augmenting effects of testosterone on muscle anabolism and metabolism,[27,28] lower testosterone in nonwestern men likely reflects an adaptive response to attenuate the metabolic costs of sexually dimorphic tissue in the face of energetic deficits resulting

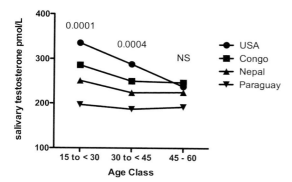

Figure 2. Population variation in salivary testosterone levels across age classes. Values are derived from the same laboratory using identical assay protocols. Although testosterone declines with age are common in industrialized societies, significant variation exists with other non-Western populations exhibiting little to no change with age. Most testosterone variation is accounted for at younger ages with populations converging after the age of 50. P-values are from one way ANOVA. NS = not significant. Drawn from data in Ellison et al.[33]

from lower caloric intake, high activity levels, and immunological challenges.[3,8,9,29]

In well-fed sedentary men, testosterone tends to decline gradually with age, especially after the age of 40.[30] Testosterone is further lowered in the presence of smoking, loss of spouse, body composition, and illness, factors that account for a greater proportion of between individual variation than age.[31] The overall effects include loss of lean body mass, increased adiposity, and possible lower libido and overall sense of well-being.[32]

However, age-related changes in testosterone are not universal with other nonwestern populations showing a modest or no decline. An analysis of testosterone associations with age in four populations revealed the greatest diminishment in American men, less so in African and Nepalese men, and no change in Ache Amerindians of Paraguay.[33] Most of the variation in testosterone levels was found at younger ages (Fig. 2). Therefore, populations with lower testosterone levels in their 20s and 30s exhibit little to no change in testosterone with age.

More subtle differences are evident in other industrialized populations. Among healthy urban Japanese men, salivary testosterone levels peak during the second decade of life, decline by the age of 40 but then remain stable well past 60.[34] Testosterone level stability with age is likely to be promoted by remaining free of chronic illness, social support, diet, and activity.[31] Circadian variation in testosterone also changes with age. Younger men tend to exhibit higher testosterone in the morning and lower levels in the evening. Older men lose this variation,[35] although forager men exhibit more modest age-related differences in circadian variation.[36]

The implications for this population variation in testosterone changes with age are that the high energy availability contexts in Western industrialized societies allow men to increase testosterone levels during early adulthood to support the metabolic needs of greater anabolism and muscle mass. As men age, the capacity to support the metabolic costs of sexually dimorphic tissue become compromised by overall somatic senescence.

E2 is also an important component of male reproductive senescence given the negative feedback effects on the hypothalamus.[37] Changes in E2 with age have been reported, however results have been inconsistent.[38] Increases in fat mass with age may promote greater adiposity through testosterone aromatization.[39] Nonetheless, no association between age and salivary E2 levels was observed among Ache men, although sample sizes were small.[40] Urinary estrone-3-glucuronide (E-3-G) levels and body mass index or fat percentage were unrelated among well-fed and undernourished Turkana men of Kenya. E-3-G levels were positively associated with age in the settled well-fed population.[41]

Follicle stimulating hormone (FSH) and LH are gonadotropins produced by the pituitary under control of gonadotropin-releasing hormone (GnRH) from the hypothalamus. FSH is involved with spermatogenesis whereas LH stimulates testosterone production.[42] Gonadotropins tend to increase with age, presumably due to the lower negative feedback effect of decreases in sex steroid production.[43] FSH and LH increases with age may be relatively stable across populations. For example, a cross sectional investigation of Ache males in Paraguay, ranging in age from 20 to over 60 revealed a significant positive association between age, serum FSH ($r = 0.75$, $P < 0.0005$), and LH ($r = 0.65$, $P < 0.01$) levels, similar to what is observed in Western industrialized males.[40]

In contrast, urinary FSH and LH levels among Turkana men of Kenya showed increases in well-fed sedentary men but not undernourished nomads.[41] Moreover, average LH and FSH levels were significantly higher among well-fed settled versus undernourished nomadic Turkana men. However, caution is warranted when interpreting urinary gonadotropin levels in response to undernutrition because urinary levels of LH and FSH increase significantly without concurrent changes in serum levels in response to fasting in men, suggesting altered renal clearance that is independent of pituitary production.[44]

Somatic investment and phenotypic plasticity
Because of greater sedentism and caloric intake, older men living in industrialized societies tend to accumulate more fat compared to those in developing regions. The rapidly growing incidence of obesity, coupled with aging, can compromise male reproductive function. Higher E2 levels, lower testosterone, and a greater risk of subfertility is evident in older obese men.[45] The effects of adiposity on testosterone, estrogen, and overall male reproductive physiology merits continued attention.

Concurrent with lifestyle, changes in body composition result from declines in testosterone levels.[46] Interestingly, older men exhibit a positive anabolic response to testosterone supplementation that is comparable to younger men, suggesting that receptor integrity and function remains intact at older ages.[47] The decline in testosterone levels and subsequent changes in body composition further suggest a loss of phenotypic plasticity and the inability to modulate muscle and fat stores. Indeed, Feinberg has suggested that the diminishment of phenotypic plasticity is an important underlying source of epigenetic disease and aging.[48] Seeing as the bulk of male reproductive effort is expressed as sexually dimorphic tissue,[3,7,9] the capacity to manage metabolic demands is attenuated with age.

Older men also lose the ability to rally energetic resources toward immunological challenges and otherwise repair damages incurred by day-to-day living.[49] Similarly, the ability to build and retain lean body mass and mobilize fat is compromised in older men.[50] A central aspect of somatic plasticity is BMR. Variation in BMR is largely dependent on lean body mass.[28] Higher BMR indicates the capacity to meet the costs of lean body mass and mobilize fat stores

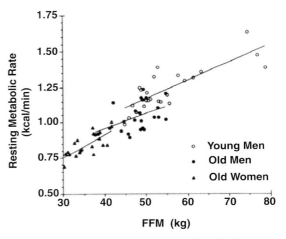

Figure 3. Associations between BMR and lean body mass in aging men and women illustrate lower fat free mass (FFM) and increased adiposity in older men. Moreover, BMR is lower in older (75 ± 1.0 SE) compared to younger men (21 ± 1.0) after controlling for FFM.[51]

when they are needed. However with age, BMR decreases in both men and women. The overall slope of the relationship between age and BMR is greater in men but nonetheless displays a significant decline with age even after controlling for lean body mass (Fig. 3).[51]

This apparent loss of plasticity may indicate an important aspect of senescence of the hypothalamic–pituitary–testicular (HPT) axis. Hypothalamic responses to acute fasting results in muted LH pulsatility in response to GnRH administration in older men (mean age 67) compared to the more robust LH responses of younger men (mean age 28) (Fig. 4),[52] implying that as males age, the HPT axis becomes less responsive to changes in energetic availability. The consequences are that somatic adjustment of lean muscle mass and fat tissue in response to variation in energy availability becomes less efficient, a compelling example of compromised phenotypic plasticity.

Male evolutionary demography
Demography illustrates the fitness outputs of the proximate mechanisms affecting lifetime reproductive effort. Coupling demographic and biological data provides a more complete image of male senescence and life histories. For example, males appear to senesce at a faster rate compared to females.[53] This may result from higher metabolic rates, investment in reproductive effort, and the

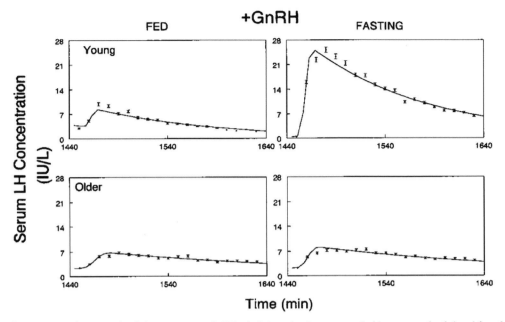

Figure 4. Luteinizing hormone levels in response to GnRH administration in young and older men under fed and fasted states. Older men fail to exhibit an expected increase in LH in response to GnRH administration during a fasted state,[52] illustrating that older males have a compromised ability to adjust HPT axis in response to energetic cues.

immunosuppressive effects of male sex hormones, just to name a few factors.[54] For an alternative perspective arguing for faster female senescence, see Graves et al.[55]

Gamete production in men is continuous throughout life. Therefore the potential for reproduction is ever-present even at older ages. Nonetheless, male fertility is not constant with age. Male fertility in industrialized populations, such as Canada, as indicated by the ratio of male age-specific fertility (ASF) and total fertility (TF), rises just before the age of 20 and peaks in dramatic fashion during the early 20s with a precipitous decline occurring thereafter. Moreover, male fertility (solid line) in industrialized populations like Canada mirror overall fertility (dashed line), with only a modest shift toward older men (Fig. 5). However, this pattern of male fertility may not have been common throughout human evolution. A comparative demographic examination of male fertility between modern industrialized societies, such as Canada, and nonindustrialized populations that experience ecological challenges that were common during human evolution, such as food limitations and high levels of activity, suggest that significant male fertility persisted beyond the age of menopause (shaded area of curves) (Fig. 5).

Tuljapurkar et al. argue that significant male fertility at older ages has important implications for the evolution of human senescence.[10] Traditional one-sex models of fertility that lead to the classic "wall of death" around the age of 50, not coincidentally matching the timing of menopause, do not accurately reflect the uniqueness of human life histories and extensive postreproductive life. That is, only through the incorporation of males into demographic models of fertility can an accurate model of human senescence be generated.[10]

Nonwestern male fertility tends to be more evenly distributed across decades, with a significant portion of fertility occurring after females' last reproduction, illustrating the commonness for men in these societies to reproduce through access to younger premenopausal women.[10] These conditions may have contributed to the extensive postreproductive life in humans. That is, the ability of males to procure mating opportunities with fertile females at older ages resulting in significant fertility later in life, selected against detrimental genes that contribute to age-related degradation.

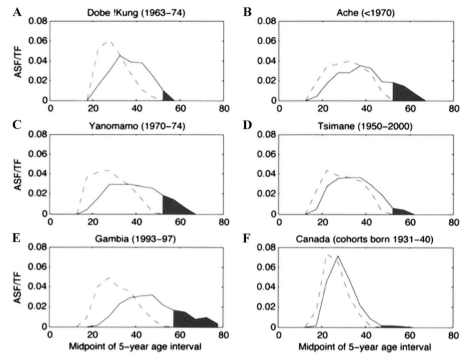

Figure 5. Observed distributions of female and male fertility. Fertility distributions (in ASF rates as a fraction of TF rate) for women (*dashed*) and men (*solid*) for (A) the hunter-gatherer Dobe !Kung of Botswana, (B) the forest-living Ache, (C) the Amazonian forager-horticulturalist Yanomamo, (D) Bolivian forager-horticulturalists the Tsimane, (E) agricultural Gambian villagers, and (F) modern Canada. The shaded area represents realized male fertility after the age of last female reproduction.[10] Non-Western men exhibit significant fertility after the age of last female reproduction, suggesting low male fertility after the age of 40 is restricted to modern industrialized populations and that reproduction by older men was more common during human evolution. Modified from Tuljapurkar et al.[10]

Consequently, males and their daughters would inherit these longevity genes resulting in extended human lifespans.[10]

But if male somatic investment and integrity declines with age, how can males procure mating opportunities with younger females? Perhaps a shift from strategies based on somatic investment to one of skill, experience, and social status provided older men with an advantage. Among Ache men, physical condition peaks during the early part of the second decade of life,[56] whereas foraging efficiency is greatest during the late 30s to early 40s.[57] Hunting ability for men is important because successful hunts promote family health, social status, and access to potential mates.[58] Perhaps human evolution was marked by a shift from a male reliance on physical strength, a characteristic that is evident in the gracile morphology of modern Homo sapiens compared to ancestral hominids, such as Homo erectus,[59] to skill and experience, a considerable benefit for older men.

Evolutionary medicine and reproductive health

Prostate cancer is the second leading cause of cancer death in American men, a risk that is especially high at older ages.[60] Rates vary widely between populations with higher incidences in more developed regions, however they are rising in less developed countries in association with development and lifestyle changes, such as higher caloric diets and sedentism.[61] Population variation intimates that lifestyle changes and environmental variability alter male physiology in a manner that increases prostate cancer risk. Moreover, the propensity of men, regardless of population,[62] to exhibit prostate disease after the age of 50 promotes the idea that lifetime, cumulative effects of endogenous androgen exposure may contribute to risk of the disease.

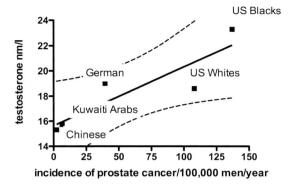

Figure 6. Linear regression of serum testosterone levels and prostate cancer risk in five populations (Kuwaiti Arab, German, US White, US Black, and Chinese) ($r^2 = 0.79$, $P = 0.04$). Data from Nigerian population excluded because of extreme variation in reported rates of prostate cancer. The authors state that the upper range may be "probably an over estimation" (p. 42). Data drawn from Table 4.[68]

The cumulative damaging effects of androgens on prostate health may reflect antagonistic pleiotropy. Traits that have a beneficial effect on fitness early in life may have a detrimental consequence on survivorship at later postreproductive ages when selection is weaker.[63] For example, testosterone supplemented dark-eyed juncos had higher lifetime extra-pair fitness but lower survivorship.[64] In human males, lower testosterone in association with infectious disease rebounds in response to illness treatment.[29,65] In the case of prostate cancer and benign prostatic hyperplasia, lifetime exposure to endogenous androgens may result in the potential for greater investment in reproduction but cause a greater propensity for prostate disease later in life.[66]

It is clear that testosterone and dihydrotestosterone promote prostate cancer cell growth and are inhibited during androgen deprivation therapy. Although testosterone variation between individuals or at any given age is not predictive of prostate cancer risk,[67] population variation in testosterone accounts for much of the variation in lifetime exposure to androgens and possibly greater susceptibility to the disease. An analysis of data compiled by Kehinde *et al.* reveals a positive association between population testosterone levels and the incidence of prostate cancer (Fig. 6),[68] similar to associations between population means in salivary progesterone and breast cancer risk.[69]

Population variation in androgen levels in association with age,[33] as well as prostate disease may involve lifestyle differences that are reflective of energetic status. GnRH, LH, and testosterone production respond to greater glucose and insulin levels.[70,71] In line with hypothalamic glucose and insulin sensitivity, population variation in prostate cancer risk is positively associated with sugar (glucose) intake (Fig. 7).[72] Among Ariaal men of Kenya, moderate to severe lower urinary tract score (LUTS) was positively related to abdominal and trunk fat. However, overall LUTS was higher than many Western populations despite poor nutritional status. Campbell suggests cultural variation in complaining and interpretation of the study questions may have affected the final results.[62]

Men who reap the potential benefits of high nutritional status and higher testosterone levels during early adulthood, such as the potential for greater anabolic activity and investment in sexually dimorphic body mass, are probably more likely to manifest prostate cancer later in life, a compelling potential example of antagonistic pleiotropy.[23] The take home message is that life history tradeoffs pervade many aspects of medical treatment strategies as well as perceptions of what is considered healthy, such as higher testosterone levels.[73] Greater awareness of these tradeoffs would aid in predicting potential detrimental effects in later life, especially in regards to hormone supplement treatment.

Conclusions and future directions

Many gaps remain in our understanding of the evolutionary biology of human male reproductive senescence. However, two important research needs can be readily addressed using available technologies and resources. First, a more detailed and comprehensive assessment of human variation is needed to determine which male reproductive traits are subject to phenotypic plasticity and which are more static. Our present understanding of human male reproductive function is largely derived from well-fed, sedentary men. Although this constituency is the primary target of the American medical community, they do not represent the majority of the human species. In addition, increased globalization has resulted in greater mobility and the placement of new immigrants in novel ecological settings. A greater awareness of this variation would not only provide a better understanding of the etiology of

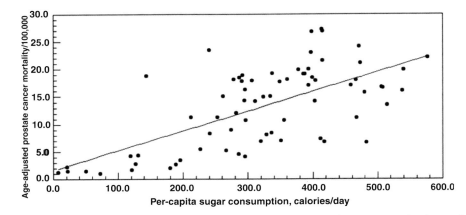

Figure 7. Positive association between sugar consumption and prostate cancer risk in 71 countries (Pearson's $R = 0.71$, $P = 0.0001$).[72] In the absence of diabetes, sugar (glucose) consumption promotes testosterone production in fasted men,[75] potentially promoting exposure to lifetime androgens and greater prostate cancer risk.

prostate cancer but also the proper dosage of testosterone supplementation when needed.

Second, a comparative perspective on male reproductive aging among chimpanzees is represented by only a handful of works.[74] Without a clear picture of how our closest evolutionary relatives respond to aging, there is no optimal comparative model for understanding human male reproductive senescence. Given the diminishing responsible access to chimpanzees in captivity for research and the endangered status of animals in the wild, quantitative assessments of chimpanzee biology should be made a priority.

Acknowledgments

The author thanks all members, past and present of the Yale Reproductive Ecology Laboratory for their insights and intellectual contributions, especially Stephanie Anestis, Beatrice Babbs, Melanie Beuerlein, Alicia Breakey, Kate Clancy, Angie Jaimez, Michael Muehlenbein, Leon Noel, Meredith Reiches, Angélica Torres, and Monica Wakefield. This manuscript was greatly improved by the insightful comments of two anonymous reviewers.

Conflicts of interest

The author declares no conflicts of interest.

References

1. Bribiescas, R.G. 2006. On the evolution of human male reproductive senescence: proximate mechanisms and life history strategies. *Evol. Anthropol.* **15:** 132–141.
2. Stearns, S.C. 1992. *The Evolution of Life Histories*. Oxford University Press. Oxford.
3. Ellison, P.T. 2003. Energetics and reproductive effort. *Am. J. Hum. Biol.* **15:** 342–351.
4. Hill, K.R. 1993. Life history theory and evolutionary anthropology. *Evol. Anthropol.* **2:** 78–88.
5. Ketterson, E.D., V. Nolan Jr., L. Wolf & C. Ziegenfus. 1992. Testosterone and avian life histories: effects of experimentally elevated testosterone on behavior and correlates of fitness in the dark-eyed junco (Junco hyemalis). *Am. Nat.* **140:** 980–999.
6. Kaplan, H., K. Hill, J. Lancaster & A.M. Hurtado. 2000. A theory of human life history evolution: diet, intelligence and longevity. *Evol. Anthropol.* **9:** 156–184.
7. Key, C. & C. Ross. 1999. Sex differences in energy expenditure in nonhuman primates. *Proc. Biol. Sci.* **266:** 2479–2485.
8. Bribiescas, R.G. 1996. Testosterone levels among Aché hunter/gatherer men: a functional interpretation of population variation among adult males. *Hum. Nat.* **7:** 163–188.
9. Bribiescas, R.G. 2001. Reproductive ecology and life history of the human male. *Yearb. Phys. Anthropol.* **Suppl. 33:** 148–176.
10. Tuljapurkar, S.D., C.O. Puleston & M.D. Gurven. 2007. Why men matter: mating patterns drive evolution of human lifespan. *PLoS One* **2:** e785.
11. Bribiescas, R.G. 2006. *Men: Evolutionary and Life History*. Harvard University Press. Cambridge.
12. de La Rochebrochard, E. *et al.* 2006. Fathers over 40 and increased failure to conceive: the lessons of in vitro fertilization in France. *Fertil. Steril.* **85:** 1420–1424.
13. Reichman, N.E. & J.O. Teitler. 2006. Paternal age as a risk factor for low birthweight. *Am. J. Public Health* **96:** 862–866.
14. Wyrobek, A.J. *et al.* 2006. Advancing age has differential effects on DNA damage, chromatin integrity, gene mutations, and aneuploidies in sperm. *Proc. Natl. Acad. Sci. USA* **103:** 9601–9606.
15. Tough, S.C., A.J. Faber, L.W. Svenson & D.W. Johnston. 2003. Is paternal age associated with an increased risk of low

15. birthweight, preterm delivery, and multiple birth? *Can. J. Public Health* **94:** 88–92.
16. Fisch, H. *et al.* 2003. The influence of paternal age on down syndrome. *J. Urol.* **169:** 2275–2278.
17. Carbone, D.J., Jr. & A.D. Seftel. 2002. Erectile dysfunction. Diagnosis and treatment in older men. *Geriatrics* **57:** 18–24.
18. Korenman, S.G. 2004. Epidemiology of erectile dysfunction. *Endocrine* **23:** 87–91.
19. Dixson, A.F. *et al.* 2003. Masculine somatotype and hirsuteness as determinants of sexual attractiveness to women. *Arch. Sex. Behav.* **32:** 29–39.
20. Pauli, E.M. *et al.* 2008. Diminished paternity and gonadal function with increasing obesity in men. *Fertil. Steril.* **90:** 346–351.
21. Moore, M.C. 1991. Application of organization-activation theory to alternative male reproductive strategies: a review. *Horm. Behav.* **25:** 154–179.
22. Finch, C.E. & M.R. Rose. 1995. Hormones and the physiological architecture of life history evolution. *Q. Rev. Biol.* **70:** 1–52.
23. Bribiescas, R.G. & P.T. Ellison. 2007. How hormones mediate trade-offs in human health and disease. In *Evolution in Health and Disease*. S.C. Stearns & J.C. Koella, Eds.: 77–93. Oxford University Press. Oxford, UK.
24. Campbell, B.C., M.T. O'Rourke & S.F. Lipson. 2003. Salivary testosterone and body composition among Ariaal males. *Am. J. Hum. Biol.* **15:** 697–708.
25. Ellison, P.T., S.F. Lipson & M.D. Meredith. 1989. Salivary testosterone levels in males from the Ituri forest of Zaire. *Am. J. Hum. Biol.* **1:** 21–24.
26. Ellison, P.T. & C. Panter-Brick. 1996. Salivary testosterone levels among Tamang and Kami males of central Nepal. *Hum. Biol.* **68:** 955–965.
27. Welle, S., R. Jozefowicz, G. Forbes & R.C. Griggs. 1992. Effect of testosterone on metabolic rate and body composition in normal men and men with muscular dystrophy. *J. Clin. Endocrinol. Metab.* **74:** 332–335.
28. Elia, M. 1992. Organ and tissue contribution to metabolic rate. In *Energy Metabolism: Tissue Determinants and Cellular Corollaries*. H.N. Tucker, Ed.: 61–79. Raven Press, Ltd. New York.
29. Muehlenbein, M.P. & R.G. Bribiescas. 2005. Testosterone-mediated immune functions and male life histories. *Am. J. Hum. Biol.* **17:** 527–558.
30. Harman, S.M. *et al.* 2001. Longitudinal effects of aging on serum total and free testosterone levels in healthy men. Baltimore Longitudinal Study of Aging. *J. Clin. Endocrinol. Metab.* **86:** 724–731.
31. Travison, T.G. *et al.* 2007. The relative contributions of aging, health, and lifestyle factors to serum testosterone decline in men. *J. Clin. Endocrinol. Metab.* **92:** 549–555.
32. Barrett-Connor, E., D.G. Von Muhlen & D. Kritz-Silverstein. 1999. Bioavailable testosterone and depressed mood in older men: the Rancho Bernardo Study. *J. Clin. Endocrinol. Metab.* **84:** 573–577.
33. Ellison, P.T. *et al.* 2002. Population variation in age-related decline in male salivary testosterone. *Hum. Reprod.* **17:** 3251–3253.
34. Uchida, A. *et al.* 2006. Age related variation of salivary testosterone values in healthy Japanese males. *Aging Male* **9:** 207–213.
35. Bremner, W.J., M.V. Vitiello & P.N. Prinz. 1983. Loss of circadian rhythmicity in blood testosterone levels with aging in normal men. *J. Clin. Endocrinol. Metab.* **56:** 1278–1281.
36. Bribiescas, R.G. & K.R. Hill. 2010. Circadian variation in salivary testosterone across age classes in Ache Amerindian males of Paraguay. *Am. J. Hum. Biol.* **22:** 216–220.
37. Hayes, F.J. *et al.* 2000. Aromatase inhibition in the human male reveals a hypothalamic site of estrogen feedback. *J. Clin. Endocrinol. Metab.* **85:** 3027–3035.
38. Orwoll, E. *et al.* 2006. Testosterone and estradiol among older men. *J. Clin. Endocrinol. Metab.* **91:** 1336–1344.
39. Zumoff, B., L.K. Miller & G.W. Strain. 2003. Reversal of the hypogonadotropic hypogonadism of obese men by administration of the aromatase inhibitor testolactone. *Metabolism* **52:** 1126–1128.
40. Bribiescas, R.G. 2005. Age-related differences in serum gonadotropin (FSH and LH), salivary testosterone, and 17-beta estradiol levels among Ache Amerindian males of Paraguay. *Am. J. Phys. Anthropol.* **127:** 114–121.
41. Campbell, B.C., P.W. Leslie & K.L. Campbell. 2006. Age-related patterns of urinary gonadotropins among Turkana men of northern Kenya. *Soc. Biol.* **53:** 31–45.
42. Griffin, J.E. & S.R. Ojeda. 2004. *Textbook of Endocrine Physiology*. Oxford University Press. Oxford; New York.
43. Feldman, H.A. *et al.* 2002. Age trends in the level of serum testosterone and other hormones in middle-aged men: longitudinal results from the Massachusetts male aging study. *J. Clin. Endocrinol. Metab.* **87:** 589–598.
44. Klibanski, A. *et al.* 1981. Reproductive function during fasting in men. *J. Clin. Endocrinol. Metab.* **53:** 258–263.
45. Chavarro, J.E. *et al.* 2009. Body mass index in relation to semen quality, sperm DNA integrity, and serum reproductive hormone levels among men attending an infertility clinic. *Fertil. Steril.* doi:10.1016/j.fertnstert.2009.01.100.
46. Wittert, G.A. *et al.* 2003. Oral testosterone supplementation increases muscle and decreases fat mass in healthy elderly males with low-normal gonadal status. *J. Gerontol. A Biol. Sci. Med. Sci.* **58:** M618–M625.
47. Bhasin, S. *et al.* 2005. Older men are as responsive as young men to the anabolic effects of graded doses of testosterone on the skeletal muscle. *J. Clin. Endocrinol. Metab.* **90:** 678–688.
48. Feinberg, A.P. 2007. Phenotypic plasticity and the epigenetics of human disease. *Nature* **447:** 433–440.
49. Reade, M.C. *et al.* 2009. Differences in immune response may explain lower survival among older men with pneumonia. *Crit. Care Med.* **37:** 1655–1662.
50. Baumgartner, R.N. *et al.* 1999. Predictors of skeletal muscle mass in elderly men and women. *Mech. Ageing Dev.* **107:** 123–136.
51. Fukagawa, N.K., L.G. Bandini & J.B. Young. 1990. Effect of age on body composition and resting metabolic rate. *Am. J. Physiol.* **259:** E233–E238.
52. Bergendahl, M. *et al.* 1998. Fasting suppresses pulsatile luteinizing hormone (LH) secretion and enhances

orderliness of LH release in young but not older men. *J. Clin. Endocrinol. Metab.* **83:** 1967–1975.
53. Rose, M.R. 1991. *Evolutionary Biology of Aging*. Oxford University Press. New York.
54. Owens, I.P. 2002. Ecology and evolution. Sex differences in mortality rate. *Science* **297:** 2008–2009.
55. Graves, B.M., M. Strand & A.R. Lindsay. 2006. A reassessment of sexual dimorphism in human senescence: theory, evidence, and causation. *Am. J. Hum. Biol.* **18:** 161–168.
56. Walker, R. & K. Hill. 2003. Modeling growth and senescence in physical performance among the ache of eastern Paraguay. *Am. J. Hum. Biol.* **15:** 196–208.
57. Walker, R., K. Hill, H. Kaplan & G. McMillan. 2002. Age-dependency in hunting ability among the Ache of eastern Paraguay. *J. Hum. Evol.* **42:** 639–657.
58. Gurven, M.D. & C. von Rueden. 2006. Hunting, social status and biological fitness. *Soc. Biol.* **53:** 81–99.
59. Klein, R.G. 1999. *The Human Career: Human Biological and Cultural Origins*. University of Chicago Press. Chicago.
60. Jemal, A. et al. 2006. Cancer Statistics, 2006. *CA Cancer J. Clin.* **56:** 106–130.
61. Quinn, M. & P. Babb. 2002. Patterns and trends in prostate cancer incidence, survival, prevalence and mortality. Part I: international comparisons. *BJU Int.* **90:** 162–173.
62. Campbell, B. 2005. High rate of prostate symptoms among Ariaal men from Northern Kenya. *Prostate* **62:** 83–90.
63. Williams, G.C. 1957. Pleiotropy, natural selection, and the evolution of senescence. *Evolution* **11:** 398–411.
64. Reed, W.L. et al. 2006. Physiological effects on demography: a long-term experimental study of testosterone's effects on fitness. *Am. Nat.* **167:** 665–681.
65. Muehlenbein, M.P. et al. 2005. The reproductive endocrine response to Plasmodium vivax infection in Hondurans. *Am. J. Trop. Med. Hyg.* **73:** 178–187.
66. Slater, S. & R.T. Oliver. 2000. Testosterone: its role in development of prostate cancer and potential risk from use as hormone replacement therapy. *Drugs Aging* **17:** 431–439.
67. Travis, R.C. et al. 2007. Serum androgens and prostate cancer among 643 cases and 643 controls in the European Prospective Investigation into Cancer and Nutrition. *Int. J. Cancer* **121:** 1331–1338.
68. Kehinde, E.O. et al. 2006. Prostate cancer risk: the significance of differences in age related changes in serum conjugated and unconjugated steroid hormone concentrations between Arab and Caucasian men. *Int. Urol. Nephrol.* **38:** 33–44.
69. Jasienska, G. & I. Thune. 2001. Lifestyle, hormones, and risk of breast cancer. *BMJ* **322:** 586–587.
70. Rojdmark, S., A. Asplund & S. Rossner. 1989. Pituitary-testicular axis in obese men during short-term fasting. *Acta Endocrinol. (Copenh)* **121:** 727–732.
71. Bruning, J.C. et al. 2000. Role of brain insulin receptor in control of body weight and reproduction. *Science* **289:** 2122–2125.
72. Colli, J.L. & A. Colli. 2006. International comparisons of prostate cancer mortality rates with dietary practices and sunlight levels. *Urol. Oncol.* **24:** 184–194.
73. Bhasin, S. & J.G. Buckwalter. 2001. Testosterone supplementation in older men: a rational idea whose time has not yet come. *J. Androl.* **22:** 718–731.
74. Seraphin, S.B., P.L. Whitten & V. Reynolds. 2008. The influence of age on fecal steroid hormone levels in male Budongo Forest chimpanzees (Pan troglodytes schweinfurthii). *Am. J. Primat.* **70:** 661–669.
75. Röjdmark, S. 1987. Influence of short-term fasting on the pituitary-testicular axis in normal men. *Horm. Res.* **25:** 140–146.

ANNALS OF THE NEW YORK ACADEMY OF SCIENCES
Issue: *Reproductive Aging*

Dynamic heterogeneity and life histories

Shripad Tuljapurkar and Ulrich K. Steiner

Department of Biology, Stanford University, Stanford, California

Address for correspondence: Shripad Tuljapurkar, Department of Biology, 454 Herrin Labs, Stanford University, Stanford, California 94305. tulja@stanford.edu

Biodemography is increasingly focused on the large and persistent differences between individuals within populations in fitness components (age at death, reproductive success) and fitness-related components (health, biomarkers) in humans and other species. To study such variation we propose the use of dynamic models of observable phenotypes of individuals. Phenotypic change in turn determines variation among individuals in their fitness components over the life course. We refer to this dynamic accumulation of fitness differences as dynamic heterogeneity and illustrate it for an animal population in which longitudinal data are studied using multistate capture-mark-recapture models. Although our approach can be applied to any characteristic, for our empirical example we use reproduction as the phenotypic character to define stages. We indicate how our stage-structured model describes the nature of the variation among individual characteristics that is generated by dynamic heterogeneity. We conclude by discussing our ongoing and planned work on animals and humans. We also discuss the connections between our work and recent work on human mortality, disability and health, and life course theory.

Keywords: individual heterogeneity; fitness; life history; life course; phenotypic variation

Introduction

An early focus of evolutionary ecology and demography was the "typical" age-pattern of mortality and reproductive status of individuals within species.[1] In recent years, increasing attention has been paid to the large and persistent differences between individuals within populations in age at death and reproductive success, as observed in humans and other species.[2–4] Demographers come at these questions by decomposing variation into contributions from demographic, socioeconomic, and biological characteristics of individuals and groups. Such decompositions now exploit results from mechanistic biology that suggest how causal relationships between genotypes, phenotypes, and environment may translate into observed variation in death and reproduction.[5,6] Evolutionary biologists have tended to focus on comparing variation among individuals within and between populations and species, and on explaining the results of such comparisons in terms of the forces of evolution,[7,8] as they play out in particular biological and environmental settings. These studies concern one or more levels of biological organization. Within a species, we are interested in describing variation among individuals and populations, the forces responsible for maintaining this variation, and in predicting the response of individuals to environmental change. Across species, we are interested in differences in both average patterns and variation in life histories, as well as species-level responses to environmental change. This paper focuses on variation among individuals, and argues that a key task of biodemography is to understand how variation among individuals is created and maintained across generations. Such an understanding is essential if we are to influence and reshape either phenotypic change at the individual level or population dynamics, in contexts as disparate as human health and ecological responses to global climate.

An influential argument says that variation in fitness-related traits, such as age at death, may be due to variation in fixed individual traits that affect relative fitness over life.[9,10] Such fixed traits are typically assumed to be fixed at birth and to affect fitness throughout life, may often be unobserved (or unobservable), and are elements of what may be called "frailty." This argument is influential in part because of an explicit or implicit assumption that such fixed differences in phenotypic traits are determined by heritable genetic differences and are being maintained by evolutionary processes. We refer to this kind of variation as "fixed heterogeneity." The assumption that frailty is the main determinant of individual fitness has been widely used to model and interpret data on human mortality[11] and also data on animal life histories.[7] We propose an alternative view of the determinants of individual fitness. This view is rooted in the fact that the observed phenotypes of individuals (including, e.g., size, energy stores, immune competence, metabolic state, health, and biomarkers) change over their lives, and that changes within individuals in their phenotypic traits over any age interval may differ among individuals. Hence any one individual follows a trajectory in a "space" of phenotypic values, and distinct individuals can follow distinct trajectories. An individual's fitness components (survival, reproduction) at any time are often largely determined by the individual's present (and past) phenotypic states, and so individuals can follow different trajectories of relative fitness over the life course. These trajectories are defined by observed values of phenotypic traits at different time points in a life course, and we describe the dynamics of such phenotypic trajectories over time by a stochastic (usually Markov) process. The distribution of alternative trajectories then determines observed variation among individuals in both phenotypic traits and fitness.[3] We refer to this kind of variation as "dynamic heterogeneity." Dynamic heterogeneity in observable phenotypes has been relatively little used to analyze and explain variation in fitness components, although models of dynamic heterogeneity are often estimated for animal populations[3,12] and some longitudinal models of human mortality are similar in spirit.[13–15]

In the next section, we explain precisely what we mean by dynamic heterogeneity and provide an illustration from an animal population in which longitudinal data are studied using multistage capture-mark-recapture models.[3,12] Following that we explain the novel features of dynamic heterogeneity in the context of a simple mathematical model. We indicate how this model describes the average pattern of life histories as well as the nature of the variation among individual characteristics that is generated by dynamic heterogeneity. We then discuss our work on other animal species and the directions for subsequent analysis. We conclude by discussing the many connections between our work and recent work on human mortality, disability and health, and life course theory[13] and our planned work on humans.

The nature of heterogeneity

Dynamic heterogeneity

We focus on one or more observable phenotypic traits of individuals. These may include, e.g., continuous traits, such as size,[16] discrete states, such as reproductive maturity or menopause,[17,18] and states, such as the level of reproduction in a year that may be treated as discrete or continuous.[3,19] More formally, every individual has a set of observable phenotypic traits $\mathbf{P}$ and individuals born at time t have a distribution $f(\mathbf{P}, 0, t)$ of these traits. As an individual ages, its phenotypic traits may change for many reasons, such as maturation, growth, reproduction, or senescence. As a result of such processes, at age a the trait distribution of a cohort born at time t changes from the distribution at birth to a new distribution $f(\mathbf{P}, a, t + a)$. We can quantify the dynamics of phenotypic change by estimating transition probabilities $p(\mathbf{P}_2, t + 1 \mid \mathbf{P}_1, t)$ from the rates at which individuals change their phenotypic state. We can also estimate the relationship between an individual's mortality and reproduction in year t and its phenotypic state $\mathbf{P}_t$ in that year. It is possible that these fitness components may depend on the individual's phenotypic trajectory and not just its current phenotype[20] but to keep things simple we assume for now that only the current state matters (i.e., the dynamics follow a Markov process[21]).

We describe a single individual by its phenotypic trajectory, i.e., a sequence of phenotypic states $\mathbf{P}_0$, $\mathbf{P}_1, \mathbf{P}_2, \ldots$ at successive times 0, 1, 2, and so on. A cohort is described by a bundle of such trajectories, one for each individual; a population is described by many bundles of trajectories, one for each cohort. Dynamic heterogeneity means that individuals who are in the same phenotypic state in one year may follow distinct phenotypic trajectories in the future, e.g., when two individuals with phenotype $\mathbf{P}_t$ in year t end up with different phenotypes $\mathbf{P}_{t+1}, \hat{\mathbf{P}}_{t+1}$ in year $t + 1$. As a result a cohort's phenotypic distribution tends to become increasingly dispersed as the cohort ages. This dispersion is balanced by intracohort selection that results from differences in mortality among different phenotypic states of individuals. Over the life course, an individual's cumulative survival probability depends on its phenotypic trajectory. In addition to survival, an individual's reproduction in any year is also a function of its current phenotypic state, and thus reproduction will vary among individuals of the same age. Lifetime reproduction is a function of the entire phenotypic trajectory and can vary even among individuals who die at the same age.

Dynamic variation of this kind is generated by a stochastic process that causes individuals to follow divergent phenotypic trajectories. To illustrate we consider data on the Mute swan, *Cygnus olor*, collected since 1976 on a well-known population in Abbotsbury, U.K. These are among the best longitudinal data on a natural population[22,23] and we have reported on these data elsewhere[3]; the latter paper provides additional detail. We classify individuals using a single discrete phenotypic trait, reproductive state. In each year each female is categorized as being in one of five reproductive stages: immature (stage 1), having a "low" clutch size of 1–4 (stage 2), having a "medium" clutch size of 5–6 (stage 3), having a "high" clutch size of 7–12 (stage 4), or being nonreproductive in that year after having reproduced at least once previously (stage 5). Capture-mark-recapture analysis[12] is used to estimate a Markov transition matrix describing the probability of moving between stages in successive years as well as age and stage specific survival rates. In the Mute swan example and other examples the results and their interpretation in terms of dynamic heterogeneity are fairly robust to using different numbers of discrete classes, or different ranges to delineate classes.[3] However, the choice of a focal phenotype (e.g., size instead of reproductive state) can lead to quite different transition dynamics. The use of discrete phenotypic classes versus continuous phenotypic values, or of just one phenotypic trait versus two or more, is driven by data quality and quantity and biological insight. When data allow, it is probably better to use continuous phenotypic values[16] rather than a large number of classes.

Figure 1A displays stage trajectories for three swans: the stars indicate actual ages of last observation and the trajectories past the stars are simulations using the estimated stage transition

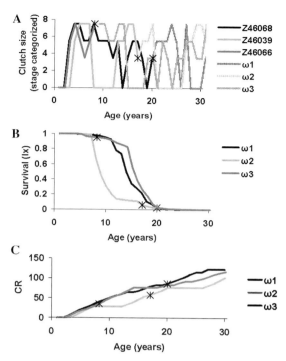

Figure 1. (A) Reproductive stage trajectories for female mute swans. Reproductive stage in each year has one of five values (see text). Observed sequences are shown for three birds and death or disappearance is indicated by a star. Past the star we have generated one possible trajectory by simulation using our estimated transition probabilities for the population. (B) Survivorship for the three trajectories shown in Panel A. A key point is that survival depends on stage and so every distinct sequence of stages (i.e., life course) determines a potentially distinct survivorship. (C) Cumulative reproduction (CR), the number of eggs laid to a given age, for the three trajectories shown in Panel A. Reproduction here varies by stage and so every distinct sequence of stages (i.e., life course) determines a potentially distinct trajectory of cumulative reproduction.

rates. Survival rates depend on stage and age and thus for each trajectory in Figure 1A we can compute a survivorship, as shown in Figure 1B. Finally Figure 1C shows the cumulative reproduction at each age. Clearly individual birds achieve rather different lifetimes and cumulative reproduction. Using the estimated model, we can also simulate a sample of birds that matches the data (in numbers and censoring) and find that we can predict nicely the variation in lifetime reproductive success in the sample; see Figure 2. Figure 2 also drives home the point that there is very large variability among individuals in lifetime reproduction, a quantity that is often used as a proxy for fitness. But in fact, these

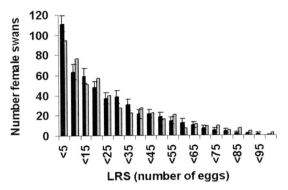

Figure 2. Distribution of Lifetime Reproductive Success (LRS) for female mute swans. The observed distribution (including potential latent factors) is shown by gray bars and the average distribution for 50 simulated synthetic populations is shown by black bars (± standard error). Strong fixed effects should lead to large deviances in distribution between the synthetic populations and the observed population. LRS is right censored for some of the observed individuals that have been taken into account for the simulations.

birds are all members of a population that has one overall growth rate. We reviewed a number of studies on different species[3] and found that large variation in fitness components is typical, not unusual. In studies that have been done over many generations, there is evidence that this large variation persists over time, suggesting that this level of variation is evolutionarily stable.

Thus, dynamic heterogeneity is generated as individuals move along the life course. Even though every individual follows the same stochastic process, the vagaries of chance produce large differences among individuals as they age. In our fitted model, differences among individuals in marginal fitness at any age, or in measures of lifetime fitness, generated by dynamic heterogeneity reflect chance, not fixed differences in "frailty" or "quality." In the real world, differences in fixed frailty could also contribute to variation among individual phenotypes or performance, but the detection of fixed differences has been difficult, as we now explain.

Fixed heterogeneity

When we observe variation in fitness components among individuals, it is tempting to argue that this variation is determined by inherent characteristics of these individuals. Suppose that the fitness components revealed in each individual's life depend on some unobserved individual trait(s), call them $\mathbf{Z}$,

that are set at birth and do not change in later life, and that an individual's mortality and fertility vary with its trait value. In demographic work such fixed traits are often called frailty and are assumed to affect mortality. Some ecological models assume two correlated components of frailty that affect mortality and reproduction, respectively. Every cohort of individuals is born with a distribution, say $\phi(\mathbf{Z})$ of trait values. Individual fitness components, such as the age at death T and lifetime reproduction M, will both vary with the trait value. A model of fixed heterogeneity attempts to infer the unknown $\phi(\mathbf{Z})$ from the observed distribution of, say, ages at death and lifetime reproduction, and then to explain the variance between individual fitness components in terms of the underlying variation in values of $\mathbf{Z}$. The dispersion of the distribution $\phi(\mathbf{Z})$ measures the amount of fixed heterogeneity in the population. Variation among individuals in age at death and lifetime reproduction is proportional to the underlying variation in $\mathbf{Z}$.

While the assumption of unobservable traits may be a useful first step, it is clearly essential to find the actual sources of fixed heterogeneity and determine whether they produce the variation we observe. But progress has been difficult since the usual suspects for the possible underlying traits (e.g., genetic variation or fetal condition) are often hard to measure and the quantification of their effects outside controlled lab environments is challenging.[24,25]

A deeper question is, what maintains variation in fixed traits? If a trait affects survival or reproduction then individuals with different trait values must differ in overall fitness and natural selection should act to eliminate variation in the trait.[26] For survival rates, previous work has suggested several possible reasons why fixed variation may persist including the continual injection of mutational variance,[27] low heritability due to a low additive genetic contribution to the trait,[28] and fluctuating selection caused by environmental variability.[29] But there is little empirical evidence to quantify the contribution of such processes to the persistence of variation, in part because our understanding of the genotype–phenotype mapping is limited.[30]

Describing variation among individual life histories

We now examine in a little more detail a model of dynamic heterogeneity that is relatively simple but still applicable to many animal populations. We assume that any individual can be in one of three states, immature (state 1), breeding with low reproduction (state 2), and breeding with high reproduction (state 3). The model requires a matrix of transition probabilities between states, conditional on survival,

$$\Psi = \begin{pmatrix} \psi_{11} & \psi_{12} & 0 \\ 0 & \psi_{22} & \psi_{23} \\ 0 & \psi_{32} & \psi_{33} \end{pmatrix}$$

Here an individual transitions to state 2 when it matures and then transitions between states 2 and 3 so long as it survives. In addition, we have survival probabilities s_i for each state i. Let the fertilities for states 2, 3 be f_2, f_3 and define a fertility matrix

$$\mathbf{F} = \begin{pmatrix} 0 & f_2 & f_3 \\ 0 & 0 & 0 \\ 0 & 0 & 0 \end{pmatrix}$$

and finally a population projection matrix

$$\mathbf{A} = \mathbf{F} + \mathbf{S}$$

where

$$\mathbf{S} = \begin{pmatrix} s_1\psi_{11} & 0 & 0 \\ s_1\psi_{12} & s_2\psi_{22} & s_3\psi_{32} \\ 0 & s_2\psi_{23} & s_3\psi_{33} \end{pmatrix}$$

It is worth noting that this matrix involves the elements of the transpose of the transition probability matrix (by sometimes confusing convention). We have no age dependence in these equations but we could easily include age dependence in the transition probabilities and survival rates. This formulation is well known in population biology as a stage-structured model.[31] Purely as a description of mortality the model above is a version of a multistage life table.[32]

Such a stage-structured model with stages described in terms of observable phenotypes is a general and powerful way to describe population dynamics. What is new about our work is that we use this model not to describe population numbers, but rather to describe dynamic heterogeneity, i.e., variation in phenotypic traits and fitness components among individuals within a population.

Variation in the age at death T measured by its mean, variance and higher moments can be computed explicitly in terms of the matrix $\mathbf{S}$, whose powers describe the probabilities of being alive in one of the three stages at any age. For example, the probability that an individual survives to age at least a and is in stage i at that age is the $(1, i)$ element of $\mathbf{S}^{a-1}$. The mean age at death, ET, and variance of age at death, $E(T^2) - (ET)^2$, for only one type of offspring being born in stage 1 can be computed[31] using

$$ET = \mathbf{e}_1^T \mathbf{I}\hat{\mathbf{N}}\mathbf{e}_1$$

and

$$ET^2 = \mathbf{e}_1^T \mathbf{I}(2\hat{\mathbf{N}} - \mathbf{I})\mathbf{I}\hat{\mathbf{N}}\mathbf{e}_1$$

Here $\mathbf{e}_1$ is a column vector of zeros except for element $i = 1$, $\mathbf{I}$ is the identity matrix, $\hat{\mathbf{N}}$ is the transpose of the fundamental matrix $\mathbf{N} = (\mathbf{I} - \mathbf{S})^{-1}$, and a superscript T indicates a transpose.

Next consider an individual's lifetime reproduction, M. The average lifetime reproduction is known[31] to be given by

$$EM = \mathbf{e}_1^T \mathbf{F}\hat{\mathbf{N}}\mathbf{e}_1$$

and we have derived a new formula for the variance (details are given elsewhere and are available on request),

$$EM^2 = \mathbf{e}_1^T \mathbf{F}(2\hat{\mathbf{N}} - \mathbf{I})\hat{\mathbf{F}}\hat{\mathbf{N}}\mathbf{e}_1$$

Here $\hat{\mathbf{F}}$ is a matrix with only diagonal elements equal to the f_i. We can also compute these moments exactly when transitions are age and stage structured, using new formulas that we report elsewhere (details are available on request).

An interesting feature of phenotypically structured models is that they generally predict late age mortality plateaus. This is a consequence of the structure of Markovian models and we find plateaus in models for plant and animal populations.[33] The plateaus generated by dynamic heterogeneity are different from those found in models with fixed frailty, a matter we return to in the last section of this paper.

An overall measure of dynamic heterogeneity is given by the entropy, H, of the matrix Ψ, which generates all possible phenotypic trajectories. Entropy is defined[34] by considering first all sequences ω of states of some length n; let the probability of sequence ω be $p(\omega)$ and define

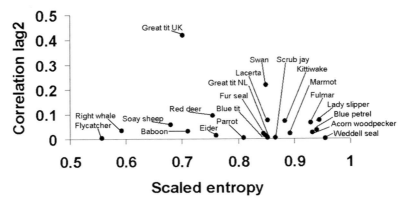

Figure 3. Dynamic heterogeneity across species. The vertical axis indicates the absolute correlation between stages two years apart, and the horizontal axis the entropy for the generating matrix of transition probabilities, scaled to a maximum value of 1.

$h(n) = -\sum_{\omega} p(\omega) \log p(\omega)$; then entropy is the limit as $n \to \infty$ of $(h(n)/n)$. This entropy describes the rate of diversification of trajectories conditional on survival and is estimated as $H = -\sum_{i=1}^{K} \sum_{j=1}^{K} \pi_i \psi_{ij} \log \psi_{ij}$.[3] The π_i are the elements of the vector describing the stable stage distribution and K is the number of stages.[3] Alternatively we can include survival weighting and compute the entropy of a matrix based on **S** that describes trajectories before death.[35] We are also interested in the persistence of an individual's current phenotypic state, i.e., if an individual is in state k at time t is it also likely to be in the same state at a later time? A useful measure of persistence is the correlation between an individual's states 2 years apart (i.e., correlation at a time lag of 2 years between states at t, $t + 2$). Figure 3 displays the entropy and correlation at lag 2 for a large number of species, estimated using reproductive success as a stage variable. In general, phenotypic state has low persistence, meaning that individuals change reproductive state frequently over the life course. Entropy varies considerably across species[3] but the level of entropy is high, meaning that individuals in all species follow diverse life course trajectories.

It is important to note that a pair of the matrices **S**, **F** constitutes a single life history phenotype. These matrices generate many alternative life course trajectories. Although individuals following distinct trajectories may have distinct marginal fitness at different stages of the life cycle, the overall phenotype's fitness is determined by all trajectories. The overall fitness $r = \log \lambda$ can also be expressed using reproductive trajectories, as the solution of the characteristic equation,

$$1 = E\left\{\sum_{a=1}^{a=T} F_a(\omega) \, e^{-r a}\right\}$$

where E indicates an expectation (i.e., average) over the probability distribution of reproductive trajectories ω, and T is age at death. This overall fitness is the population growth rate. Evolutionary change comes about when the elements of these two matrices change, e.g., if a mutation produces a phenotype in which the probability to remain immature ψ_{11} increases so that individuals on average take a longer time to mature. Thus, dynamic heterogeneity is intrinsic to the life history phenotype and does not require arguments for its maintenance.

Analyzing variation and its causes

Although we have emphasized the difference between fixed and dynamic heterogeneity, we certainly expect to find both kinds of heterogeneity in nature. In our picture, we expect to find fixed and heritable variation among individuals in phenotypic transition probabilities and mortality. In looking for such variation, our current work focuses on several questions. In natural populations, how much fixed and dynamic heterogeneity can we detect among individuals using multistage models in which individuals also differ in fixed traits? What do statistical models tell us about the interaction between fixed and dynamic heterogeneity, and do they support the hypothesis that tradeoffs act via fixed latent factors? Is dynamic heterogeneity associated with or predicted by individual covariates, environmental factors, and temporal variability? Is there evidence for the heritability of traits that determine fixed and dynamic heterogeneity?

A different analytical strategy that we are using for phenotypic models is to work with the change over time and age of the distribution $f(\mathbf{P}, a, t)$ of phenotypic traits. A powerful way to study this distribution is to use a generalization of the Price equation that we recently derived and applied to longitudinal data on natural populations of mammals.[36] This generalized Price equation enables us to study the evolution of phenotypic means and variances, and to exactly decompose changes in these moments into contributions from selection, phenotypic plasticity, and parent–offspring transmission. There is a natural connection between this approach and one based on estimating stage-structured models, and we aim to explore this in forthcoming work.

Researchers working on human mortality and health have long used longitudinal models that are similar in spirit to our stage structured approach. Recent studies have used an explicit analysis of trajectories of observed or self-reported health state,[14] and entry into and exit from disability[37]; these studies also use a Markov process to predict health status as a phenotypic trait, but have not yet examined the nature of dynamic heterogeneity. There is also a close connection between our view of variation as generated along the life course and work on life course health development.[13] The latter work marches with ours in focusing attention on the transition probabilities between stages of health and functioning as being critical targets for study and possible intervention.

Much of our work uses data from longitudinal studies of natural populations of animals and plants. We have also begun work on human populations, including a study by Michael Gurven of health, biomarkers, and demography among the Tsimane.[38] Our approach aims to study a broad set of phenotypic traits, analyze transition rates between them, and then map phenotypic trajectories onto health states. In our perspective, some phenotypic states will map onto high risks of mortality or morbidity, and we can estimate age and state dependent probabilities that an individual will transition into and out of such states. The population-level age-pattern of mortality reflects the distribution of individual phenotypes between low-risk and high-risk states. Old-age mortality plateaus are generated because individuals continue to make transitions among risk states at all ages, and as a cohort ages it displays a stable distribution of individuals across different risk states. These plateaus are not the result of the differential survival of hardy individuals. The power of our approach lies in quantifying the forces that determine phenotypic transitions and phenotypic trajectories. Transition rates can be related to mechanistic arguments about physiology, environment, and genes. They provide the right setting in which to examine the consequences of interventions that aim to improve health over the life course. Finally, the use of phenotypic trajectories will make it possible to examine how much variation among individuals may be due to fixed effects, such as genotypes, rather than simply to dynamic heterogeneity.

Acknowledgments

We thank the U.S. National Institute on Aging and National Science Foundation for support. For the data on the swans at Abbotsbury, we are grateful to Mrs. Charlotte Townshend for allowing the study to be made, to Prof. C.M. Perrins and the late Dr. R.H. McCleery for collating the data, and to David Wheeler and Steve Groves at the Swannery and the many volunteers who have helped to collect the data. For comments on a previous version of this manuscript we thank Tim Coulson and an anonymous referee. The study is funded by NIH.

Conflicts of interest

The authors declare no conflicts of interest.

References

1. Stearns, S.C. 1976. Life-history tactics: a review of the ideas. *Q. Rev. Biol.* **51:** 3–47.
2. Edwards, R.D. & S. Tuljapurkar. 2005. Inequality in life spans and a new perspective on mortality convergence across industrialized countries. *Popul. Dev. Rev.* **31:** 645–674.
3. Tuljapurkar, S., U.K. Steiner & S.H. Orzack. 2009. Dynamic heterogeneity in life histories. *Ecol. Lett.* **12:** 93–106.
4. Lenormand, T., D. Roze & F. Rousset. 2009. Stochasticity in evolution. *Trends Ecol. Evol.* **24:** 157–165.
5. Crimmins, E.M. & C.E. Finch. 2006. Infection, inflammation, height, and longevity. *Proc. Natl. Acad. Sci. USA* **103:** 498–503.
6. Seplaki, C.L., N. Goldman, M. Weinstein & Y.H. Lin. 2004. How are biomarkers related to physical and mental well-being? *J. Gerontol. A Biol. Sci. Med. Sci.* **59:** 201–217.
7. Cam, E. *et al.* 2002. Individual covariation in life-history traits: seeing the trees despite the forest. *Am. Nat.* **159:** 96–105.

8. Jones, O.R. et al. 2008. Senescence rates are determined by ranking on the fast-slow life-history continuum. *Ecol. Lett.* **11:** 664–673.
9. Vaupel, J.W., K.G. Manton & E. Stallard. 1979. Impact of heterogeneity in individual frailty on the dynamics of mortality. *Demography* **16:** 439–454.
10. Vaupel, J.W. 1988. Inherited frailty and longevity. *Demography* **25:** 277–287.
11. Yashin, A.I. et al. 2007. Model of hidden heterogeneity in longitudinal data. *Theor. Popul. Biol.* **73:** 1–10.
12. Nichols, J.D. & W.L. Kendall. 1995. The use of multi-state capture-recapture models to address questions in evolutionary ecology. *J. Appl. Stat.* **22:** 835–846.
13. Halfon, N. & M. Hochstein. 2002. Life course health development: an integrated framework for developing health, policy, and research. *Milbank Q.* **80:** 433–479.
14. Sacker, A., R.D. Wiggins, M. Bartley & P. McDonough. 2007. Self-rated health trajectories in the United States and the United Kingdom: a comparative study. *Am. J. Public Health* **97:** 812–818.
15. Næss, O., F. Hernes & D. Blane. 2006. Life-course influences on mortality at older ages: evidence from the Oslo Mortality Study. *Soc. Sci. Med.* **62:** 329–336.
16. Ellner, S.P. & M. Rees. 2006. Integral projection models for species with complex demography. *Am. Nat.* **167:** 410–428.
17. Tuljapurkar, S. & C. Horvitz. 2006. From stage to age in variable environments: life expectancy and survivorship. *Ecology* **87:** 1497–1509.
18. Coulson, T. et al. 1997. Population substructure, local density, and calf winter survival in red deer (*Cervus elaphus*). *Ecology* **78:** 852–863.
19. Cam, E. & J. Monnat. 2000. Stratification based on reproductive state reveals contrasting patterns of age-related variation in demographic parameters in the kittiwake. *Oikos* **90:** 560–574.
20. Brownie, C. et al. 1993. Capture-recapture studies for multiple strata including non-markovian transitions. *Biometrics* **49:** 1173–1187.
21. Seneta, E. 2006. *Non-negative Matrices and Markov Chains*. Springer. New York.
22. McCleery, R.H., C. Perrins, D. Wheeler & S. Groves. 2002. Population structure, survival rates and productivity of mute swans breeding in a colony at Abbotsbury, Dorset, England. *Waterbirds* **25:** 192–201.
23. Charmantier, A., C. Perrins, R.H. McCleery & B.C. Sheldon. 2006. Evolutionary response to selection on clutch size in a long-term study of the Mute swan. *Am. Nat.* **167:** 453–465.
24. Harshman, L.G. 2003. Life span estimation of drosophila melanogaster: genetic and population studies. In *Life Span: Evolutionary, Ecological and Developmental Perspectives*. J.R. Carey & S. Tuljapurkar, Eds.: Population and Development Review. Vol. 29 (Suppl.): 99–126. Population Council. New York.
25. Horiuchi, S. 2003. Interspecies differences in the life span distribution: humans versus invertebrates. In *Life Span: Evolutionary, Ecological and Developmental Perspectives*. J.R. Carey & S. Tuljapurkar, Eds.: Population and Development Review. Vol. 29 (Suppl.):127–191. Population Council. New York.
26. Charlesworth, B. 1994. *Evolution in Age-Structured Populations*. Cambridge University Press. Cambridge.
27. Charlesworth, B. 1990. Optimization models, quantitative genetics, and mutation. *Evolution* **44:** 520–538.
28. Vaupel, J. 1988. Inherited frailty and longevity. *Demography* **25:** 277–287.
29. Tuljapurkar, S. 1997. The evolution of senescence. In *Between Zeus and the Salmon*. K.W. Wachter, C.E. Finch, Eds.: 65–77. National Academy of Sciences. Washington, DC.
30. Flint, J. & T.F.C. Mackay. 2009. Genetic architecture of quantitative traits in mice, flies, and humans. *Genome Res.* **19:** 723–733.
31. Caswell, H. 2001. *Matrix Population Models: Construction, Analysis and Interpretation*. 2nd edn. Sinauer Associates Inc. Sunderland, Mass.
32. Schoen, R. 1988. *Modeling Multigroup Populations*. Plenum Press. New York.
33. Horvitz, C.C. & S. Tuljapurkar. 2008. Stage dynamics, period survival and mortality plateaus. *Am. Nat.* **172:** 203–215.
34. Khinchin, A. I. 1957. *Mathematical Foundations of Information Theory*. Dover Publications. New York.
35. Matthews, J. 1970. A central limit theorem for absorbing Markov chains. *Biometrika* **57:** 129–139.
36. Coulson, T. & S. Tuljapurkar. 2008. The dynamics of a quantitative trait in an age-structured population living in a variable environment. *Am. Nat.* **172:** 599–612.
37. Gill, T.M., H.G. Allore, T.R. Holford & Z. Guo. 2004. Hospitalization, restricted activity, and the development of disability among older persons. *JAMA* **292:** 2115–2124.
38. Gurven, M., H. Kaplan & A.Z. Supa. 2007. Mortality experience of Tsimane amerindians of Bolivia: regional variation and temporal trends. *Am. J. Hum. Biol.* **19:** 376–398.

Mechanisms of reproductive aging: conserved mechanisms and environmental factors

Mary Ann Ottinger

Department of Animal and Avian Sciences, University of Maryland, College Park, Maryland

Address for correspondence: Mary Ann Ottinger, Department of Animal and Avian Sciences, University of Maryland, 3115 Animal Science Building, College Park, Maryland 20742. maottinger@umresearch.umd.edu

The interplay of neuroendocrine processes and gonadal function is exquisitely expressed during aging. In females, loss of ovarian function results in decreased circulating estradiol. As a result, estrogen-dependent endocrine and behavioral responses decline, including impaired cognitive function reflecting the impact of declining estrogen on the hippocampus circuits, and decreased metabolic endocrine function. Concurrently, age-related changes in neuroendocrine response also contribute to the declining reproductive function. Our session considered key mechanisms in reproductive aging including the roles of ovarian function (Finch and Holmes) and the hypothalamic median eminence (Yin and Gore) with an associated age-related cognitive decline that accompanies estrogen loss (Morrison and colleagues). Effects of smoking, obesity, and insulin resistance (Sowers and colleagues) impact the timing of the perimenopause transition in women. Animal models provide excellent insights into conserved mechanisms and key overarching events that bring about endocrine and behavioral aging. Environmental factors are key triggers in timing endocrine aging with implications for eventual disease. Session presentations will be considered in the context of the broader topic of indices and predictors of aging-related change.

Keywords: neuroendocrine aging; negligent senescence; perimenopause transition; healthy aging; insulin resistance; environmental factors

Introduction

The timing of the age-related decline in reproduction function depends on a complex interplay of interacting changes at all levels of the hypothalamic–pituitary–gonad (HPG) axis. In women, the perimenopause transition may be relatively short or persist over many years with few clues as to the triggers for the timing of these changes in an individual. Despite vast differences in lifetime reproductive strategies across vertebrates, there are interesting commonalities in aging processes. Comparison of patterns in reproductive aging across vertebrate species reveal that key elements and the underlying mechanisms that bring about age-related changes are conserved; that is, the fundamental sequence and elements of events that occur during aging are common across vertebrates. Comparative animal models and data from nonhuman primates provide valuable insights into the functional changes that accompany the age-related reproductive decline and how these changes relate to aging of other physiological and endocrine systems. Furthermore, there are interesting examples of species that have surprisingly long lifespan compared with those that are short-lived; for example, seabirds and tortoises have an extraordinarily long lifespan, whereas Japanese quail and mice are relatively short-lived. Early exposure to external environmental factors as well as lifetime habits, such as diet and disease, are likely critical triggers in the timing of reproductive decline and eventual senescence. These environmental factors may be proximate, such as smoking or lifestyle, or they may be inadvertent, such as exposure to contaminants. For example, recent data strongly supports the role of environmental chemicals that have the ability to mimic hormones as contributors to age-related demise in physiological function and neural disease.[1] Taken together, there are complex interactions directing the progression of aging processes, with the onset and time course of events potentially modified by environmental factors according to the target and mechanisms of effects associated with exposure.

Progression of endocrine events during aging

Age-related changes in the reproductive endocrine axis (HPG axis) follow individually dictated timelines. That is, a population and species have general characteristics in the timing and sequence of age-related decline in endocrine function; an individual may experience earlier, average, or later onset in the aging process. In general, males experience declining reproductive function in a gradual manner with significant reduction in circulating androgen levels relatively late in life. Reduction *in libido* and other age-related changes in metabolic endocrine and physiology gradually emerge, often preceding loss of spermatogenesis. These outcomes may be exacerbated by the use of pharmaceuticals, especially those that impact neural systems that also regulate endocrine and behavioral components of reproduction. Females experience a sharper decline in reproductive function, due to a combination of progressively erratic ovulations and eventual ovarian failure. The loss of ovarian function is accompanied by the loss of cycles in reproductive hormones. There is remarkable similarity in the pattern of ovarian decline across a variety of species, which provides a powerful comparison to understand basic mechanisms in aging processes (see the chapter by Finch and Holmes[2]). And although there is controversy regarding the extent and even occurrence of postreproductive lifespan, studies, especially in long-lived species, provide supporting data for life following cessation in reproductive function. For example, if there is a social structure that incorporates postreproductive roles for individuals in caring for young (grandmother hypothesis), then the function and presence of postreproductive individuals have been well documented. Despite the discussion about postreproductive lifespan, the events leading to the ultimate cessation of reproduction have been the source of extensive research, especially in rodent models. These studies and comparative data collected from primates and nonmammalian models point to the critical role of ovarian cycling and accompanying ovarian steroidogenesis in the maintenance of female reproduction. However, assessing the status of ovarian function during the aging process has proven difficult. Ovarian reserve tests have proven to be good indicators of ovarian follicular content and response.[3] A strong positive response to an ovarian reserve test can serve as a key indicator for continued ovarian cycling, fertility, and potential success with the use of assisted reproductive technologies. The responsiveness of ovarian follicles, especially related to steroid production, diminishes during the perimenopause transition leading to reduced estradiol production and lower circulating estradiol levels, which in turn translate into blunted stimulation of the preovulatory luteinizing hormone (LH) surge. This "domino effect" continues; the consequence of insufficient LH release is failure to ovulate, at least for some cycles. Ovulation becomes increasingly erratic accompanied by loss of progesterone production, which would have been produced postovulation. This general sequence of events occurs across females of various species and vertebrate classes.[4,5] Comparison of ovarian function in short- and long-lived species affords the opportunity to study reproductive aging processes over differing time scales. As such, aging progresses more slowly in long-lived organisms and there are coincident differences in lifetime reproductive strategies. These comparisons also provide a compelling case for insights from studies in the comparative biology of aging.

In primates, menopause is the signal of the complete collapse of ovarian function. The demise of ovarian function is accompanied by the loss of circulating estradiol and progesterone, and increased pituitary gonadotropin levels. The perimenopause transition may be quite extended, with declining fertility as early as the 30s and certainly in the 40s for women. Markers of the perimenopause transition have become much more precise at giving information about individual status relative to this transition. Decreasing ovarian follicular reserve is particularly revealing because it is associated with decreased steroid synthesis, reduced inhibin β and anti-mullerian hormone production.[3,6–10] As discussed earlier, decreased ovarian function has several ramifications including reduced steroid feedback and signaling to hypothalamic areas that modulate gonadotropin releasing hormone (GnRH) production and release. Concurrently, there is altered hypothalamic signaling to the pituitary gland against a background of reduced ovarian feedback, resulting in elevated follicle-stimulating hormone (FSH) and a transient increase in LH. As ovarian reserve decreases, ovulation becomes more sporadic with a coincident increase in variability in cycle length. During this transition, circulating steroids shift from cycle associated peak estradiol in the follicular phase and peak progesterone levels in the luteal phase, to unopposed elevated estradiol owing to the absence of ovulation. The unopposed estradiol production associated with stimulation of follicular development is exacerbated by increasing FSH and LH levels in response to diminished negative feedback due to loss of progesterone and inhibin β. This cascade is soon followed by ovarian failure and loss of estradiol as well as progesterone. These changes in the menstrual cycle of primates are shown in Figure 1, which shows cycles in the rhesus macaque. The figures present data from our studies which illustrate these shifting hormone levels during the perimenopause transition in the rhesus macaque in which regular cycles (Fig. 1A) are followed with irregular cycles and intermittent ovulation (Fig. 1B) and finally to menopause with loss of circulating steroid hormones (Fig. 1C).[3,8,11,12] It is interesting to note that a phase of unopposed high-circulating estrogen prior to ovarian failure also occurs in rodent models, in which females enter a phase of constant estrus associated with high levels of estradiol. However, this phase may persist for a longer period in rodents compared with primates.

Although transient, these changes in circulating steroid hormone levels represent a roller coaster in hormonal background that have unclear consequences for hormone-dependent systems, including brain regions responsible for cognition. As discussed in the chapter by Morrison and colleagues,[13] estradiol has neuroprotective actions as well as stimulates hippocampal neuronal spines; whereas, loss of estradiol is accompanied by diminished cognitive function and altered neuroplasticity.[14,15] Hormone replacement therapy (HRT) remains controversial for postmenopausal women. In some women, HRT effectively blunts the effects of these dramatic changes in circulating steroid hormone levels including diminished cognitive function. Risks associated with the use of HRT have led to additional studies which seek to resolve the controversy about the merits of HRT, including the use of native hormones versus conjugated steroids.[15]

Triggers that signal and direct the timing of the perimenopause transition in an individual are modulated by both internal and external factors. Because these are complex interacting factors, studies in nonhuman primates provide the most precise insights into the complexities of these interactions for biomedical applications and human populations.[10–12,16] Moreover, the applicability of a variety of possible interventions have been a rich source of options for aging populations as we seek to diminish the effects of age-related maladies and disease. Some of the most effective interventions have included dietary paradigms, such as caloric restriction and healthy diets.[4,12,16,17] Conversely, as reviewed in the chapter by Sowers and colleagues,[18] factors such as smoking, obesity, and insulin resistance can have adverse effects on the HPG axis, ultimately resulting in earlier and perhaps premature occurrence of the final cycle. It is also clear that there is a great deal of individual variation in both the response to these factors and the extent to which an individual may be affected. This means that the timing of menopause may be premature if the individual's ovarian function is severely impaired. However, in the event that ovarian processes are muted by environmental factors, there may be a less dramatic outcome, meaning that there may not be premature menopause but there may be more irregular cycling or prolonged perimenopause transition. Further, physiological status and disease also contribute to the overall health of the individual, making it difficult to discern clear causal factors; the mechanisms by which this interaction may occur remain complex and multifactor. Although the adrenal axis and stress-related effects are generally suspected as exacerbating symptoms of the perimenopause transition, much more information is needed. Males also exhibit age-related reproductive decline albeit more gradual than in females.[19] Moreover, there is evidence for benefits of dietary interventions including caloric restriction.[12,20,21] A key element in the aging process is increasing disruption in biological rhythms; in the case of reproductive endocrine function there are altered circadian rhythms that occur even at the level of the pituitary gland.[22] This breakdown in circadian rhythms also appears to be a fundamental aspect of the process of aging for physiological rhythms in general and is expressed in sleep–wake cycles and other behavioral responses.

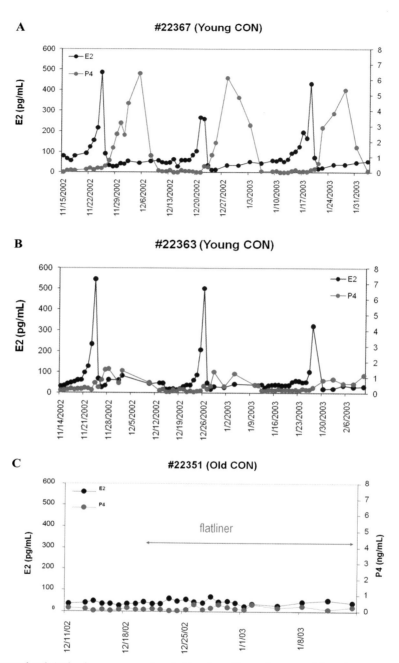

Figure 1. The menstrual cycle in the rhesus macaque female (data from dissertation research by Julie Wu, Ph.D.). Panels A, B, and C show the progression of hormone cycles over the reproductive lifespan of the female with a transition from (A) regular cycles to (B) irregular and anovulatory cycles and (C) collapse of ovarian function in the postmenopausal female (from Wu et al.[8]).

Neuroendocrine systems and aging

Although declining ovarian reserve and the associated decrease in steroid production are critical triggers in reproductive aging, neural systems also show a concomitant deterioration in functional response. The GnRH system is the primary hypothalamic regulator of reproduction. Despite extensive study to understand the role of the GnRH system in age-related reproductive failure, it has been difficult to discern clear changes in cellular morphology or integrity of the function response of neural systems that occur in correlation with diminished ovarian function and the perimenopause transition.[3] Recent data revealed altered functional responses of the GnRH system that contribute to fundamental biological changes that occur in the reproductive axis during aging.[10–12] The difficulties in identifying and characterizing these age-related changes are captured eloquently in the chapter by Yin and Gore.[23] They also describe the background provided by studies that measured GnRH release using push–pull perfusion. These studies documented different patterns in aging female rats and rhesus macaques by characterizing the neural mechanisms that regulate GnRH release at the level of the median eminence. As pointed out by the authors, there are differences in age-related changes, which likely reflect different phases in the aging process. These include the transitional phase that occurs during the loss of ovarian function and concomitant decline in ovarian steroids. This phase is followed by the next phase as ovarian steroid feedback to the hypothalamus diminishes; and neural systems attempt to compensate for the loss of estradiol by increasing amplitude and frequency of GnRH release. As Yin and Gore[23] have observed, the glial cells that surround the GnRH axonal terminals have the capacity to modulate release of GnRH. The significance of an age-related change in the interaction of GnRH axonal terminals and the surrounding tanycytes in the median represent a fundamental piece of the puzzle in terms of the functional elements of hypothalamic systems as they become altered in aging female rats. Ultimately, these changing interactions and altered modulation of GnRH combine with changing signals from the ovarian steroids to produce an age-related functional response leading to eventual reproductive failure.

Parallel findings from comparative studies in the quail model

We have conducted studies on reproductive endocrine and behavioral aging in a short-lived avian model, the Japanese quail. Our initial studies revealed that the process of aging follows a chronological pattern similar to that observed in mammals. That is, male Japanese quail undergo a gradual loss in the frequency and intensity of sexual behavior. The regulation and age-related decline mirror processes observed in mammals, especially those observed in primates that transition from regular cycles to irregular cycling and finally ovarian failure. As pointed out by Finch and Holmes,[2] the age-related loss of ovarian reserve parallels that observed in mammals. Similarly, there is a period of unopposed estrogen exposure during which ovulation does not reliably occur followed by collapse of ovarian function and reproductive failure.

A number of molecular forms of GnRH have been identified in other species and classes of vertebrates. In quail, of the two forms of GnRH, GnRH-I is the reproductively active form secreted under the modulation of the neural system that regulates reproductive endocrine function. Neurons containing GnRH-I originate from the preoptic-septal region with release at the median eminence, similar to mammals.[11] In addition to conserved anatomical location of the GnRH system, we have shown an age-related decrease in GnRH-I release *in vitro* from hypothalamic slices taken from young and from reproductively senescent male and female quail (Fig. 2).[5] As shown in Figure 2A and 2B, young females and males show strong response to norepinephrine (NE) release (denoted by the bars on the figures); whereas reproductively senescent females and males show reduced response to NE stimulation and a more erratic release response. When averaged over males and females, both baseline and NE induced GnRH-I release diminish with age (Fig. 2C). These data show similar GnRH-I release patterns to those reported in reproductively senescent mammals, reinforcing the contention that many of the age-related changes are conserved across vertebrate classes.[4,19–21]

A remarkable attribute observed in some non-mammalian species is the plasticity of neural systems throughout the life cycle and the ability to regenerate select tissues. The latter has been demonstrated in phylogenetically lower species, such as lizards, that regenerate appendages and of course invertebrates that can regenerate major portions of their bodies. The neural plasticity observed in a number of species has been of great interest, especially when comparing reproductive aging in short- or long-lived species. Although not entirely surprising, given extensive studies on birth of new cells in the central nervous system throughout life, neural plasticity enables the individual to continue

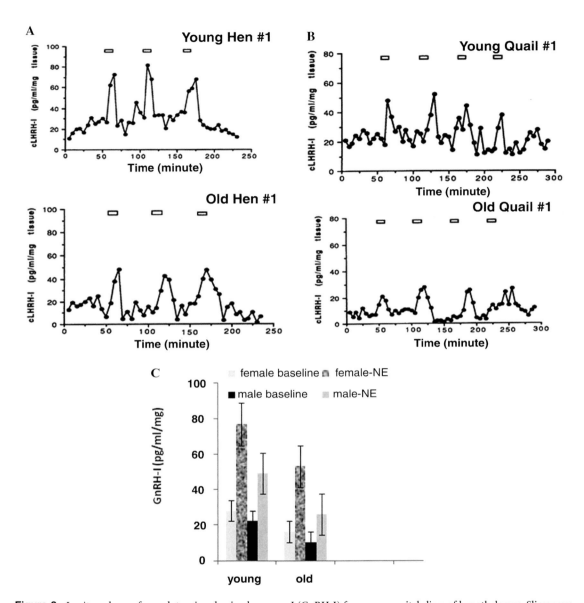

Figure 2. *In vitro* release of gonadotropin releasing hormone-I (GnRH-I) from parasaggital slices of hypothalamus. Slices were taken from young reproductive and old senescent males and females. An age-related diminution in pulse height occurred in response to norepinephrine (NE) stimulation, which was administered as a single injection made at three to four time points (denoted by bars) in both females (A) and males (B). Both baseline and NE induced GnRH-I differed with aging in males and females as shown in the bar graph (based on Ottinger[5] and Ottinger et al.[1]).

to reproduce during aging. A specific example is the Japanese quail, which is a short-lived bird that undergoes age-related reproductive decline. As noted earlier, in males, similar to mammals, plasma androgen levels decline gradually accompanied by loss of reproductive behavior, including courtship and mating behaviors. Albeit the loss of behavior is not directly correlated to declining circulating androgen levels in aging mammals or birds, testosterone replacement restores male sexual behavior and the associated neural systems that modulate these behaviors.[5,19,24] Study of these resilient

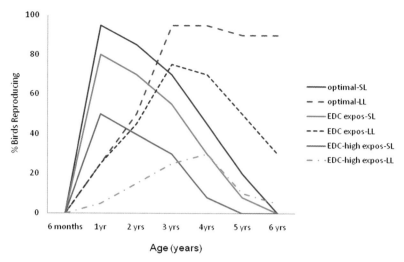

Figure 3. Schematic diagram showing the potential effects of exposure to endocrine disrupting chemicals (EDCs) over the lifespan of short-lived (SL) or over part of the lifespan for long-lived (LL) birds. The schematic is hypothetical and is based on information about longevity data from birds and the impact of EDCs on birds, both in field observations and from laboratory studies. In birds, reproduction peaks quickly in SL birds and declines quickly with the short lifespan. In these species, intermittent exposure to EDCs would be predicted to impair reproduction, which may impact SL species more drastically than LL birds depending on the timing of exposure during the life cycle. High exposure would be deleterious for both SL and LL birds resulting in reduced fitness and possibly shortened reproductive lifespan (based on Ottinger et al.[1] and Gore[25]).

neural systems and the mechanisms by which they are restored in aging individuals will augment our understanding of age-related alterations leading to reproductive senescence.

Environmental factors: endocrine disruption and potential effects on aging

In recent years, increasing attention has been paid to the effects of endocrine disrupting chemicals (EDCs) that have endocrine activity in wildlife, domestic species, and humans. Because we all carry a body burden of environmental chemicals and EDCs, it is important to determine the potential impact of lifelong exposure and discern differences in vulnerability to exposure. There is also an issue of variable sensitivity according to the phase of the life cycle. Although EDCs are generally not lethal at field relevant concentrations, it has become clear that there are both species differences in sensitivity to EDCs as well as individual differences in response to EDC exposure.[1] The question of early exposure to EDCs and effects on later reproductive lifespan may be a critical aspect of aging processes, especially if exposure attenuates reproductive function prematurely. In addition, it is important to consider lifetime exposure leading to a body burden of resident EDCs, especially in lipids stored in fat cells or liver. In birds and other wildlife, EDCs may alter sexual differentiation and impair reproductive success in adults.[1] This is shown schematically for a species of migratory bird in Figure 3, in which optimal conditions foster maximal breeding success and higher exposure is predicted to result in reduced reproductive success in breeding pairs and potentially fewer birds reproducing over their lifespan. There are also compelling data showing that EDCs can directly impact the GnRH system as well as reproductive behavior, including courtship, mating, and parental behavior.[1,25,26] EDC effects are likely to impact the sexual differentiation of neuroendocrine systems that modulate reproductive behavior during embryonic and early post hatch development. It is important to note that this relative exposure to estradiol and androgen is permuted by additional steroid or EDCs that interact with endogenous systems during this critical organizational period. Rather, it is a precise balance in exposure to steroids that dictate the organizational processes, making the avian system vulnerable to added steroid-like effects from EDCs. It is also becoming increasingly clear that precocial birds are more vulnerable to these organizational effects of EDCs during embryonic development;

whereas, altricial species may suffer less effect from embryonic EDC exposure but greater lifetime impacts from EDCs.[1] Therefore, there is evidence for neural effects of EDCs in both mammals and birds, which are likely to impact reproductive function in adults.

Conclusions

In summary, there is evidence for conserved mechanisms in the process of age-related reproductive decline. Ovarian follicular reserve declines across species resulting in decreasing steroid production. As ovarian function declines, there is a period of elevated estrogen levels unopposed by ovarian progesterone owing to intermittent ovulation. This is followed by loss of circulating estradiol with the collapse of ovarian function and associated impacts on cognitive function. Coincidently, the hypothalamic GnRH system shows functional alterations attributable in part to the glial cells that modulate GnRH release from the axonal terminals. Studies in comparative models of aging support the contention that many of these processes are common among vertebrate species. Finally, environmental factors including smoking, diet, and EDCs are likely to alter reproductive function and the timing of reproductive senescence in individuals.

Acknowledgments

This paper was prepared as a contribution to the Workshop on Reproductive Aging sponsored by the National Institute on Aging and Georgetown University and hosted by the Center for Population and Health at Georgetown University. The author would like to acknowledge Dr. Patricia Goldman, who encouraged me to enter into the world of aging research and Dr. Caleb Finch, who enabled many to pursue studies in the comparative biology of aging. The author thanks many other wonderful collaborators, especially at the NIA and ONPRC, who continue to travel with me in studies on aging and whose work is also cited in this paper. Special thanks to Dr. Julie Wu, who persisted in understanding the endocrine changes in aging nonhuman primates and to Dr. Mary Zelinski, a premier researcher and wonderful collaborator in our studies on aging in primates. Research reviewed here was supported in part by the Intramural Research Program of the National Institute on Aging, NIH; NIH #U01-AG21380-03 (MZ and MAO); USEPA STAR grant (MAO); and NIH #1 R21-AG031387 (MAO).

Conflicts of interest

The author declares no conflicts of interest; NIH.

References

1. Ottinger, M.A., E. Lavoie, K. Dean, *et al.* 2010. Endocrine disruption and senescence: consequences for reproductive endocrine and neuroendocrine systems, behavioral responses and immune function. In *Neurotoxicology*, 3rd ed. H. Tilson & J. Harry, Eds.: 138–155. Informa Healthcare. New York, NY.
2. Finch, C.E. & D.J. Holmes. 2010. Ovarian aging in developmental and evolutionary context. *Ann. N.Y. Acad. Sci.* **1204:** 82–94.
3. Wu, J.M., D.L. Takahashi, D.K. Ingram, *et al.* 2010. Ovarian reserve tests and their utility in predicting response to controlled ovarian stimulation in rhesus monkeys. *Am. J. Primatol.* Accepted.
4. Ottinger, M.A., M. Mobarak, M.A. Abdelnabi, *et al.* 2005. Effects of caloric restriction on reproductive and adrenal systems in Japanese quail: are responses similar to mammals, particularly primates? *Mech. Ageing Dev.* **126:** 967–975.
5. Ottinger, M.A. 2007. Neuroendocrine aging in birds: comparing lifespan differences and conserved mechanisms. Special issue of *Aging Research Reviews* entitled: Genetics of Aging in Vertebrates. **6:** 46–53.
6. Klein, N.A., B.S. Houmard, K.R. Hansen, *et al.* 2004. Age-related analysis of inhibin A, inhibin B, and activin a relative to the intercycle monotropic follicle-stimulating hormone rise in normal ovulatory women. *J. Clin. Endocrinol. Metab.* **89:** 2977–2981.
7. van Rooij I.A., F.J. Broekmans, E.R. te Velde, *et al.* 2002. Serum anti-Mullerian hormone levels: a novel measure of ovarian reserve. *Hum. Reprod.* **17:** 3065–3071.
8. Wu, J.M., M. Zelinski-Wooten, D.K. Ingram & M.A. Ottinger. 2005. Ovarian aging and menopause: current theories and models. *Exp. Biol. Med.* **230:** 818–828.
9. Tremellen, K.P., M. Kolo, A. Gilmore & D.N. Lekamge. 2005. Anti-mullerian hormone as a marker of ovarian reserve. *Aust. N. Z. J. Obstet. Gynaecol.* **45:** 20–24.
10. Downs, J.L. & H.F. Urbanski. 2006. Neuroendocrine changes in the aging reproductive axis of female rhesus macaques (*Macaca mulatta*). *Biol. Reprod.* **75:** 539–546.
11. Ottinger, M.A., J.A. Mattison, M.B. Zelinski, *et al.* 2006. The rhesus macaque: a biomedical model for human health issues, aging, and cognition. In *Handbook of Models for Human Aging*. P. Michael Conn, Ed.: 457–468. Amsterdam. Elsevier.
12. Roth, G.S., J.A. Mattison, M.A. Ottinger, *et al.* 2004. Rhesus monkeys: relevance to human health interventions. *Science* **305:** 1423–1426.
13. Dumitriu, D., P.R. Rapp, B.S. McEwen & J.H. Morrison. 2010. Estrogen and the aging brain: an elixir for the weary cortical network? *Ann. N.Y. Acad. Sci.* **1204:** 104–112.
14. Wise, P.M., D.B. Dubal, M.E. Wilson, *et al.* 2001. Estradiol is a protective factor in the adult and aging brain: understanding

of mechanisms derived from in vivo and in vitro studies. *Brain Res. Rev.* **37:** 313–319.
15. Rapp, S.R., M.A. Espeland, S.A. Shumaker, *et al.* 2003. WHIMS Investigators. Effect of estrogen plus progestin on global cognitive function in postmenopausal women: the Women's Health Initiative Memory Study: a randomized controlled trial. *J. Am. Med. Assoc.* **289:** 2663–2672.
16. Downs, J.L., J.A. Mattison, D.K. Ingram & H.F. Urbanski. 2008. Effect of age and caloric restriction on circadian adrenal steroid rhythms in rhesus macaques. *Neurobiol. Aging* **29:** 1412–1422.
17. McShane, T.M. & P.M. Wise. 1996. Life-long moderate caloric restriction prolongs reproductive life span in rats without interrupting estrous cyclicity: effects on the gonadotropin-releasing hormone/luteinizing hormone axis. *Biol. Reprod.* **54:** 70–75.
18. Sowers, M.F.R., D. McConnell, M. Yosef, *et al.* 2010. Relating smoking, obesity, insulin resistance, and ovarian biomarker changes to the final menstrual period. *Ann. N.Y. Acad. Sci.* **1204:** 95–103.
19. Ottinger, M.A. 1998. Male reproduction: testosterone, gonadotropins, and aging. In *Functional Endocrinology of Aging*, Vol. 29. C.V. Mobbs & P.R. Hof, Eds.: 105–126. Karger Press. Basel.
20. Sitzmann, B.D., H.F. Urbanski & M.A. Ottinger. 2008. Aging in male primates: reproductive decline, effects of calorie restriction and future research potential. *AGE* **30:** 157–168.
21. Sitzmann, B.D., J.A. Mattison, D.K. Ingram, *et al.* 2009. Impact of moderate calorie restriction on the reproductive neuroendocrine axis of male rhesus macaques. *Open Longevity Sci.* **3:** 38–47.
22. Sitzmann, B.D., J.A. Mattison, D.K. Ingram, *et al.* 2010. Effect of moderate calorie restriction on pituitary gland gene expression in the male rhesus macaque (*Macaca mulatta*). *Neurobiol. Aging* **31:** 696–705.
23. Yin, W. & A.C. Gore. 2010. The hypothalamic median eminence and its role in reproductive aging. *Ann. N.Y. Acad. Sci.* **1204:** 113–122.
24. Ottinger, M.A., I.C.T. Nisbet & C.E. Finch. 1995. Aging and reproduction: comparative endocrinology of the common tern and Japanese quail. *Amer. Zoologist* **35:** 299–306.
25. Gore, A.C. 2008. Neuroendocrine systems as targets for environmental endocrine disrupting chemicals. *Fertil. Steril.* **89**(2 Suppl.): e103–e108.
26. Ottinger, M.A., E.T. Lavoie, N. Thompson, *et al.* 2009. Is the gonadotropin releasing hormone vulnerable to endocrine disruption in birds? *Gen. Comp. Endocrinol.* **163:** 104–108.

ANNALS OF THE NEW YORK ACADEMY OF SCIENCES
Issue: *Reproductive Aging*

Ovarian aging in developmental and evolutionary contexts

Caleb E. Finch[1] and Donna J. Holmes[2]

[1]Ethel Percy Andrus Gerontology Center, Department of Biological Sciences, University of Southern California, Los Angeles, California. [2]School of Biological Sciences, Center for Reproductive Biology, Washington State University, Pullman, Washington

Address for correspondence: Donna J. Holmes, School of Biological Sciences, Center for Reproductive Biology, P.O. Box 644236, Washington State University, Pullman, Washington 99164-4236. djholmes@wsu.edu

Evolutionary theory predicts that aging-related fertility declines result from tradeoffs between reproduction and somatic maintenance. Developmental programs for oogenesis also contribute to variation in aging-related reproductive declines among female vertebrates. Documented reproductive aging patterns in female vertebrates, including humans, are consistent with canonical aging patterns determined developmentally and require no special adaptive explanation. Here we discuss patterns of aging-related ovarian decline in diverse female vertebrates, and place human ovarian aging in comparative context. Depletion of finite oocyte stores accompanied by fertility loss occurs in a variety of nonhuman mammals and vertebrates, including short-lived rodents, birds, and some fishes; moreover, postreproductive lifespans of considerable length clearly are not limited to long-lived, social species with well-developed kin networks. We argue for a more rigorous comparative approach for understanding the evolutionary and developmental bases of ovarian aging in vertebrates with a wider range of aging patterns and social structures.

Keywords: follicular reserve; oocytes; ovarian aging; ovarian development; menopause; vertebrates

Introduction

Evolutionary aging and life-history theory predict that natural selection will not actively promote somatic integrity beyond the reproductive period. Successful reproduction is the only real evolutionary currency, and under most circumstances selection is expected to eliminate germline mutations that decrease reproductive success or are otherwise detrimental to genetic fitness. Aging-related fertility declines generally can be viewed as a result of evolutionary, genetic, and physiological tradeoffs between investment in reproduction and long-term somatic maintenance, or between early-life payoffs and late-life costs.[1–4] Variation in developmental programs is associated with variation among species in life-history patterns and fitness tradeoffs, including tradeoffs between patterns of lifetime reproductive investment, somatic maintenance, and longevity. Variation in developmental programs is also expected to shape differences among female vertebrates in other aging-related physiological declines and the accumulation of damage thought to be responsible for clinical aging syndromes and related disease states.[2,5,6] From this perspective, reproductive aging patterns in most female vertebrates for which data exist—including humans—resemble typical or "canonical" aging syndromes reflecting loss of adaptive function in other physiological systems,[6–8] and hence they do not require any special evolutionary explanation.

This chapter addresses aging-related patterns of ovarian decline and other aspects of reproductive aging in female vertebrates. We survey current information about ovarian aging in major vertebrate groups, and place patterns of aging-related fertility declines in human females in comparative context, particularly in relation to extensive studies of reproductive aging in female laboratory rodents. We show that human menopause and postreproductive lifespans (PRLSs) do not constitute an unusual evolutionary case. Nor do they require that we invoke special adaptive explanations, like kin selection or intergenerational resource transfer, in which

Table 1. Documented evidence of reproductive aging and PRLS consistent with progressive depletion of finite oocyte stores in major vertebrate groups

Vertebrate group					Reproductive aging patterns/available information on female reproductive aging	Documented female PRLSs
Gnathostomes	Anamniotes		Agnatha	Jawless fishes (hagfish and lampreys)	---	---
				Cartilaginous fishes (sharks, rays, etc.)	---	---
				Bony fishes	Iteroparity with gradual senescence in some (guppies, platyfish); extremely rapid fertility loss with semelparity in some (salmonids); no or extremely slow female reproductive aging in some (rockfish); data generally sparse[11,12,21]	Guppies: yes; platyfish: no; rockfish (2 spp.): no
	Tetrapods			Amphibians (frogs, salamanders, etc.)	---	---
		Amniotes		Reptiles (lizards, snakes, alligators, turtles)	Turtles: increasing reproductive success with age[9,20] (Bronikowski unpublished data)	Turtles: no Garter snakes: no
				Birds	Galliforms; passerines[16,19]	Quail, chicken: yes Some passerine songbird species in captivity: yes
				Mammals	Monotremes (platypus, echidna): no data	---
					Marsupials (opossums, kangaroos, etc.): semelparity in some marsupial mice (*Antechinus* spp.)[10]; gradual fertility loss in some New World opossums[22] (Holmes unpublished); data sparse	New World opossum species: yes
					Placentals: Data available from captive or wild representatives showing gradual fertility loss with oocyte depletion or PRLSs in 8-9 orders (see Fig. 2)[14]	Yes: eight additional lineages (see Fig. 2)

Classification scheme from Ref. 47. Additional references cited in text.

females sacrifice their own continued reproductive success or forgo further reproductive risk for the benefit of daughters and other close kin. We argue that human menopause is physiologically similar to patterns of fertility loss that typify reproductive senescence in many other mammals and nonmammalian vertebrates. When placed in this comparative context, ovarian aging syndromes in women and females of many other species appear to be more consistent with waning natural selection with age or a failure of developmental programs than with expectations for adaptive physiological systems shaped by natural selection to function well during postreproductive life.

Female reproductive aging in phylogenetic context

Vertebrate animals exhibit a striking variety of reproductive aging patterns,[9–22] ranging from the dramatic and rapid declines in fertility seen in semelparous species, like migratory salmon and some mouse-sized marsupials, to the extremely slow or even negligible decreases in reproductive success documented for some seabirds, some reptiles (e.g., turtles), and some fishes (e.g., rockfish) (Table 1). Markers of physiological senescence—including gonadal changes—are now documented in natural populations of a variety of birds and mammals, and there is demographic evidence consistent with aging-related reproductive declines in females from a much wider range of vertebrate taxa than generally recognized, including some fishes. This evidence provides an important comparative perspective on human ovarian aging. The ovarian and neuroendocrine correlates of female reproductive aging, however, have been well studied in only a few nonhuman mammalian models—primarily laboratory strains of mice and rats, as well as several species of primates. For nonmammalian taxa (birds, reptiles, amphibians, and fish), physiological or histological data are even sparser.

Demographic data from a wide cross-section of captive and wild homeotherms ("warm-blooded" mammals and birds) suggest that after peak

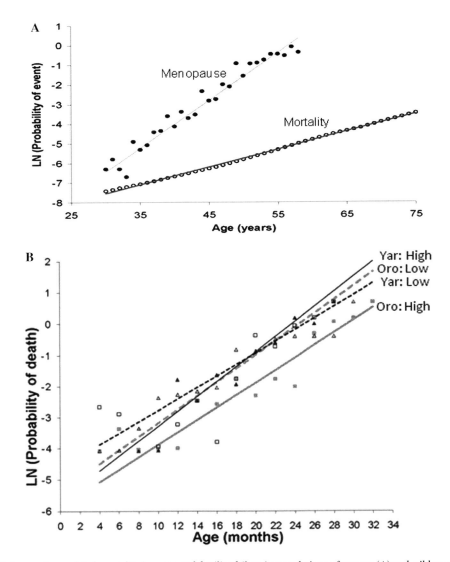

Figure 1. Patterns of age-related mortality increase and fertility failure in populations of women (A) and wild guppies (*Poecilia guttata*) (B, C) (figure continued next page). These are both consistent with canonical aging patterns in female animals that undergo gradual loss of ovarian reserve or capacity to produce new oogonia. Guppies (B, C) are derived from two populations (Oropuche and Yarra) with different histories of natural selection by predation (high and low). In C, data are for guppies maintained in captivity for their entire lifespans; a–d: the proportion of the population still reproducing at a given age (*y*-axis) versus age at last reproduction (*x*-axis); high predation (solid line), low predation (stippled line); e–h: the natural log of negative natural log of the percent still reproducing (*y*-axis) versus age at last reproduction (natural log-transformed, *x*-axis); high predation (filled triangles); low predation (open triangles); Oropuche, high food (a, e); Oropuche, low food (b, f); Yarra, high food (c, g); Yarra, low food (d, h). (A and B reprinted from Ref. 68 with permission of the author and Elsevier Press; C reprinted from Refs. 58, 59 with permission from the authors and Nature Publishing Group.)

reproductive maturity is reached, subsequent declines in female fecundity are the rule rather than the exception, and these declines may be considered to be among the canonical patterns of aging,[10] in the sense that they typify physiological and organismal senescence across a range of animal taxa (see, e.g., Refs. 14 and 17). Gradual, monotonic aging-related declines in female reproductive success generally resembling those in women and nonhuman primates are well documented in captive vertebrate species as

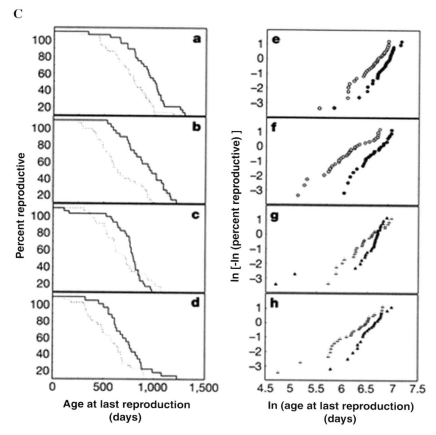

Figure 1. *Continued.*

diverse as laboratory mice, domestic Japanese quail, and wild guppies (Fig. 1).

Human female reproductive aging—midlife cessation of ovulation with a subsequent postreproductive period of up to half of the total lifespan—has been suggested to be unusually sudden, taxonomically rare or otherwise exceptional among mammals, and hence to require special evolutionary explanation in the form of kin selection.[14,23] In this scenario, females forgo their own continued reproduction (or the increasing risks it may entail) in exchange for contributing to the health and reproductive success of their close kin.[24,25] But a thorough understanding of the physiological constraints, phylogenetic origins, and evolutionary correlates of human female reproductive aging requires that these patterns be placed in a broader and more rigorous comparative context, both for nonhuman mammals and among vertebrates generally.

Mammals

Data from captive and wild populations of mammals of a range of orders show steady declines in female fecundity variables (e.g., live births and weaning rates) that are consistent with gradual exhaustion of finite oocyte stores from birth onward over the course of the reproductive lifespan (Fig. 2).[11,12,14,17] Moreover, PRLSs of considerable length ($\geq$20–30% of maximum documented species lifespan) have been documented in at least nine phylogenetically diverse mammalian lineages. These lineages include some short-lived, nongregarious marsupials and rodents, as well as a variety of longer-lived, social species, such as primates and some whale species. Hence PRLSs are not limited to—nor do they occur consistently in—long-lived, social species with well-developed kin networks, as would be predicted by kin selection theories for the evolution of menopause.[13,14,26,27] On the other hand, the longest documented female mammalian

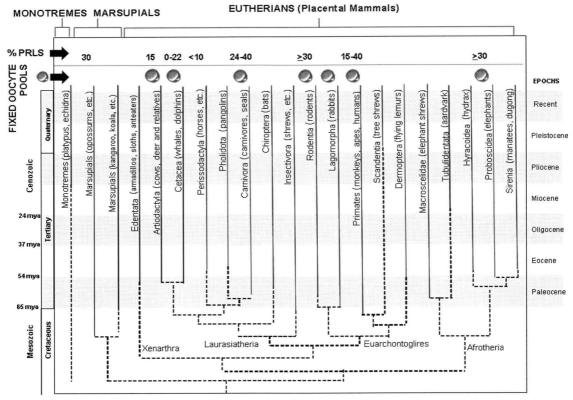

Figure 2. Phylogram summarizing evidence of reproductive aging consistent either with depletion of finite oocyte pools (symbol) or documented PRLSs (as percent of documented maximum species lifespan) of significant length in nine mammalian orders. These lineages have been phylogenetically distinct for at least 50 million years, suggesting that ovarian aging patterns in mammals are evolutionarily conservative. Redrawn from Ref. 47; see also Ref. 48.

postreproductive periods (25–50% of total documented lifespan) occur in females of highly social, long-lived species with extended periods of maternal care: women, including women both in industrialized societies and some traditional unindustrialized settings (e.g., the !Kung, Ache, and Hadza peoples) and at least two species of long-lived whales (e.g., short-finned pilot whale, *Globicephalus macrorhyncus* and killer whale, *Orcinus orcus*) living freely in nature.

Aging-related depletion of a finite ovarian follicular pool. In laboratory rodent models, the physiological correlates of female reproductive senescence have been explored in considerable depth. Rodent studies have included intensive analysis of ovarian and neuroendocrine changes across the entire reproductive lifespan, including aging-related changes in the hypothalamic-pituitary-ovarian axis (Yin and Gore chapter).[28–31] In mice, as in humans, the pool of primary follicles and oocytes declines exponentially during prenatal development, with less than 10% of the original embryonic follicular pool surviving to reproductive maturity (Fig. 3). The vast majority of primary follicles and oocytes are formed during embryonic development (for recent discussions of the controversy surrounding the possibility of *de novo* oogenesis in adult mammals, see Tilly *et al.* [pro][32] and Begum *et al.* [con].[33] The initial size of the pool of oocytes is a major determinant of the time of onset of reproductive senescence, as has been shown by elegant experiments by Nelson and Felicio involving partial removal of ovarian tissue in mice[34] (Fig. 4). The timing of onset of aging-related ovarian acyclicity is precisely related to the mass of ovarian tissue removed: when the depletion of follicles reaches a species-specific threshold, reproductive cycles cease. The onset of fetal

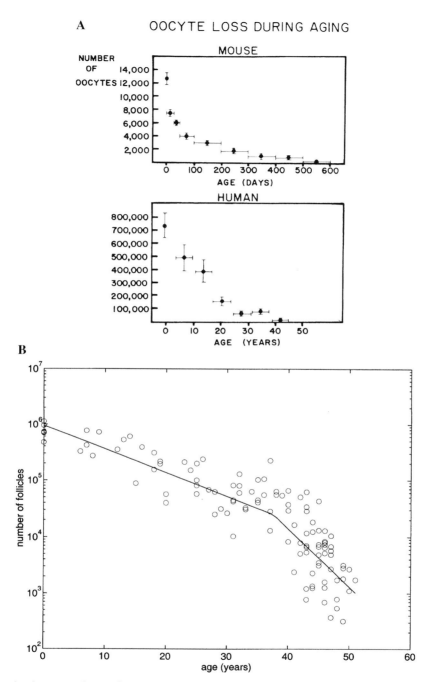

Figure 3. Age-related patterns of oocyte loss in A-strain laboratory mice and women (A); changes in rate of oocyte loss with age in women (B). In (B), open circles represent observed total follicle counts and lines represent results of "broken-stick" regression analysis, showing a significant increase in the rate of follicle loss at around 38 years of age. (A) Reprinted from Ref. 10 with permission of the author and University of Chicago Press (original data from Ref. 79; (B) reprinted from Ref. 80 with permission of the author and publisher.)

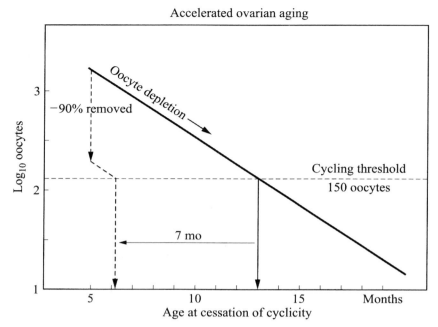

Figure 4. Regression model predicting effects of partial ovariectomy on timing of reproductive aging in mice. Nelson and Felicio calculated that surgical reduction of 90% of follicles should accelerate age of cessation of estrous cycles by 7 months; this prediction was confirmed by monitoring vaginal cytology. (Reprinted from Ref. 36 with permission of the author and The Oxford University Press; original redrawn from Ref. 34.)

aneuploidy, an additional marker of reproductive senescence that precedes acyclicity, is also accelerated by hemi-ovariectomy in rodents.[35] Such experimental manipulations of the ovarian follicular pool in rodents provide models for premature reproductive senescence syndromes arising from partial ovarian dysgenesis in humans. These studies are also relevant to the controversy surrounding the possibility of adult *de novo* oogenesis, since they have shown no indication of compensation for experimentally induced follicular deficits with the formation of new oocytes.

Individual and chance variation in ovarian aging. Major variations of follicular number are present at birth in inbred mice and general human populations (Fig. 5),[36] suggesting a critical role for individual differences in the embryonic follicular endowment in the individual duration of fertility. This may be relevant to the exceptional PRLS seen in some primates: individuals in natural-fertility populations of chimpanzees and humans that continue childbearing to unusually advanced ages[37,38] may have had exceptionally large follicular endowments initially. Conversely, premature reproductive senescence could arise from deficits in follicular endowment, as well as from specific pathological conditions. In inbred mice, variation in the size of the follicular endowment among individual females can be seen throughout the period of embryonic development of the ovary. Primordial germ cells from which oocytes arise show substantial stochastic variation in proliferative capacity, as shown in *in vitro* studies in which cells were grown on soft agar.[36,39] Future studies are needed to tease apart the complex regulatory interactions between the aging ovary and the hypothalamic-pituitary system. A systems engineering approach may ultimately reveal important sources of developmental variation at many levels besides the ovary.[29,30,36,39]

PRLSs: a general mammalian trait? From an evolutionary standpoint, reproductive aging with a definitive PRLS is more likely to be seen in captive colonies or in other situations where animals are protected from natural sources of mortality or, alternatively, where natural levels of adult mortality are low and very stable.[40] In a survey of female

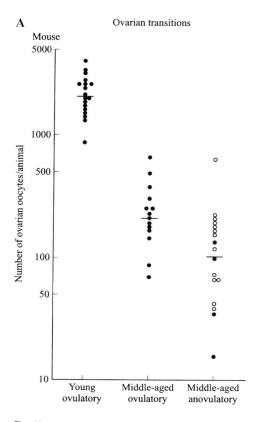

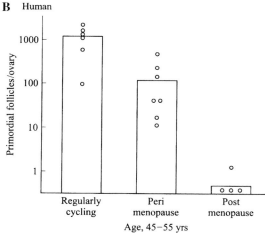

Figure 5. Chance variation in numbers of oocytes in C57BL/6J mice and women at different ages. (A) By middle age, oocyte numbers vary from zero in prematurely senescent ovaries in mice to >1,000 in mice of the same age and strain. (B) Premenopausal women show a similar range (>1,000-fold) of variation. (Reprinted from Ref. 36 with permission from author and The Oxford University Press; original data from Ref. 81.)

reproductive aging in mammals from approximately 40 species, including captive and wild populations, Cohen[14] found evidence of PRLSs ranging from 14 to over 50% of species' maximum lifespans in seven of eight phylogenetically distinct mammalian orders. These groups include a disproportionate number of long-lived species with well-developed social behavior and kin networks, and which have been domesticated or studied in captive colonies (e.g., nonhuman primates; cats; dogs; marine mammals). We would emphasize, however, that PRLSs also are typical of short-lived rodents[12,41] and some nongregarious marsupials (e.g., *Monodelphis* and *Didelphis* New World opossum spp.) (Holmes, unpublished data) in captivity. Thus, the occurrence of PRLS clearly does not depend on the existence of kin networks, since these are either absent or rudimentary in nongregarious species that engage in only short periods of parental care, and in which nonreproductive females would be expected to transfer social or genetic benefits to close kin. On the other hand, some highly social mammals, including long-lived whales in natural populations and naked mole-rats in captivity, have shown no appreciable decline in reproduction and no evidence of a PRLS, even when records for hundreds of individuals are examined (R. Buffenstein, personal communication).[14,26,42–44] Moreover, among highly social species, the occurrence of a PRLS extends to those lacking clear matrilineal kin-support networks. The golden-lion tamarin, for example, has documented postreproductive periods in captivity of over 25% of total lifespan (unpublished data, National Zoo Studbook, courtesy of J.D. Ballou). Females of some marmoset species display anovulatory cycles associated with documented depletion of follicular stores at older ages closer to the maximum lifespan.[45] This pattern of female reproductive aging is particularly interesting in view of the fact that marmoset fathers provide a substantial portion of parental care, and older, dominant females may actively inhibit reproduction by daughters.[46]

In sum, a gradual pattern of female reproductive aging, with substantial postreproductive periods consistent with expectations for mammals with fixed oocyte pools, appears to be a conservative mammalian trait seen in a wide range of orders with their last common ancestors occurring at least 50 million years ago (Fig. 2).[47,48] Aside from domestic rodents, some monkeys and humans, however, for most mammalian species with documented female reproductive declines there are few data on dynamics of follicular depletion or other physiological variables. The question remains whether the declines in fecundity seen in diverse groups of mammals can be attributed to physiologically homologous patterns of ovarian decline; a great deal more comparative data are needed to address this issue.

Birds

In general, birds age considerably more slowly than mammals of equivalent size. However, most documented patterns of avian reproductive aging are characterized by gradual to slow senescence, versus the extremely slow to negligible fertility loss seen in some fishes and reptiles.[16,18,19,49] While some data are available for wild birds maintained in captivity, most information on avian reproductive senescence come from banding and monitoring studies of wild bird populations. Much less is known for birds than mammals about the comparative prevalence or length of avian PRLS.

Surveys of the literature on aging-related changes in avian demographic variables have shown statistically reliable changes in some measures of reproductive success in over 35 species of wild birds from a wide range of orders.[17,50] Declines have been documented in many species for specific "female" reproductive variables, including number of eggs laid, clutch size, and hatching and fledging rates. Brunet-Rossinni and Austad,[17] in a recent review by of the literature on avian aging in the wild, identified diverse avian species in which older females laid eggs reduced in size or volume; older females in some of these species also laid fewer eggs per clutch at older ages. Lower hatching rates or other measures of fertility or fecundity (indirect indicators of ovulatory rates or egg quality) have also been documented for older female birds. These kinds of changes are consistent with, but not conclusive evidence of, ovarian aging, since male aging effects and additional behavioral factors can not be distinguished from female physiological variables.

Direct evidence of limited follicular reserves is available for very few avian species.[16,51] For aged chickens and domestic quail, there is both endocrinological and functional evidence of depletion of the ovarian reserve.[52,53] Random counts of primordial follicles in Japanese quail showed significant declines (to below 35% of original numbers of primordial follicles) in the ovarian reserve in newly postreproductive Japanese quail hens,[54] relative to young (peak egg production) and middle-aged layers (Holmes, Wu, and Ottinger, unpublished data). This is consistent with the mammalian pattern of exhaustion of fixed oocyte pools. However, few studies of ovarian, endocrine, or neuroendocrine changes have been conducted in wild or nonpoultry avian species. In a study of wild common terns (*Sterna hirundo*), the oldest females showed little decline in reproductive success, and no aging-related changes in plasma steroid hormones.[55] Histological examination of ovaries of birds from this tern population again showed declining numbers of primordial and primary follicles (Holmes and Nisbet, unpublished data).

Like mammals, birds are homeothermic amniotes. However, birds are oviparous and produce yolky eggs, in contrast with most mammals. Presumably, female birds also produce the vast majority of oogonia before sexual maturity.[56] Studies of ovarian aging in a wider range of bird orders with different reproductive aging patterns are needed to clarify the role of a finite ovarian reserve in reproductive aging; it remains unknown whether declines in fixed oocyte pools or other aspects of ovarian aging seen in representatives of these vertebrate groups are the result of shared phylogeny, or arise from separate evolutionary trajectories.

Nonhomeothermic vertebrates: turtles and fishes

Limited data are available on female reproductive aging in nonhomeothermic vertebrates, including a few species of turtles and fishes. Most of these species are iteroparous spawners, and some appear to have very slow or negligible changes in reproductive function. In extensive longitudinal studies of the long-lived Blanding's tortoise (*Emydoidea blandingii*), the oldest females (≥ 60 years) showed no reliable declines in reproductive success consistent with ovarian aging.[9] On the contrary, older Blanding's females exhibited the highest rates of survivorship, the most frequent reproductive episodes, and significantly larger clutches, even when possible effect of larger body size on clutch size were statistically controlled for. Although egg hatch rates were somewhat reduced in older females, this difference was not statistically significant. A shorter-lived species, the painted turtle (*Chrysemys picta*), also appears to maintain reproductive success at older ages.[54] Aging-related changes in follicular reserve, however, have not been examined systematically for any reptile species.

A bit more is known about reproductive aging in fishes than in reptiles. Unlike mammals and birds, some seasonally spawning fishes presumably produce new oogonia throughout their reproductive lifespans. Some fish species (e.g., rockfish) are extremely long-lived (>100 years), with correspondingly long reproductive lives. Other, more short-lived fish species, however, undergo reproductive aging resembling that in short-lived mammals and birds. Wild guppy strains (*P. reticulata*) maintained in captivity clearly show gradual reproductive declines and exhibit definitive PRLSs (Fig. 1).[57,58] Studies by Reznick *et al.*[59] have shown that the lengths of PRLS in representatives of two guppy populations maintained in captivity were uncorrelated with their previous histories of natural selection in the wild (Fig. 1C). This finding is consistent with a lack of selection for PRLS in this fish—not surprising, given the lack of parental care or kin networks in this live-bearing species. The ovarian and neuroendocrine changes underlying these patterns of functional reproductive decline remain unstudied for guppies, as well as most other teleost fishes.

Ovarian histological data from a couple of other fish species show evidence of sustained *de novo* oogenesis even in adults of advanced ages. In two species of long-lived Pacific rockfish (*Sebastes aleutianus and S. alutus*), the oldest fish examined (75–80 years) produced new primary oocytes seasonally, and showed no indications of ovarian oocyte depletion.[21] Many other species of long-lived fish appear to maintain fertility at advanced ages.[54] In the zebrafish, which lives 4–5 years, the adult ovary also exhibits continuous *de novo* oogenesis, and expresses *nanos*1, a stem cell marker which regulates translation of specific mRNAs and is required to maintain oogenesis in adult females.[60] An important question remains as to whether or not oocyte quality declines with aging in fishes, as in mammals; this question could be addressed using experimental approaches, such as *in vitro* fertilization or ovarian transplant techniques. Additionally, we note that at least one group of cartilaginous fishes, the lampreys, have embryonically fixed oocyte pools, as do mammals.[10] In sum, it seems likely that the fishes—an exceptionally old (450 million years) and diverse (some 24,000 species) group of vertebrates—exhibit a commensurate diversity of reproductive developmental styles and reproductive aging patterns.

Human menopause in comparative context

Human menopause, in which the cessation of fertile ovulatory cycles can be followed by a PRLS of up to three decades or more, is considered by many researchers to be a phenomenon requiring special evolutionary explanation.[14,26,27,61,62] But as we have emphasized here, human ovarian aging differs little in its general outlines from that seen in a variety of other mammalian species. Furthermore, the reproductive lifespan in women does not end abruptly in midlife, as many authors have suggested. Instead, fertility declines gradually from the late 20s onward, and the perimenopausal transitional period, characterized by increasing irregularity of ovulatory cycles, precedes menopause by 4–7 years.[63,64] In industrialized societies with long lifespans like the United States, the timing of menopause (end of ovulation) is accompanied by sharply accelerating rates of death from cardiovascular disease, the leading aging-related cause of mortality.[65,66] This excess mortality stands in direct contrast to the trend expected if nonreproductive physiological functions were optimized by natural selection (including kin selection) in older women during the postreproductive years.

The physiological underpinnings and developmental tradeoffs associated with female reproductive aging are still not thoroughly understood in most nonhuman animals. In laboratory rodents, however, declines in primordial oocytes in relation to failure of ovarian cyclicity are well studied, resemble those seen in women, are correlated with aging-related increases in mortality (Fig. 1A and B), and are associated with a range of other aging-related physiological declines, as well as increasing incidence of disease.[6,10] Human reproductive aging, moreover, is similar demographically and functionally in many respects to patterns of fertility decline seen in a diverse array of other female vertebrates, ranging from solitary mammals to Japanese quail and guppies. In a wide variety (but not all) animals for which data are available, a sustained, accelerating loss of fertility (versus a sudden, precipitous loss) is often followed by a significant postreproductive period—particularly in captivity or under regimes of very stable, low adult mortality.[26,67,68] This pattern is typical of senescent deterioration of physiological systems, and not the kind of pattern that would be expected if physiological function were optimized by natural selection.[25,59,67,69,70]

Researchers from diverse fields, including evolutionary biology, anthropology, and biogerontology, have endorsed various versions of kin selection theory to explain this pattern of reproductive aging in women. Often referred to under the rubric of the "grandmother hypothesis," this type of evolutionary scenario posits that social, nutritional, or reproductive benefits to close kin compensate for midlife loss of a female's individual reproductive capacity. This style of explanation for the timing of reproductive aging in women has gathered a great deal of popular and scientific momentum, and a substantial body of evidence has been presented in the literature that is consistent with it. Attempts are rarely undertaken to falsify this evolutionary model for human menopause using conventional evolutionary or comparative approaches, however. As we have shown here, this pattern of fertility loss associated with oocyte exhaustion is not limited to species with extensive kin networks, extended periods of female parental care, or significant opportunities for intergenerational resource transfer. These observations argue strongly against the idea that human menopause requires a special evolutionary or social explanation.

In evolutionary terms, the most parsimonious explanation for aging-related fertility loss is that it occurs by default, as a result of absent or weakening natural selection for a longer reproductive lifespan. Rather than being physiologically or evolutionarily "programmed" in a direct sense, we propose that aging-related declines in reproduction and survivorship are expected to arise as an *indirect* result of the *failure* of adaptive genetic, developmental, and physiological tradeoffs for optimizing organismal fitness and reproductive success. Reproduction and somatic maintenance are expected to be sustained physiologically only for the period that organisms have a reasonable probability of survival in nature.

Most would agree that women's PRLS can reach 25–50% of maximum recorded female lifespan in long-lived, modern populations. The modern PRLS is thus significantly longer than that documented for any other vertebrate, except for some whale species. Human PRLS could, in part, represent an artifact of relatively recent increases in human lifespans; this remains an intriguing, often contentious possibility.[23,26,42,71,72] But invoking a kin selection hypothesis or other alternative to a more straightforward, parsimonious evolutionary scenario, in which aging occurs as a result of failure of selection on individuals for longer healthy lifespans, calls for rigorous analysis and falsification using mainstream evolutionary and comparative approaches,[25,73,74] demographic and analytical techniques from biogerontology[1,67–70] and reproductive physiology.[11,12,64,75–78] Ultimately, a rigorous comparative approach to understanding the evolutionary and developmental bases of ovarian aging will include careful examination of basic aspects of female reproductive aging—demographic, developmental, anatomical and physiological—and consideration of functional patterns of senescence in broad phylogenetic context, in a wide range of vertebrates with divergent lifespans and social capacities.

Acknowledgments

C. Finch is grateful for the support of the National Institute on Aging (Grants AG14571 and AG026752) and the Ellison Medical Foundation. D. Holmes would like to acknowledge S. Austad, R. Gosden, and D. Sherry for intellectual support of this work, and to thank M.A. Ottinger, J. Wu, and I. Nisbet for their generous assistance with the quail and tern ovary work. Financial support for the bird ovary studies was provided by an NIH INBRE grant to D. Holmes through the University of Idaho, and by the Center for Reproductive Biology at Washington State University.

Conflicts of interest

The authors declare no conflicts of interest.

References

1. Kirkwood, T.B.L. & S.N. Austad. 2000. Why do we age? *Nature* **408**: 233–238.
2. Partridge, L., D. Gems & D.J. Withers. 2005. Sex and death: what is the connection? *Cell* **120**: 461–472.
3. Partridge, L. & N.H. Barton. 1993. Optimality, mutation and the evolution of ageing. *Nature* **362**: 305–311.
4. Promislow, D.E.L., K.M. Fedorka & J.M.S. Burger. 2006. Evolutionary biology of aging: future directions. In *Handbook of the Biology of Aging*. E.J. Masoro & S.N. Austad, Eds.: 217–242. Academic Press. New York.
5. de Magalhães, J.P. & G.M. Church. 2005. Genomes optimize reproduction: aging as a consequence of the developmental program. *Physiology (Bethesda)* **20**: 252–259.

6. Finch, C.E. 2007. *The Biology of Human Longevity: Inflammation, Nutrition, and Aging in the Evolution of Lifespans.* Academic (Elsevier). Amsterdam.
7. Finch, C.E., M.C. Pike & M. Witten. 1990. Slow mortality rate accelerations during aging in some animals approximate that of humans. *Science* **249:** 902–905.
8. Finch, C.E. & M.R. Rose. 1995. Hormones and the physiological architecture of life history evolution. *Q. Rev. Biol.* **70:** 1–52.
9. Congdon, J., R.D. Nagle, O.M. Kinney & R.C. van Loben Sels. 2001. Hypotheses of aging in a long-lived vertebrate (Blanding's turtle, *Emydoidea blandingii*). *Exp. Gerontol.* **36:** 813–827.
10. Finch, C.E. 1990. *Longevity, Senescence, and the Genome.* University of Chicago Press. Chicago, IL.
11. vom Saal, F.S., C.E. Finch & J.F. Nelson. 1994. Natural history and mechanisms of reproductive aging in humans, laboratory rodents, and other selected vertebrates. In *The Physiology of Reproduction.* E. Knobil & J.D. Neill, Eds.: 1213–1314. Raven Press. New York.
12. Gosden, R.F. & C.E. Finch. 2000. Definition and character of reproductive aging and senescence. In *Female Reproductive Aging.* E. Te Velde, P. Pearson & F. Broekmans, Eds.: 11–25. Parthenon. New York.
13. Ricklefs, R.E., A. Scheuerlein & A. Cohen. 2003. Age-related patterns of fertility in captive populations of birds and mammals. *Exp. Gerontol.* **38:** 741–745.
14. Cohen, A.A. 2004. Female post-reproductive lifespan: a general mammalian trait. *Biol. Rev. Camb. Philos. Soc.* **79:** 733–750.
15. Holmes, D.J. 2003. Aging in birds. In *Aging in Organisms.* H. Osiewacz, Ed.: 201–219. Kluwer Press. Amsterdam.
16. Holmes, D. & M.A. Ottinger. 2006. Domestic and wild bird models for the study of aging. In *Handbook of Models for Human Aging.* P. Conn, Ed.: 351–365. Academic Press. New York.
17. Brunet-Rossinni, A.K. & S.N. Austad. 2006. Senescence in wild populations of mammals and birds. In *Handbook of the Biology of Aging.* E.J. Masoro & S.N. Austad, Eds.: 243–266. Academic Press. Burlington, MA.
18. Ricklefs, R. 2008. The evolution of senescence from a comparative perspective. *Funct. Ecol.* **22:** 379–392.
19. Holmes, D. & K. Martin. 2009. A bird's-eye view of aging: what's in it for ornithologists? *Auk* **126:** 1–23.
20. Patnaik, B. 1994. Ageing in reptiles. *Gerontology* **40:** 200–220.
21. deBruin, J.-P., R.G. Gosden, C.E. Finch & B.M. Leaman. 2004. Ovarian aging in two species of long-lived rockfish, *Sebastes aleutianus* and *S. alutus. Biol. Reprod.* **71:** 1036–1042.
22. Austad, S.N. 1993. Retarded senescence in an insular population of Virginia opossums (*Didelphis virginiana*). *J. Zool.* **229:** 695–708.
23. Voland, E., A. Chasiotis & W. Schiefenhovel. 2005. *Grandmotherhood: The Evolutionary Significance of the Second Half of Life.* Rutgers University Press. New Brunswick, NJ.
24. Hamilton, W.D. 1964. The genetical evolution of social behavior, I and II. *J. Theor. Biol.* **7:** 1–52.
25. Futuyma, D. 2005. *Evolution.* Sinauer. Sunderland, MA.
26. Austad, S.N. 1994. Menopause: an evolutionary perspective. *Exp. Gerontol.* **29:** 255–263.
27. Packer, C., M. Tatar & A. Collins. 1998. Reproductive cessation in female mammals. *Nature* **392:** 807–811.
28. Yin, W. & A.C. Gore. 2010. The hypothalamic median eminence and its role in reproductive aging. *Ann. N.Y. Acad. Sci.* **1204:** 113–122.
29. Finch, C.E., L.S. Felicio, C.V. Mobbs & J.F. Nelson. 1984. Ovarian and steroidal influences on neuroendocrine aging processes in female rodents. *Endocr. Rev.* **5:** 467–497.
30. Nelson, J.F., L.S. Felicio, H.H. Osterburg & C.E. Finch. 1992. Differential contributions of ovarian and extraovarian factors to age-related reductions in plasma estradiol and progesterone during the estrous cycle of C57BL/6J mice. *Endocrinology* **130:** 805–810.
31. Downs, J.L. & P.M. Wise. 2009. The role of the brain in female reproductive aging. *Mol. Cell. Endocrinol.* **299:** 32–38.
32. Tilly, J.L., Y. Niikura & B.R. Rueda. 2009. The current status of evidence for and against postnatal oogenesis in mammals: a case of ovarian optimism versus pessimism? *Biol. Reprod.* **80:** 2–12.
33. Begum, S., V.E. Papaioannou & R.G. Gosden. 2008. The oocyte population is not renewed in transplanted or irradiated adult ovaries. *Hum. Reprod.* **23:** 2326–2330.
34. Nelson, J.F. & L.S. Felicio. 1986. Radical ovarian resection advances the onset of persistent vaginal cornification but only transiently disrupts hypothalamic-pituitary regulation of cyclicity in C57BL/6J mice. *Biol. Reprod.* **35:** 957–964.
35. Brook, J.D., R.G. Gosden & A.C. Chandley. 1984. Maternal ageing and aneuploid embryos—evidence from the mouse that biological and not chronological age is the important influence. *Hum. Genet.* **66:** 41–45.
36. Finch, C.E. & T.B.L. Kirkwood. 2000. *Chance, Development, and Aging.* Oxford University Press. New York.
37. Hawkes, K. & K.R. Smith. 2010. Do women stop early? Similarities in fertility decline in humans and chimpanzees. *Ann. N.Y. Acad. Sci.* **1204:** 43–53.
38. Kaplan, H. *et al.* 2010. Learning, menopause, and the human adaptive complex. *Ann. N.Y. Acad. Sci.* **1204:** 30–42.
39. Ohkubo, Y., Y. Shirayoshi & N. Nakatsuji. 1996. Autonomous regulation of proliferation and growth arrest in mouse primordial germ cells studied by mixed and clonal cultures. *Exp. Cell Res.* **222:** 291–297.
40. Kirkwood, T.B. 2002. Evolution of ageing. *Mech. Ageing Dev.* **123:** 737–745.
41. Faddy, M.J., R.G. Gosden & R.G. Edwards. 1983. Ovarian follicular dynamics in mice: a comparative study of three inbred strains and an F1 hybrid. *J. Endocrinol.* **96:** 23–33.
42. Austad, S.N. 1997. Postreproductive survival. In *Between Zeus and the Salmon: The Biodemography of Longevity.* K.W. Wachter & C.E. Finch, Eds.: 161–174. National Research Council, National Academy Press. Washington, DC.
43. Haenel, N.J. 1986. General notes on the behavioral ontogeny of Puget Sound killer whales and the occurrence of allomaternal behavior. In *Behavioral Biology of Killer Whales.* B.C. Kirkevold & J.S. Lockard, Eds.: 285–300. Alan R. Liss. New York.
44. Olesiuk, P.F., M.A. Bigg & G.M. Ellis. 1990. Life history and population dynamics of resident killer whales (*Orcinus*

orca) in the coastal waters of British Columbia and Washington State. Report of the International Whaling Commission (Special Issue 2). International Whaling Commission. Cambridge, UK.

45. Harrison, M.L. & S.D. Tardif. 1988. Kin preference in marmosets and tamarins: *Saguinus oedipus* and *Callithrix jacchus* (Callitrichidae, primates). *Am. J. Phys. Anthropol.* **77:** 377–384.

46. Saltzman, W., L. Digby & D. Abbott. 2009. Reproductive skew in female common marmosets: what can proximate mechanisms tell us about ultimate causes? *Proc. R. Soc. Lond. B Biol. Sci.* **276:** 389–399.

47. Kardong, K. 2009. *Vertebrates: Comparative Anatomy, Function, Evolution*. McGraw Hill. New York.

48. Murphy, W. J. *et al*. 2001. Resolution of the early placental mammal radiation using Bayesian phylogenetics. *Science* **294:** 2348–2351.

49. Holmes, D.J. & S.N. Austad. 1995. Birds as animal models for the comparative biology of aging: a prospectus. *J. Gerontol. Biol. Sci.* **50A:** 59–66.

50. Holmes, D.J., R. Flückiger & S. N. Austad. 2001. Comparative biology of aging in birds: an update. *Exp. Gerontol.* **36:** 869–883.

51. Holmes, D.J., S.L. Thomson, J. Wu & M.A. Ottinger. 2003. Reproductive aging in female birds. *Exp. Gerontol.* **38:** 751–756.

52. Bahr, J.M. & S.S. Palmer. 1989. The influence of aging on ovarian function. *Crit. Rev. Poult. Biol.* **2:** 103–110.

53. Woodard, A.E. & H. Abplanalp. 1971. Longevity and reproduction in Japanese quail maintained under stimulatory lighting. *Poult. Sci.* **50:** 688–692.

54. Finch, C.E. 2009. Update on slow aging and negligible senescence—a mini-review. *Gerontology* **55:** 307–313.

55. Nisbet, I. *et al*. 1999. Endocrine patterns during aging in the common tern (*Sterna hirundo*). *Gen. Comp. Endocrinol.* **114:** 279–286.

56. Johnson, A.L. & D.C. Woods. 2007. Ovarian dynamics and follicle development. In *Reproductive Biology and Phylogeny of Aves*. B.G.M. Jamieson, Ed.: 243–277. Science Publishers, Inc. Plymouth, UK.

57. Comfort, A. 1964. *Ageing: The Biology of Senescence*. Routledge & Kegan Paul. London, UK.

58. Reznick, D.N. *et al*. 2004. Effect of extrinsic mortality on the evolution of senescence in guppies. *Nature* **431:** 1095–1099.

59. Reznick, D., M. Bryant & D. Holmes. 2006. The evolution of senescence and post-reproductive lifespan in guppies (*Poecilia reticulata*). *PLoS Biol.* **4:** e7.

60. Draper, B.W., C.M. McCallum & C.B. Moens. 2007. Nanos1 is required to maintain oocyte production in adult zebrafish. *Dev. Biol.* **305:** 589–598.

61. Hawkes, K. 2003. Grandmothers and the evolution of human longevity. *Am. J. Hum. Biol.* **15:** 380–400.

62. Hawkes, K. 2004. Human longevity: the grandmother effect. *Nature* **428:** 128–129.

63. Santoro, N. 2005. The menopausal transition. *Am. J. Med.* **118**(Suppl. 12B): 8–13.

64. Gosden, R.G. 1985. *The Biology of Menopause: The Causes and Consequences of Ovarian Ageing*. Academic Press. London.

65. Parker, S.L., T. Tong, S. Bolden & P. Wingo. 1997. Cancer statistics, 1997. *Ca. Cancer J. Clin.* **47:** 5–27.

66. Heron, M. *et al*. 2009. Deaths: final data for 2006. In *National Vital Statistics Reports*, vol. **57:** 1–128. U.S. Department of Health and Human Services, Centers for Disease Control and Prevention. National Center for Health Statistics. Washington, DC.

67. Wood, J.W., D.J. Holman & K. O'Connor. 2000. Did menopause evolve by antagonistic pleiotropy? *Homo* **4:** 483–490.

68. Bronikowski, A.M. & D.E.L. Promislow. 2005. Testing evolutionary theories of aging in wild populations. *Trends Ecol. Evol.* **20:** 271–273.

69. Wood, J.W. *et al*. 2001. The evolution of menopause by antagonistic pleiotropy. Working Paper 01-04 Center for Studies in Demography & Ecology, University of Washington.

70. Rose, M.R. 1991. *Evolutionary Biology of Aging*. Oxford University Press. New York.

71. Hill, K. & A.M. Hurtado. 1991. The evolution of premature reproductive senescence and menopause in human females: an evaluation of the "grandmother hypothesis." *Hum. Nat.* **2:** 313–350.

72. Kaplan, H. 1997. The evolution of the human life course. In *Between Zeus and the Salmon*. K.W. Wachter, Ed.: 175–211. National Academy Press. Washington, DC.

73. Gould, S.J. & R.C. Lewontin. 1979. The spandrels of San Marco and the Panglossian paradigm: a critique of the adaptationist programme. *Proc. R. Soc. Lond. B Biol. Sci.* **205:** 581–598.

74. Austad, S.N. & D.J. Holmes. 1999. Evolutionary approaches to probing aging mechanisms. In *Methods in Aging Research*. B.P. Yu, Ed.: 437–452. CRC Press. Boca Raton, FL.

75. Nelson, J.F. & L.S. Felicio. 1985. Reproductive aging in the female: an etiological perspective. *Rev. Biol. Res. Aging.* **2:** 251–314.

76. Faddy, M.J. *et al*. 1992. Accelerated disappearance of ovarian follicles in mid-life: implications for forecasting menopause. *Hum. Reprod.* **7:** 1342–1346.

77. Faddy, M. & R. Gosden. 2007. Numbers of ovarian follicles and testing germ line renewal in the postnatal ovary: facts and fallacies. *Cell Cycle* **6:** 1951–1952.

78. Jones, K.P., *et al*. 2007. Depletion of ovarian follicles with age in chimpanzees: similarities to humans. *Biol. Reprod.* **77:** 247–251.

79. Jones, E.C. & P.L. Krohn. 1960. The effect of unilateral ovariectomy on the reproductive lifespan of mice. *J. Endocrinol.* **20:** 129–134.

80. Faddy, M.J. 2000. Follicle dynamics during ovarian ageing. *Mol. Cell Endocrinol.* **163:** 43–48.

81. Richardson, S., V. Senikas & J.F. Nelson. 1987. Follicular depletion during the menopausal transition: evidence for accelerated loss and ultimate exhaustion. *J. Clin. Endocrinol. Metab.* **65:** 1231–1237.

ANNALS OF THE NEW YORK ACADEMY OF SCIENCES
Issue: *Reproductive Aging*

Relating smoking, obesity, insulin resistance, and ovarian biomarker changes to the final menstrual period

MaryFran R. Sowers,[1,2] Daniel McConnell,[1] Matheos Yosef,[1] Mary L. Jannausch,[1] Sioban D. Harlow,[1] and John F. Randolph, Jr.[2]

[1]Department of Epidemiology, School of Public Health, University of Michigan, Ann Arbor, Michigan. [2]Department of Obstetrics and Gynecology, University of Michigan Health Sciences System, Ann Arbor, Michigan

Address for correspondence: MaryFran Sowers, Ph.D., Department of Epidemiology, School of Public Health, University of Michigan, 1415 Washington Heights, Rm 1846, Ann Arbor, Michigan 48109. mfsowers@umich.edu

To determine if smoking, obesity, and insulin resistance mediated age at final menstrual period (FMP), we examined anti-Müllerian hormone (AMH), inhibin B, and follicle-stimulating hormone (FSH) as biomarkers of changing follicle status and ovarian aging. We performed a longitudinal data analysis from a cohort of premenopausal women followed to their FMP. Our results found that smokers had an earlier age at FMP ($P < 0.003$) and a more rapid decline in their AMH slope relative to age at FMP ($P < 0.002$). Smokers had a lower baseline inhibin B level relative to age at the FMP than nonsmokers ($P = 0.002$). Increasing insulin resistance was associated with a shorter time to FMP ($P < 0.003$) and associations of obesity and time to FMP were observed ($P = 0.004$, in model with FSH). Change in ovarian biomarkers did not mediate the time to FMP. We found that smoking was associated with age at FMP and modified associations of AMH and inhibin B with age at FMP. Insulin resistance was associated with shorter time to FMP independent of the biomarkers. Interventions targeting smoking and insulin resistance could curtail the undue advancement of reproductive aging.

Keywords: obesity; insulin resistance; smoking; anti-Müllerian hormone; inhibin B; menopause

Introduction

Epidemiological studies have reported that cigarette smoking leads to reduced ovarian function and fertility and an earlier age at menopause,[1] and both body size and insulin resistance have been variably associated with measures of ovarian function.[2] The degree to which environmental factors, such as smoking, obesity, and insulin resistance, may impact follicle number and quality is an important question as the nature and quantity of ovarian reserve is indicative of reproductive capacity and the time remaining during which conception can occur prior to the menopause. Increasing evidence suggests that measured anti-Müllerian hormone (AMH) and inhibin B in conjunction with follicle-stimulating hormone (FSH) could reflect the shrinking ovarian reserve over time and provide a useful means of investigating the impact of environmental factors.

In women, AMH (Müllerian Inhibiting Substance [MIS]) from the granulosa cells of ovarian follicles reflects the transition of resting primordial follicles into growing primary and secondary follicles and the subsequent recruitment of FSH-sensitive follicles in the early antral stage.[3–5] Because AMH is produced only in growing ovarian follicles, serum AMH levels are regarded as a direct indicator of ovarian reserve, representing the quantity and quality of the recruitable ovarian follicle pool.[6] We have identified a linear decline of $_{\log}$AMH to low or nondetectable levels 5 years prior to the natural final menstrual period (FMP).[7]

Inhibin B is the primary inhibin produced by the small antral follicles; its levels have been interpreted as indicating growth of the antral follicle cohort.[8] Produced by granulosa cells, inhibin B suppresses FSH secretion through direct negative feedback to the pituitary.[9–11] There is a curvilinear decline of follicular-phase $_{\log}$inhibin B to undetectable levels 4–5 years prior to the natural FMP.[7]

Reproductive aging is also characterized by a progressive rise in serum FSH levels and reduced levels of ovarian steroids.[12–14] This FSH rise, a central endocrine feature of the perimenopause, was described by Sherman and Korenman in 1975,[15] and has been confirmed in subsequent epidemiological studies of the menopausal transition.[16–18] Globally, the FSH rise is associated with a progressive loss of ovulatory function.[19]

The degree to which environmental factors, such as smoking, obesity, and insulin resistance, impact the association of these ovarian markers in their relation to time to FMP and age at FMP is the subject of this research. The goal was to determine if smoking behavior, obesity, and HOMA-IR, an indicator of insulin resistance among nondiabetics, were associated with time to FMP independent of age and age at FMP. Further, we evaluated if women who smoked or were more insulin resistant had different AMH, inhibin B, and FSH profiles assuming that these biomarkers reflected ovarian aging.

Methods and procedures

Population

This report is based on data from 50 Michigan Bone Health and Metabolism study (MBHMS) enrollees of a possible 629 women. The women were pre- and early perimenopausal at their baseline evaluation. Archival serum specimens from the initial six consecutive annual visits were assayed for AMH, inhibin B, and FSH. In addition, archival information from physical measurements and interviews were used to identify obesity and smoking behavior at these visits. MBHMS enrollees were followed annually, allowing the subsequent documentation of their FMP.

The organization of the population-based MBHMS cohort has been described.[20] It was organized in 1992 from two sampling sources including a list of the female offspring, aged 24–44 years, from the community-based Tecumseh (Michigan) Community Health Study (TCHS) enrollees and Kohl's Directory, a listing of community female residents (also aged 24–44) whose parents had not participated in the TCHS. This report includes data collected during the 15-year period from 1992/1993 through 2006/2007, excluding the 18- and 14-month lapses in funding in 1997 and 2003, respectively. Written informed consent has been obtained from all participants; this study has been approved by the University of Michigan Institutional Review Board. Since study inception, the annual cohort visits have included interviews about health status, menstrual bleeding patterns and health-related behaviors, and phlebotomy to provide serum and urine specimens for assay of hormones, metabolic measures, and repository storage.

To develop this substudy, which has been reported,[7] we selected and assayed repository specimens to correspond in time to 6 yearly measures, beginning in 1993, when regularly cycling (nine or more menstrual cycles per year) women were in their pre- and early menopause stages. This was to assure that we could examine both level and rate of change in the ovarian markers as women entered the menopause transition. Therefore, women were eligible for inclusion in this substudy only if their FSH values in 1992/1993 or 1993/1994 were ≤ 14 mIU/mL. Then, women were selected for study if they had a naturally occurring FMP by the 2006 annual visit (within the 13-year period after the baseline) and if their age at FMP was more than 41 years and therefore not reflective of factors thought to contribute to premature ovarian aging. Further, these women had no exposure to hormone therapy use or gynecological surgeries during the 13-year study period. It was important to be able to examine rates of change in the ovarian markers to address our hypotheses that smoking, insulin resistance and obesity altered ovarian aging and were thereby associated with menopause characteristics.

Data from 50 women meeting these eligibility criteria were selected. They were, on average, 4 years older than women not selected (a mean 1992 baseline age of 41 [SD = 2.6] years versus 38 years [SD = 5.0]). Initial body mass index (BMI), insulin resistance index, and smoking frequency were not statistically significantly different in selected versus nonselected women.

Specimens and assays

Specimens were collected fasting in days 2–7 of the follicular phase of the menstrual cycle, aliquoted and stored at $-80°$C without thaw until assay. A commercially available enzyme-linked immunosorbent assay (ELISA) from Diagnostic Systems Laboratories (Beckman Coulter, DSL,

Webster, TX) was used for the *in vitro* quantitative measurement of Müllerian Inhibiting Substance/Anti-Müllerian Hormone in human serum. This ELISA is a direct competitive immunoassay without sample extraction or hydrolysis. The detection system consisted of a biotinylated secondary antibody and streptavidin-labeled horseradish peroxidase. Samples were assayed in duplicate. There is no detectable cross-reactivity with closely related compounds. The assay measured AMH concentrations ranging from 0.017 to 10 ng/mL with an assay range (standard curve) of 0.05 to 10 ng/mL. Manufacturer-specified interassay coefficients of variation (CV) were 8.0% at 0.15 ng/mL, 4.8% at 0.85 ng/mL, and 6.7% at 4.28 ng/mL (mean = 6.5%); intra-assay CVs were 4.6% at 0.14 ng/mL, 2.4% at 0.84 ng/mL, and 3.3% at 4.41 ng/mL (mean = 4.0%). The level of assay detection was 0.05 ng/mL.

Serum inhibin B concentrations were measured in duplicate with an α-ßB dimeric ELISA (DSL-10–84100, Diagnostic Systems Laboratories) and referenced to a standard of human inhibin B preparation isolated from human follicular fluid provided by Nigel Groome (Oxford Brookes University, Oxford, UK). The within- and between-assay variations were 11.7% and 15.6%, respectively. The assay lower limit of detection (lowest standard curve point) was 10.0 pg/mL. Assays for both AMH and inhibin B were measured in a single time period and kits came from single lots.

FSH concentrations were measured in annual batches across time using an in-house (CLASS laboratory, University of Michigan) two-site chemiluminoscence immunoassay directed to different regions on the beta subunit.[21] It has an interassay CV of 12.0% and 6.0% and a lower limit of detection of 1.05 mIU/mL. The intra-assay CV at five locations along the standard curve were as follows: 7.8% (3.3 mIU/mL), 3.2% (9.9 mIU/mL), 5.1% (18.2 mIU/mL), 4.4% (22 mIU/mL), and 3.3% (60.8 mIU/mL).

Serum insulin was measured using Radioimmuneassay (DPC Coat-a-count, Los Angeles, CA). Glucose was measured with a hexokinase-coupled reaction on a Hitachi 747–200 (Boehringer Mannheim Diagnostics, Indianapolis, IN). The hemostatic model-based insulin resistance index (HOMA-IR) was calculated as ([fasting insulin × fasting glucose]/22.5).[22] Serum glucose and insulin values were not available from the 1992/1993 assessment, but available at the subsequent five data points. HOMA-IR was treated as a time-varying covariate.

Other measures

Height (cm) and weight (kg) were measured at annual study visits with a calibrated stadiometer and balance-beam scale and these data used to calculate BMI as weight (in kilograms) divided by the square of height (in meters). Obesity was defined by dichotomizing BMI at 30 kg/m^2. Smoking status at study entry was included as an ever versus never dichotomous variable.

Statistical analysis

Variable distributions were examined for normality, the presence of nonplausible outliers, and/or changing variability over time. Scatter plots and box plots were used to determine whether transformations of outcome measures were necessary for satisfying model assumptions.

Biomarkers ($_{log}$AMH, $_{log}$inhibin B, and $_{log}$FSH levels) as independent variables were related to time to FMP and to age at FMP as the outcomes of interest. First, the six annual biomarker values for each woman were decomposed into subject-specific intercept and slope. Then, these were incorporated as random independent variables and related to the outcome measure, age at FMP. When values below assay detection were present (for the biomarkers AMH and inhibin B), we used the nonlinear mixed model procedure, Proc NLMixed (in SAS) to address those below-detection values as in Thiébaut and Jacqmin-Gadda.[23] Otherwise, modeling was undertaken using Generalized Estimating Equations.

Baseline variables for age, smoking, or BMI were added to the basic models as independent main effects. In addition, and in separate models, HOMA-IR was treated as a continuous, time-varying covariate. Then, interaction terms were included in the models to test the hypotheses that in women who smoked, were obese or insulin resistant, these environmental factors modified the relationships of the measured biomarkers with FMP.

Appropriateness of model fitting was assessed both graphically and using residual analyses. SAS 9.1 and Macro facilities (SAS Institute, Cary, NC) were used to perform the statistical analyses.

Results

The baseline hormone biomarker characteristics of the 50 women included mean (SD) AMH of 0.66 ± 0.50 ng/mL; mean (SD) inhibin B of 72 ± 44 pg/mL; and mean (SD) FSH of 8.0 ± 2.4 mIU/mL. Table 1 shows the biomarker values across the six time points and the number of values that were below the AMH and inhibin B assay detections. The profiles of AMH and inhibin B changed significantly over the 6-year interval ($P < 0.0001$, respectively). Increasingly over time, the proportion of values below the levels of assay detection increased ($P < 0.0001$, respectively). The change in AMH and inhibin B profiles predated the rise in the follicular-phase FSH profile. The mean age in 1993 of the 50 women was 41.5 years and the mean (SD) BMI was 27.0 ± 5.9 kg/m^2. Twenty-four percent of women were obese and 18% of women smoked at baseline. Mean age at FMP was 51 years.

Age at FMP, smoking, and ovarian biomarkers

As seen in Table 2, there were statistically significant and important associations between the ovarian markers and age at FMP, but the nature of the association varied according to smoking status. The rate of change of AMH is related to age at FMP. However, women who smoked were likely to have an earlier age at FMP and the slope of their AMH levels in relation to age at FMP was steeper (that is, AMH levels declined faster in women who smoked, which was, in turn, associated with an earlier age at FMP). As seen in the second model of Table 2, smoking was associated with an earlier age at FMP and smoking modified the level at which inhibin B was associated with the earlier age at menopause in that it happened at a higher level of baseline inhibin B ($P = 0.002$). Obesity and HOMA-IR were not associated with age at FMP. FSH not associated with age at FMP and there was no interaction between FSH and smoking in relation to age at FMP.

Time to FMP, AMH, and lifestyle modifiers

$_{\log}$AMH was significantly associated with time to FMP, with a decline of one unit in observed $_{\log}$AMH being associated with 1.75 earlier years to FMP ($P < 0.0001$). When AMH became nondetectable, it was also significantly associated with time to FMP. An increasingly higher HOMA-IR was significantly associated with a closer time to FMP ($P = 0.009$), apart from AMH. However, neither baseline smoking nor obesity measures were significantly associated with time to FMP (see Table 3). There were no statistically significant interactions between AMH and smoking, obesity, or HOMA-IR in relation to time to FMP, indicating that the relationship between AMH and the time to FMP was not significantly different in women who smoked, were obese, or had greater insulin resistance (models not shown).

Table 1. Sample characteristics from 50 pre- and perimenopausal women at six points across time—MBHMS

	1993/1994 Mean ± SD	1994/1995 Mean ± SD	1995/1996 Mean ± SD	1997/1998 Mean ± SD	1998/1999 Mean ± SD	1999/2000 Mean ± SD
Age	41.5 ± 2.67	42.5 ± 2.67	43.4 ± 2.61	46.3 ± 2.60	47.3 ± 2.67	48.3 ± 2.56
AMH (ng/mL)						
Observed values[a]	0.66 ± 0.50	0.51 ± 0.36	0.39 ± 0.30	0.16 ± 0.11	0.19 ± 0.11	0.15 ± 0.10
% ↓ detection	3 (6%)	5 (10%)	8 (16%)	26 (52%)	27 (54%)	32 (64%)
Inhibin B (pg/mL)						
Observed values[a]	72 ± 44	56 ± 43.5	59 ± 35	38 ± 21	47 ± 72	34 ± 30
% ↓ detection	2 (4%)	2 (4%)	9 (18%)	18 (36%)	23 (46%)	23 (46%)
FSH (mIU/mL)	8.0 ± 2.4	7.6 ± 3.7	7.7 ± 4.9	16.4 ± 15.7	18.3 ± 15.2	21.5 ± 18.3
HOMA-IR	N/A[b]	1.68 ± 1.05	1.84 ± 1.74	2.48 ± 2.36	2.35 ± 2.10	2.60 ± 2.97
BMI (kg/m^2)	27.0 ± 5.9	27.6 ± 6.1	27.6 ± 5.9	28.3 ± 6.5	28.5 ± 6.3	28.8 ± 6.4

[a]Data are for AMH and inhibin B that are observed above level of assay detection.
[b]HOMA-IR data are not available from this annual visit.

Table 2. Association of age at final menstrual period with $_{\log}$AMH and $_{\log}$inhibin B profiles according to smoking status

Covariates	Beta ± SE	P-value
	Model of AMH and smoking	
$_{\log}$AMH intercept	0.12 ± 0.39	0.76
$_{\log}$AMH slope	**11.49 ± 4.14**	**0.008**
Smoked at baseline	**−7.31 ± 2.36**	**0.003**
$_{\log}$AMHintcpt*smoking at baseline	**1.92 ± 0.65**	**0.005**
$_{\log}$AMHslope*smoking at baseline	**−19.27 ± 5.84**	**0.002**
	Model of inhibin B and smoking	
$_{\log}$Inhibin B intercept	−2.75 ± 2.19	0.22
$_{\log}$Inhibin B slope	3.92 ± 4.02	0.33
Smoked at baseline	**−44.32 ± 12.69**	**0.001**
$_{\log}$Inhibin B intcpt*smoking at baseline	**10.20 ± 3.12**	**0.002**
$_{\log}$Inhibin B slope*smoking at baseline	−7.99 ± 6.23	0.21

Bolded variables are statistically significant.

Time to FMP, $_{log}$inhibin B, and lifestyle modifiers

A decline of one unit in observed $_{\log}$inhibin B was associated with 1.58 earlier years to FMP. Inhibin B below detection was also associated with time to FMP. As seen in Table 4, insulin resistance, expressed with HOMA-IR, was significantly independently associated with time to FMP ($P = 0.003$); the association with obesity, defined as a BMI >30 kg/m^2, was of borderline statistical significance ($P = 0.06$). Women who smoked did not have a statistically different time to FMP than women who did not smoke. There were no statistically significant interactions with smoking behavior, obesity classification, or HOMA-IR and inhibin B in relation to time to FMP (data not shown).

Time to FMP and FSH

Though the linear component of $_{\log}$FSH was not associated with time to the FMP ($P = 0.13$), the curvilinear rate of change in $_{\log}$FSH was associated with the time to the FMP ($P = 0.0005$). As reported in Table 5, smoking was not associated with time to FMP ($P = 0.46$). In addition to the curvilinear rise in FSH, obesity ($P = 0.004$) and HOMA-IR ($P = 0.0003$) were independently associated with time to FMP in separate models. There were no significant interactions with smoking, obesity, or HOMA-IR and FSH in relation to time to FMP (data not shown).

Discussion

The ovarian aging concept incorporates the timing of reproductive events, including the beginning of subfertility, the transition from menstrual cycle regularity to irregularity, absolute infertility, and the final menstrual period.[24] Identifying lifestyle factors that influence the rate of ovarian aging is of interest, both clinically and from a public health perspective. After including 6-year levels and changes in AMH, inhibin B, and FSH as biomarkers of the progressive decline in functional ovarian cells, we considered whether three factors, smoking, obesity, and insulin resistance, might be associated with the rate of ovarian aging in their relation to both age at FMP and time to FMP independent of age.

Role of smoking

We identified a marked difference between smokers and nonsmokers in the relation of both baseline $_{\log}$AMH and change in $_{\log}$AMH with age at FMP. There was an earlier age at FMP and a more rapid decline in AMH levels among women who were smokers, suggesting that smoking behavior may lead to either fewer oocytes or an earlier decline in oocyte number. In addition, we found that smoking was associated with a lower initial level of inhibin B but not associated with the rate of change in inhibin B levels. This led us to conclude that smoking might be associated with an earlier onset of processes associated

Table 3. The associations of time to FMP and AMH incorporating smoking, obesity (defined at baseline), or HOMA-IR (time-varying) as main effects

	Beta ± SE	P-value
	Model of AMH and smoking	
$_{\log}$observed AMH	−1.75 ± 0.14	**<0.0001**
AMH below detection	**6.57 ± 0.31**	**<0.0001**
Smoked at baseline (dummy variable)	−0.22 ± 0.38	0.57
	Model of AMH and obesity (dichotomized)	
$_{\log}$observed AMH	−1.75 ± 0.14	**<0.0001**
AMH below detection	**6.57 ± 0.32**	**<0.0001**
Obesity > 30 kg/m^2	0.03 ± 0.44	0.94
	Model of AMH with time-varying HOMA-IR	
$_{\log}$observed AMH	−1.65 ± 0.16	**<0.0001**
AMH below detection	**6.06 ± 0.36**	**<0.0001**
HOMA-IR (time-varying)	**0.16 ± 0.06**	**0.009**

Bolded variables are statistically significant. Smokers were designated 1 while nonsmokers were designated with 0.

with ovarian aging, but not necessarily a disruption in the sequence of activities and that inhibin B and FSH are better indicators of the sequence of activities leading to the selection of a dominant follicle as compared to being indicators of the size of the follicle pool.

Multiple epidemiological studies have reported that cigarette smoking leads to reduced ovarian function and fertility and an earlier age at menopause,[25] suggesting that smoking impairs ovarian function. The mechanism of tobacco's toxic effect on the ovary is unclear but may be due to effects on oocyte quantity,[26] oocyte quality, or disruption of endocrine function.[27,28] Cotinine, a long-lived metabolite of nicotine, can be detected in ovarian follicular fluid, indicating that toxic constituents of cigarette smoke including nicotine and cadmium have access to the follicular environment and could affect ovarian function.[29–32]

The sample size of this report, limited to data from 50 women, precluded our investigation of important related public health questions, such as whether quitting smoking altered these associations and whether the patterns varied according to the time of initiation, duration, or total exposure to tobacco use.

Role of obesity and HOMA-IR

We identified that there were stronger and more frequent associations of insulin resistance with the biomarkers of ovarian aging or their change with time than with obesity when using a widely used cutpoint for obesity (or with BMI treated as a continuous variable, data not reported). Data on the association of AMH and obesity are scarce. A recent study reported that AMH levels were significantly lower in obese women as compared to nonobese women in the late reproductive years; however, that study did not report measures of insulin resistance.[33] There are reports of an inverse association of obesity with inhibin B levels in women with polycystic ovary syndrome (PCOS) and in anovulatory premenopausal women.[34,35] Thus, there is limited data in healthy reproductively aged women suggesting a significant association between decreasing inhibin B levels and greater BMI but mechanisms for the association have not been established.[36]

We identified that AMH and insulin resistance were predictive of time to FMP. Importantly, obesity and insulin resistance should not be treated as synonymous as there are obese women without insulin resistance and lean insulin resistant women.[37] Published studies of insulin resistance and ovarian biomarkers are largely limited to investigations of PCOS.[38] Several studies in women with PCOS have identified that elevated circulating levels of AMH are correlated with increased numbers of small antral follicles in the ovary,[39] potentially reflecting alterations granulosa cell function and a role not necessarily associated with body size.[40,41] If insulin

Table 4. The associations of time to FMP with $_{\log}$inhibin B incorporating smoking, obesity (defined at baseline), or HOMA-IR (time-varying) as main effects

Covariates	Beta ± SE	P-value
	Model of inhibin B with smoking	
$_{\log}$Inhibin B	−1.58 ± 0.28	**<0.0001**
Inhibin B below assay detection	**−2.52 ± 1.08**	**0.02**
Smoked at baseline	0.33 ± 1.81	0.86
	Model of inhibin B with obesity	
$_{\log}$Inhibin B	−1.58 ± 0.22	**<0.0001**
Inhibin B below detection	**−2.43 ± 0.88**	**0.0065**
Obesity > 30 kg/m^2	0.77 ± 0.41	0.06
	Model of inhibin B with time-varying HOMA-IR	
$_{\log}$Inhibin B	−1.27 ± 0.24	**<0.0001**
Inhibin B below detection	−1.81 ± 0.92	0.052
HOMA-IR (time-varying)	**0.22 ± 0.07**	**0.003**

Bolded variables are statistically significant.

resistance has an impact on granulosa or theca cell functioning, this would be consistent with our findings. As the perception of PCOS changes from that of a reproductive disorder to one of a metabolic syndrome with reproductive implications, greater focus is being placed on dysfunction of the hypothalamic–pituitary axis and primary defects of insulin activity as contributory causes to the syndrome.[42] Although the underlying cause of the ovulatory dysfunction associated with PCOS is unknown, it is thought to be associated with the dysregulation of thecal steroidogenesis.[43] The resulting hyperandrogenism plays a role in the arrest of folliculogenesis,[44,45] a mechanism that has not been evaluated in normal ovarian aging. Alternatively, studies in knockout models suggest considering other mechanisms including insulin receptor insufficiency or GLUT4 dysregulation.[46]

This report includes strengths and limitations. The data reflect appropriately collected specimens without thaw-refreeze prior to assay. Data were obtained from a substudy nested in a population-based study that includes documentation of the natural progression of women through stages of the menopause transition to the FMP and subsequent postmenopause. This allowed the selection of archival specimens for assay that represented the late reproductive period. Simultaneously, information about smoking behavior, obesity status, and insulin resistance were obtained concurrently with the specimens assayed for AMH, inhibin B, and FSH and did not require women to engage in interviews that required long-term recall. The study design allowed us to consider change in biomarkers of ovarian aging and how environmental factors were related to these changes in relation to menopause characteristics. However, this is an epidemiological study without access to certain measures of follicle status, such as antral follicle count. The population is Caucasian so findings may not be generalizable to non-Caucasian women, although the study addresses factors that are often disproportionately associated with non-Caucasian populations including smoking, insulin resistance, and obesity. The major deficit is that the size of the sample evaluated may be too small to detect important interactions in relation to the biomarkers and time to FMP. The sample size was too limited to detect three-way interaction patterns (insulin resistance in smokers who were obese versus insulin resistance in smokers who were not obese) that may characterize important subgroups that could be targeted for intervention.

In summary, smoking was not only associated with age at FMP but also modified the associations of two biomarkers, AMH and inhibin B, with age at FMP. These biochemical markers for oocyte quality and quantity may provide an indication of the potential mechanisms whereby smoking is associated

Table 5. The associations of time to FMP with $_{\log}$FSH and considering smoking, obesity (defined at baseline), or HOMA-IR (time-varying) as main effects

Covariates	Beta ± SE	P-value
	Model of FSH with smoking	
$_{\log}$FSH	−1.61 ± 1.05	0.13
$_{\log}$FSH *$_{\log}$FSH	**0.72 ± 0.20**	**0.0005**
Smoked at baseline	−0.29 ± 0.40	0.46
	Model of FSH with obesity (dichotomized)	
$_{\log}$FSH	−1.62 ± 1.04	0.12
$_{\log}$FSH *$_{\log}$FSH	**0.73 ± 0.20**	**0.0004**
Obesity (dichotomized) at baseline	**1.22 ± 0.42**	**0.004**
	Model of FSH with time-varying HOMA-IR	
$_{\log}$FSH	−0.37 ± 0.99	0.71
$_{\log}$FSH *$_{\log}$FSH	**0.43 ± 0.19**	**0.03**
HOMA-IR (time-varying)	**0.27 ± 0.07**	**0.0003**

Bolded variables are statistically significant.

with the earlier age at FMP, but does not necessarily disrupt progression in the timing of events leading to the FMP. The absence of significant interactions between insulin resistance with AMH and inhibin B in relation to time to FMP may indicate that other mechanisms in ovarian and adrenal steroidogenesis should be considered that are not limited to antral follicle recruitment. Although future studies will more fully reveal these relationships, both smoking behavior and greater insulin resistance are likely to accelerate reproductive aging. This report contributes additional evidence that targeting smoking and insulin resistance would be important practices to moderate in women as they appear to generate undue advancement in reproductive aging.

Acknowledgments

Grants supporting the data collection and writing of this manuscript: AR051384 (Sowers, PI), AR040888 (Sowers, PI), and AR20557 (Sowers, PI).

Conflicts of interest

The authors declare no conflicts of interest.

References

1. Sowers, M.F. & M. La Pietra. 1995. Menopause: its epidemiology and potential association with chronic diseases. *Epidemiol. Rev.* **17**: 287–302.
2. Gracia, C.R. et al. 2005. The relationship between obesity and race on inhibin B during the menopause transition. *Menopause* **12**: 559–566.
3. Vigier, B. et al. 1984. Production of anti-Müllerian hormone: another homology between Sertoli and granulosa cells. *Endocrinology* **114**: 1315–1320.
4. Durlinger, A.L. et al. 1999. Control of primordial follicle recruitment by anti-Müllerian hormone in the mouse ovary. *Endocrinology* **140**: 5789–5796.
5. Durlinger, A.L. et al. 2001. Anti-müllerian hormone attenuates the effects of FSH on follicle development in the mouse ovary. *Endocrinology* **142**: 4891–4899.
6. te Velde, E.R. & P.L. Pearson. 2002. The variability of female reproductive aging. *Hum. Reprod. Update* **8**: 141–154.
7. Sowers, M.F. et al. 2008. Anti-Mullerian hormone (AMH) and inhibin-B in the definition of ovarian aging and the menopause transition. *J. Clin. Endocrinol. Metab.* **93**: 3478–3483.
8. Welt, C.K. 2004. Regulation and function of inhibins in the normal menstrual cycle. *Semin. Reprod. Med.* **22**: 187–193.
9. Vale, W., A. Hsueh, C. Rivier & J. Yu. 1991. The inhibin/activin family of growth factors. In *Peptide Growth Factors and their Receptors, Handbook of Experimental Pharmacology*. M.A. Sporn & A.B. Roberts, Eds.: 211–248. Springer-Verlag. Heidelberg.
10. Pierson, T.M. et al. 2000. Regulable expression of inhibin a in wild-type and inhibin alpha null mice. *Molecul. Endocrinol.* **14**: 1075–1085.
11. Welt, C., Y. Sidis, H. Kneutmann & A. Schneyer. 2002. Activins, inhibins, and follistatins: from endocrinology to signaling. A paradigm for the new millennium. *Exp. Biol. Med. (Maywood)* **227**: 724–752.
12. Burger, H. et al. 1999. Prospectively measured levels of serum follicle-stimulating hormone, estradiol and the dimeric inhibins during the menopausal transition in a population based cohort of women. *J. Clin. Endocrinol. Metab.* **84**: 4025–4030.
13. Sowers, M.F. et al. 2008. Follicle stimulating hormone (FSH) and its rate of change to define menopause transition stages. *J. Clin. Endocrinol. Metab.* **93**: 3958–3964.

14. Burger, H. et al. 1999. Prospectively measured levels of serum follicle-stimulating hormone, estradiol and the dimeric inhibins during the menopausal transition in a population based cohort of women. *J. Clin. Endocrinol. Metab.* **84:** 4025–4030.
15. Sherman, B.M. & S.G. Korenman. 1975. Hormonal characteristics of the human menstrual cycle throughout reproductive life. *J. Clin. Invest.* **55:** 699–706.
16. Lenton, E.A., L. Sexton, S. Lee & I.D. Cooke. 1988. Progressive changes in LH and FSH and LH: FSH ratio in women throughout reproductive life. *Maturitas* **10:** 35–43.
17. Longcope, C. et al. 1986. Steroid and gonadotropin levels in women during the peri-menopausal years. *Maturitas* **8:** 189–196.
18. Rannevik, G. et al. 1995. A longitudinal study of the perimenopausal transition: altered profiles of steroid and pituitary hormones, SHBG and bone mineral density. *Maturitas* **21:** 103–113.
19. Metcalf, M.G. 1986. Incidence of ovulation from the menarche to the menopause: observations of 622 New Zealand women. *N.Z. Med. J.* **96:** 645–648.
20. Sowers, M.F. et al. 2008. Anti-Mullerian hormone (AMH) and inhibin-B in the definition of ovarian aging and the menopause transition. *J. Clin. Endocrinol. Metab.* **93:** 3478–3483.
21. Randolph, J.F. Jr. et al. 2003. Reproductive hormones in the early menopausal transition: relationship to ethnicity, body size, and menopausal status. *J. Clin. Endocrinol. Metab.* **88:** 1516–1522.
22. Haffner, S.M. 1997. Progress in population analyses of insulin resistance syndrome. *Ann. N.Y. Acad. Sci.* **827:** 1–12.
23. Thiébaut, R. & H. Jacqmin-Gadda. 2004. Mixed models for longitudinal left-censored repeated measures. *Comput. Methods Prog. Biomed.* **74:** 255–260.
24. te Velde, E.R. 1998. Ovarian ageing and postponement of childbearing. *Maturitas* **30:** 103–104.
25. Sowers, M.F. & M. La Pietra. 1995. Menopause: its epidemiology and potential association with chronic diseases. *Epidemiol. Rev.* **17:** 287–302.
26. Mattison, D.R. & S.S. Thorgeirsson. 1978. Smoking and industrial pollution, and their effects on menopause and ovarian cancer. *Lancet* **1:** 187–188.
27. Paszkowski, T., R.N. Clarke & M.D. Hornstein. 2002. Smoking induces oxidative stress inside the Graafian follicle. *Hum. Reprod.* **17:** 921–925.
28. Valdez, K.E. & B.K. Petroff. 2004. Potential roles of the aryl hydrocarbon receptor in female reproductive senescence. *Reprod. Biol.* **4:** 243–258.
29. Weiss, T. & A. Eckert. 1989. Cotinine levels in follicular fluid and serum of IVF patients: effect of granulosa-luteal cell function in bitro. *Hum. Reprod.* **4:** 482–485.
30. Barbieri, R.L., P.M. McShane & K.H. Ryan. 1986. Constituents of cigarette smoke inhibit human granulosa cell aromatase. *Fertil. Steril.* **46:** 232–236.
31. Zenzes, M.T. et al. 1995. Cadmium accumulation in follicular fluid of women in invitro fertilization-embryo transfer is higher in smokers. *Fertil. Steril.* **64:** 599–603.
32. Varga, B. et al. 1993. Age-dependent accumulation of cadmium in the human ovary. *Reprod. Toxicol.* **7:** 225–258.
33. Freeman, E.W. et al. 2005. Follicular phase hormone levels and menstrual bleeding status in the approach to menopause. *Fertil. Steril.* **83:** 383–392.
34. Pigny, P. et al. 2000. Serum levels of inhibins are differentially altered in patients with polycystic ovary syndrome: effects of being overweight and relevance to hyperandrogenism. *Fertil. Steril.* **73:** 972–997.
35. Cortet-Rudelli, C. et al. 2002. Obesity and serum luteinizing hormone level have an independent and opposite effect on the serum inhibin B level in patients with polycystic ovary syndrome. *Fertil. Steril.* **77:** 281–287.
36. Gracia, C.R. et al. 2005. The relationship between obesity and race on inhibin B during the menopause transition. *Menopause* **12:** 559–566.
37. Gerald Reaven, G., F. Abbasi & T. McLaughlin. 2004. Obesity, insulin resistance, and cardiovascular disease. *Recent Prog. Horm. Res.* **59:** 207–223.
38. Eyvazzadeh, A.D. et al. 2009. The role of the endogenous opioid system in polycystic ovary syndrome. *Fertil. Steril.* **92:** 1–12.
39. Piltonen, T. et al. 2005. Serum anti-Müllerian hormone levels remain high until late reproductive age and decrease during metformin therapy in women with polycystic ovary syndrome. *Hum. Reprod.* **20:** 1820–1826.
40. Spandorfer, S.D. et al. 2004. Obesity and in vitro fertilization: negative influences on outcome. *J. Reprod. Med.* **49:** 973–977.
41. Pigny, P. et al. 2003. Elevated serum level of anti-Müllerian hormone in patients with polycystic ovary syndrome: relationship to the ovarian follicle excess and to the follicular arrest. *J. Clin. Endocrinol. Metab.* **88:** 5957–5962.
42. Chang, R.J., R.M. Nakamura, H.L. Judd & S.A. Kaplan. 1983. Insulin resistance in nonobese patients with polycystic ovarian disease. *J. Clin. Endocrinol. Metab.* **57:** 356–359.
43. Catteau-Jonard, S. et al. 2008. Anti-Mullerian hormone, its receptor, FSH receptor, and androgen receptor genes are overexpressed by granulosa cells from stimulated follicles in women with polycystic ovary syndrome. *J. Clin. Endocrinol. Metab.* **93:** 4456–4461.
44. Jonard, S. & D. Dewailly. 2004. The follicular excess in polycystic ovaries, due to intra-ovarian hyperandrogenism, may be the main culprit for the follicular arrest. *Hum. Reprod. Update* **10:** 107–117.
45. Hughesdon, P.E. 1982. Morphology and morphogenesis of the Stein-Leventhal ovary and of so-called hyperthecosis. *Obstet. Gynecol. Surv.* **37:** 59–77.
46. Kadowaki, T. 2000. Insights into insulin resistance and type 2 diabetes from knockout mouse models. *J. Clin. Invest.* **106:** 459–465.

ANNALS OF THE NEW YORK ACADEMY OF SCIENCES
Issue: *Reproductive Aging*

Estrogen and the aging brain: an elixir for the weary cortical network

Dani Dumitriu,[1] Peter R. Rapp,[2] Bruce S. McEwen,[3] and John H. Morrison[1,4]

[1]Department of Neuroscience, Mount Sinai School of Medicine, One Gustave L. Levy Place, New York, New York. [2]Laboratory of Experimental Gerontology, National Institute of Aging, Baltimore, Maryland. [3]Harold and Margaret Milliken Hatch Laboratory of Neuroendocrinology, The Rockefeller University, New York, New York. [4]Brookdale Department of Geriatrics and Palliative Medicine, Mount Sinai School of Medicine, New York, New York

Address for correspondence: John H. Morrison, Department of Neuroscience, Mount Sinai School of Medicine, 1425 Madison Avenue, P.O. Box 1065, New York, New York 10029. john.morrison@mssm.edu

The surprising discovery in 1990 that estrogen modulates hippocampal structural plasticity launched a whole new field of scientific inquiry. Over the past two decades, estrogen-induced spinogenesis has been described in several brain areas involved in cognition in a number of species, in both sexes and on multiple time scales. Exploration into the interaction between estrogen and aging has illuminated some of the hormone's neuroprotective effects, most notably on age-related cognitive decline in nonhuman primates. Although there is still much to be learned about the mechanisms by which estrogen exerts its actions, key components of the signal transduction pathways are beginning to be elucidated and nongenomic actions via membrane bound estrogen receptors are of particular interest. Future studies are focused on identifying the most clinically relevant hormone treatment, as well as the potential identification of new therapeutics that can prevent or reverse age-related cognitive impairment by intercepting specific signal transduction pathways initiated by estrogen.

Keywords: estrogen; aging; hippocampus; prefrontal cortex; dendritic spines; cognition

Key experiments on female rats in the McEwen laboratory in the early 1990s showed that the dendritic spine density on the neurons in the CA1 region of the hippocampus varies over the course of the ovarian cycle (Fig. 1),[1] and surgical ovariectomy (OVX) leads to a 30% loss in spine density that can be rescued by estrogen replacement.[2] The estrogen-induced spinogenesis is mirrored by an equal increase in synapses,[3] pointing to potential integration of the new spines into the hippocampal network. The effects of fluctuating hormone levels, OVX, and estrogen replacement on spine density have now been extended to the somatosensory and motor cortex of rats.[4]

Classically, it has been thought that the induction of spine formation happened over 1–2 days. Recently however, induction of both spinogenesis[5] and synaptogenesis[6] has been demonstrated to also occur within half an hour of a subcutaneous injection of estrogen. Because induction of the immediate early gene c-Fos (a measure of neuronal activation) following estrogen administration is biphasic, with one peak at 1 h and a second peak at 24 h,[7] and because both the spinogenesis and the synaptogenesis are reduced at 2 and 4 h post injection, respectively, it is possible that estrogen induces two distinct phenomena over the two different time courses. In a powerful parallel to the structural plasticity body of work, estrogen has been shown to enhance hippocampal-dependent memory both over the course of days[8] and rapidly, over the course of minutes.[9]

The existence of nongenomic mechanisms of actions for estrogen were postulated early on, following the initial discovery of estrogen's effect on hippocampal electrophysiology[10] and the first observation of rapid spinogenesis in cultured hippocampal cells.[11] Both of these phenomena occurred on the order of seconds to minutes and thus precluded the possibility of being mediated

doi: 10.1111/j.1749-6632.2010.05529.x

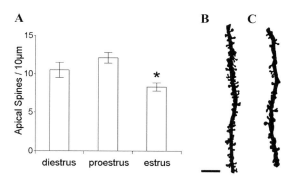

Figure 1. Spine density fluctuations in the CA1 region of the hippocampus from naturally cycling female rats. (A) Manual quantification of spine density from micrographs of Golgi-impregnated CA1 pyramidal cells. Values represent mean + SEM; asterisk denotes significant difference from proestrus ($P < 0.01$). (B, C) Representative Camera Lucida drawings of apical dendrites from proestrus and estrus females, respectively. Scale bar = 10 μm. Adapted from Woolley et al.[1]

via gene transcription and translation. In fact, the first evidence of estrogen's ability to bind plasma membranes came as early as 1983,[12] but it would take almost two more decades before the receptors would begin to be described.

It is now clear that estrogen receptor, (ER)-α and ERβ as well as more recently described receptors, such as ER-X and G protein-coupled ER (GPR30), are present in multiple neuronal compartments in hippocampus, including synapses.[13–16] Through these receptors estrogen activates multiple signaling pathways and cellular processes via both genomic and nongenomic actions.[17] The McEwen lab has been characterizing multiple pathways linked to the synaptic effects of estrogen in hippocampus, particularly those linked to phosphoinositide-3-kinase (PI3K) (Fig. 2) (see Spencer et al.[18] for review). For example, we have shown that estrogen rapidly activates dendritic and synaptic Akt that leads to the stimulation of local translation of PSD95, a key scaffolding protein required for synaptic expansion[19] (Fig. 2). We have also been particularly interested in the serine/threonine kinase LIM kinase (LIMK), which plays a key role in actin polymerization through phosphorylation of cofilin.[20] LIMK is present in CA1 synapses,[21] and critically important to the dynamics of actin polymerization required for spine formation[18] (Fig. 2). Further characterization of these rapidly activated signaling cascades will be required to reveal the mechanisms underlying the nongenomic effects of estrogen on synapse formation and synaptic plasticity linked to learning and memory.

Estrogen and the aging hippocampus: lessons learned from the rat model

Convincing evidence that estrogen modulates synaptic plasticity in young female rats naturally led us to the clinically relevant question of estrogen's role in age-related cognitive decline. Our initial task was simply to examine if the synaptic remodeling observed in response to estrogen in young animals would also occur in the hippocampus of aged OVX rats. For this purpose, we used electron microscopy to quantify the axospinous synapse density in young (3–4 months) and old (23–24 months) OVX rats in the presence or absence of estrogen. Although we were able to replicate previous findings[2] of a 30% increase in synapse density in the young group, estrogen treatment failed to increase synapse density in the CA1 region of aged OVX rats (Fig. 3A).[22] Interestingly, when we quantified synaptic levels of the obligate NMDA receptor subunit NR1 in the same animals, we found that estrogen increased the synaptic pool of NR1 (defined as being located within 30 nm of the postsynaptic membrane) in the aged but not the young cohort.[22] Consistently, we later found a trend ($P = 0.06$) of higher representation of the NMDA receptor subunit NR2A in the same synaptic region and a significant increase in the representation of the NR2B subunit in the lateral portion of the postsynaptic density of aged animals, again with no estrogen-induced change in the young group.[23] These important findings indicate that the aged synapse is not entirely insensitive to the circulating estrogen. Although estrogen-induced hippocampal structural plasticity is restricted to young animals, the estrogen-induced molecular plasticity with respect to NMDA receptor composition might allow the aged network to partly compensate for the loss of youthful connectivity.

An interesting parallel could be drawn between the age-dependent estrogen-induced molecular adaptations in NMDA receptor profile described earlier and electrophysiological findings by Thompson and colleagues on the interactions between aging, stress, and estrogen.[24] The group used acute slices from male rats, an *in vitro* model that preserves local neuronal connections and allows for detailed assessment of changes in long-term potentiation

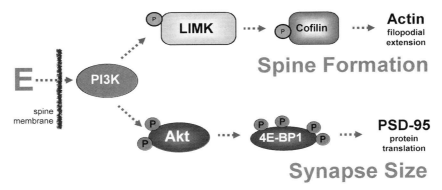

Figure 2. Non-genomic effects of estrogen. Estrogen initiates a complex set of signal transduction pathways in the hippocampal neuron via several membrane-bound receptors. Above are two examples of estrogen-initiated signal transduction leading to spinogenesis and changes in synapse size. Rapid activation of Akt (protein kinase B) via PI3K is thought to be mediated by ERα. Subsequently, activated Akt initiates translation of PSD-95 by removing the repression of the initiation factor 4E-binding protein 1 (4E-BP1). Estradiol-mediated phosphorylation of cofilin has been shown to occur via activation of LIMK. Cofilin is an actin depolymerization factor and it is inactivated by phosphorylation. Therefore, in the presence of estrogen, cofilin repression of actin polymerization is removed, resulting in an increase in filopodial density. The signal transduction pathways illustrated here are an oversimplification of a large body of work done in an *in vitro* cell line.

(LTP) and long-term depression (LTD), two types of synaptic strength modification commonly thought to underlie learning and memory processes.[25] For example, LTD is enhanced by age and this age-induced enhancement is blocked by estrogen. Furthermore, estrogen is protective against changes in both LTP and LTD following stress in aged animals. Specifically, estrogen ameliorates the stress-induced decrease in LTP and blocks the stress-induced enhancement in LTD.[24] Both LTP and LTD are NMDA receptor-dependent forms of plasticity. Estrogen exerts its effect within seconds to minutes, indicating that genomic mechanisms can also be excluded here. One potential nongenomic mechanism underlying these functions is the phosphorylation of the NR2 subunit via the src/MAPK pathway.[26] It is interesting to note that this phosphorylation of NR2 is diminished in aging,[27] which could explain why estrogen confers stronger protection in LTD than LTP changes in slices from aged stressed animals.[24] In summary, this is another model in which, by altering NMDA receptor function, estrogen is able to maintain select network properties in the aged hippocampus.

We next investigated some of the possible mechanisms underlying the loss of structural plasticity in the aged hippocampus by quantitative ultrastructural analysis on tissue from the same animals in which we conducted our initial study. We first decided to look at the distribution of ERα because it had recently been found to localize in dendritic spines in CA1[13] and therefore was ideally positioned to induce local, nongenomic structural changes. We found that only about half as many synapses contain ERα in aged versus young animals (Fig. 3B and C), irrespective of estrogen treatment. In addition, although subtle, an estrogen-induced decrease in synaptic ERα levels was observed in both axon terminals and dendritic spines selectively in the young cohort, once again pointing to a loss of responsivity by the neurons in the aged CA1 subfield.[28]

As discussed earlier, we have been particularly interested in phosphorylated LIMK (pLIMK) located within the CA1 synapses activated by estrogen, due to its critically important role in regulation of actin dynamics linked to spine formation. As in the case of ERα, we found selective responsivity of synaptic pLIMK to estrogen in the young cohort, where we observed a 30% increase in the number of synapses containing pLIMK in the presence of estrogen[21] (Fig. 3D and E). In contrast, the aged animals had decreased synaptic pLIMK levels in CA1 and this decrease was not reversed by estrogen treatment. Taken together, the following model emerges:

(1) *Young animals*: estrogen replacement induces an increase in pLIMK via ERα, which then leads to phosphorylation of cofilin. Phosphorylation of cofilin inhibits its binding to actin,

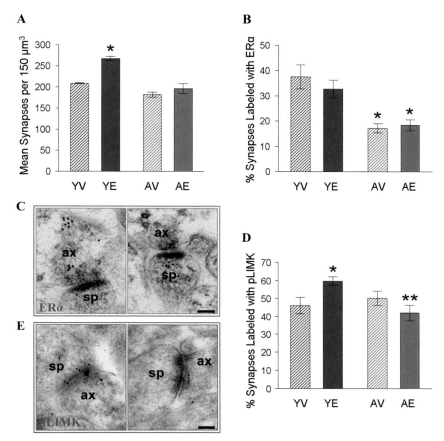

Figure 3. Interaction between estrogen and aging in stratum radiatum of the CA1 subfield of the hippocampus from ovariectomized rats. (A) Synapse density in young and aged animals treated with either estrogen or vehicle. For unbiased quantitative ultrastructural analysis, we took 15 pairs of adjacent electron micrographs from each animal and counted synapses using the dissector method. A significant increase was observed in the young estrogen treated as compared to vehicle treated animals (30%, $P < 0.005$). No synapse density changes occurred in the aged cohort. Adapted from Adams et al.[22] (B) Percentage of ERα-immunoreactively labeled synapses. Our data illustrate that ERα is significantly reduced with age ($P < 0.0001$) but does not respond to estrogen treatment. (C) Sample electron micrographs showing the distribution of ERα in the synapse. Gold particles were observed to be localized both pre- and postsynaptically. Scale bar = 100 nm. Panels (B) and (C) are adapted from Adams et al.[28] (D) Percentage of synapses immunoreactive to pLIMK. The graph illustrates that the percentage of pLIMK containing synapses increases with estrogen treatment in the young cohort (30%, $P = 0.02$) and decreases with age in the presence of estrogen (43%, $P = 0.04$). (E) Representative electron micrographs of pLIMK immunoreactivity. Gold particles are observed both pre- and postsynaptically and are often affiliated with the postsynaptic density. Scale bar = 100 nm. Panels (D) and (E) are adapted from Yildirim et al.[21] *denotes $P < 0.05$, **denotes $P < 0.01$. YV, young vehicle treated; YE, young estrogen treated; AV, aged vehicle treated; AE, aged estrogen treated; ax, axon; sp, spine.

promoting actin polymerization required for the formation of new spines.

(2) *Aged animals*: estrogen is unable to increase levels of pLIMK, possibly due to the low levels of synaptic ERα. This results in decreased deactivation of cofilin and ultimately a disruption of actin polymerization required for formation of new spines.

Interestingly, the abovementioned collection of age-related losses of responsivity to estrogen with respect to structural plasticity, molecular rearrangement, and biochemical function correlate with an age-related loss of estrogen-induced memory enhancement. In young, but not aged OVX rats, estrogen rescues the learning deficits associated with experimentally induced cholinergic impairment.

When given the drug scopolamine, rats perform worse on a T maze active avoidance task. Estrogen remains protective against scopolamine during middle age (as defined by irregular cycling), but loses its effect at more advanced stages in reproductive aging.[29]

Estrogen and the aging prefrontal cortex: lessons learned from the nonhuman primate model

Menopause is an important physiological and psychological milestone in the lives of half of our population. Although some clinical studies have shown that hormone treatment (HT) following menopause can protect against cognitive decline, other studies have shown an increased risk for neurologic complications with HT (see Sherwin and Henry[30] for review). The inconsistencies within the clinical studies likely result from multiple differences in design, primary among them, differences in the timing of treatment relative to the menopausal transition in women.[30] A further impediment to the emergence of a consensus on HT for women has been the disconnect between basic science and the medical reality of menopause.[31] Although it is beyond the scope of this review to discuss the implications of our cellular and biochemical findings for cognitive function, this is a topic at the forefront of current dialogue in the field (see Brinton[32] for review). Until recently, the investigation into estrogen's role on brain function has largely been restricted to young OVX rats. Extending the model into the aged rat was an important first step toward increasing the clinical relevance of the animal work. The translational power of the animal studies has now been enhanced further by the design of experiments on NHPs modeled after our findings in the rat described earlier.[31]

The NHP model is ideal for preclinical studies on age-related cognitive decline and HT for two main reasons. First, unlike the rat which experiences a chronic high estrogen state at reproductive senescence,[33] the NHP menopause is very similar to humans,[34] characterized by loss of circulating estrogen. Second, NHPs are an excellent model to investigate age-dependent cognitive impairment linked to the dorsolateral prefrontal cortex (dlPFC), which is comparable to that of humans.[35] To look at the interaction between aging and estrogen, both young (∼10 years) and aged (∼22 years) female rhesus monkeys had surgical OVX and were given an injection of either 100 μg estradiol cypionate (E) or vehicle (V) every 21 days. Note that this is a cyclical treatment, designed to mimic the physiological peak in estrogen at ovulation and is therefore very different from the HT that women are currently prescribed by their physicians. At various stages pre- and postoperatively, the monkeys are cognitively assessed on multiple tasks, including delayed response (DR), a task that is dependent on dlPFC; and a delayed nonmatching to sample (DNMS) task that is more closely tied to the medial temporal lobe system, which includes the hippocampus. In this review, we will focus our discussion on our dlPFC findings, a brain area that is particularly vulnerable to cognitive aging in both NHPs[35] and humans.[36]

Our first important finding is that estrogen restores performance on the DR task in the aged cohort,[37] but has no effect in the young monkeys, where resilience in cognitive function is observed even in the absence of estrogen (Fig. 4A).[38] Thus, the only group that experiences cognitive decline is the group that models perimenopausal women, and we provide evidence that cyclic treatment with estrogen initiated shortly after OVX restores cognitive function to levels seen in young monkeys. To examine the underlying neural basis for this observation, we turned our attention to the layer III neurons in area 46 of the dlPFC, some of the key neurons that execute the brain computations underlying a monkey's performance in the DR task.[39] Detailed quantitative morphometric analyses were performed by microinjection of individual cells with the fluorescent dye Lucifer Yellow followed by three-dimensional tracing and high-resolution confocal imaging of dendritic spines at systematic distances from the soma (Fig. 4B).[38] Interestingly, although we found that estrogen induces spinogenesis in both young and aged monkeys (Fig. 4C), there were two key age-dependent differences that can account for the selective decrease in performance in DR in the aged vehicle group. First, the decrease in spine density in the young OVX+V group compared to OVX+E occurred against a background of an adaptive increase in dendritic length in the young V-treated animals. This implies that, in spite of the higher spine density in the E-treated group, overall there is no difference in the number of spines per neuron in the young cohort. Second, the estrogen-induced spinogenesis in the aged cohort occurred against the background

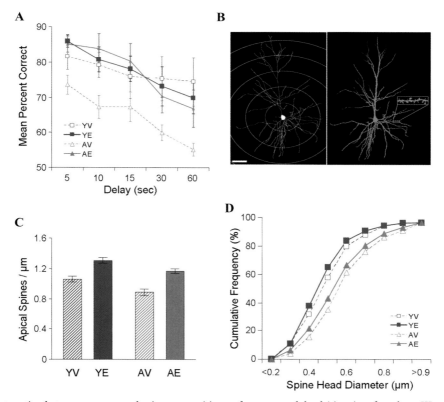

Figure 4. Interaction between estrogen and aging on cognitive performance and dendritic spines from layer III neurons in area 46 of the monkey dlPFC. Young and aged NHPs were OVXed and after an average period of 30 weeks were given either estrogen or vehicle injections every 3 weeks for a period of approximately 2.5 years. The monkeys were sacrificed and perfused 24 h after final treatment. Neuropsychological testing was initiated two days after the second estrogen or vehicle injection. (A) Performance on a delayed response task. Although young animals perform well on the task irrespective of their estrogen status, aged vehicle treated animals perform significantly worse than all other groups. (B) Representative Lucifer Yellow-loaded pyramidal neuron. To ensure systematic sampling for spine density analysis, neurons were first traced in three-dimensions and concentric circles were drawn at multiples of 60 μm from the cell body. Dendritic analysis was subsequently restricted to segments intersecting the circles. Scale bar = 100 μm. (C) Apical spine density of layer III pyramidal neurons from the dlPFC. Dendritic segments were imaged at high resolution using confocal microscopy, followed by manual, blind counts of all dendritic protrusions. Significant treatment effects were observed in both young and aged animals. In addition, there was an age-related decrease in spine density. (D) Cumulative frequency analysis of spine head diameter. A significant increase in spine head is observed with aging, whereas estrogen shifts the distribution in favor of spines with smaller heads. Adapted from Hao et al.[38] YV, young vehicle treated; YE, young estrogen treated; AV, aged vehicle treated; AE, aged estrogen treated.

of an age-related "second hit" to the spine density of the aged V-treated group. Thus, the vulnerability to cognitive decline in the aged V-treated group can be explained in terms of a "double hit" to the connectivity in the dlPFC of these animals: an age-induced loss of spines, coupled with an estrogen deficiency-induced loss of spines (Fig. 5).

Because spine size is highly correlated with both synapse size as well as glutamate receptor number,[40] we were very interested in examining dlPFC spine morphology in our model. We found that estrogen shifts the distribution of spine head diameter toward smaller size in both the young and aged animals (Fig. 4D). However, most remarkably, we found that aging dramatically reduced the representation of spines with small heads (Fig. 4D) and long necks.[38] This age-related selective loss of small spines (and the partial recovery with estrogen) fits in nicely with a developing framework in neurobiology of the essential role that small spines play in learning and memory. Both Kasai et al.[40] and Bourne and Harris[41] have recently suggested that thin spines are

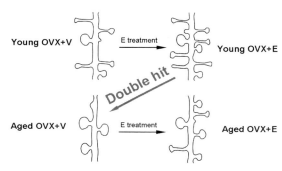

Figure 5. "Double-hit" to the dlPFC connectivity of aged vehicle treated NHPs. Our observations of estrogen-induced spinogenesis and age-related spine loss have led us to the hypothesis that the cognitive impairment observed in the aged vehicle-treated monkeys could be explained by a double-hit to their neural connectivity in the dlPFC. The diagram illustrates that this group lacks the threshold density of small spines seemingly necessary for cognitive function. OVX, ovariectomized; V, vehicle; E, estrogen.

"learning" whereas big, mushroom-type spines represent "memory" traces. Spontaneous spinogenesis visualized by two-photon time-lapse microscopy has shown that new spines are small, highly plastic, responsive to external manipulations, and form synapses within days of appearing.[42] In addition, small spines can enlarge and stabilize in response to LTP,[43] which has been suggested as a potential mechanism for learning.[41] Therefore, the "double hit" to the aged V-treated group in our model (Fig. 5) might likely be most detrimental for the animal's cognitive performance precisely because small, plastic spines are missing in the dlPFC. When estrogen replacement is provided to the aged animals, their DR performance matches the performance of young animals (Fig. 4A) in spite of an overall smaller spine density than the young E-treated animals (Fig. 4C). This could indicate that a modest increase in small spines goes a long way in providing neurobiological resilience.

Future directions: bench science with the potential to alter clinical practice

We have developed an NHP model for the age-related cognitive decline experienced by postmenopausal women and have shown that long-term cyclic treatment with estrogen is protective against both cognitive impairment as well as some of the structural changes that occur with aging and low hormone status. At present, it is unclear whether the same mechanisms underlie both the dendritic spine plasticity found in the rat hippocampus as well as that found in monkey dlPFC. However, given that we observe similar morphological changes, and given that we have found ERα to be present in spines both in the rat hippocampus[28] as well as the monkey dlPFC,[44] this is a logical inference and we plan to pursue this idea in more depth in our future studies on the signaling mechanisms mediating estrogen-induced spinogenesis.

The development of the NHP model is an important first step in the long-term goal of identifying the best HT in the clinic. Currently, we are investigating the difference between continuous and cyclic estrogen treatment, as well as the effects of estrogen with progesterone. Following the identification of the best HT, we plan to test the idea that there is a "window of opportunity" within which treatment must be started to retain beneficial effects. This is a question of enormous clinical relevance in light of the recent findings from the Women's Health Initiative (WHI) that HT does not confer any protection against age-related cognitive decline in women when the onset of therapy lags by many years the transition into menopause. The idea of a "window of opportunity" has already been suggested from work in the rat model, where it has been shown that HT initiated immediately after surgery but not at 10-months post-OVX is effective at preventing cognitive decline.[45] A reduction in estrogen-induced spinogenesis in CA1 following long-term estrogen deprivation has also been shown.[46]

The entire body of work on the effects of estrogen on the brain has necessarily come from OVX animals. It is of enormous experimental difficulty to design well-controlled experiments on NHPs that enter menopause naturally. Therefore, although we acknowledge that at present we do not know if the effects of estrogen are the same in surgically intact aged animals, we believe that basic science must continue to use OVX animals until the best HT is found, and only then move toward the much harder experiment of evaluating the effects of the HT in intact aged NHPs.

Another important clinical question is the role of locally synthesized estrogen. The identification of the presence of the enzymatic machinery capable of synthesizing estrogen from cholesterol in the brain[47] has brought a spotlight onto the interaction between the two sources of estrogen in the female. It is possible, for example, that the observed resilience in our young V-treated monkeys is a result of local compensatory production of estrogen within the cortical network following removal of circulating estrogen by OVX. In addition, an investigation into the role of local estrogen has the potential to extend our findings to males, as mouse male brains have been found to contain estrogen levels similar to females, presumably from local synthesis.[48] Although estrogen is not thought to affect spine density in the hippocampus of males,[49,50] it does have a spinogenic effect in male rat prefrontal cortex.[51]

Finally, we plan to continue to probe for molecular targets in the rat model for subsequent evaluation of their relevance in our NHP model. Given the large number of signaling cascades activated by estrogen in neurons, therapeutic strategies based on naturally occurring hormones will continue to be a major focus. However, we will also seek to identify new pharmaceutical targets within the signal transduction pathways initiated by estrogen. Such refined analysis of the molecular mechanisms and signaling pathways activated by estrogen in brain regions linked to cognition will likely lead to new treatments that could also be used to protect against age-related cognitive impairment in males.

Conflicts of Interest

The authors declare no conflicts of interest.

References

1. Woolley, C.S., E. Gould, M. Frankfurt & B.S. McEwen. 1990. Naturally occurring fluctuation in dendritic spine density on adult hippocampal pyramidal neurons. *J. Neurosci.* **10:** 4035–4039.
2. Gould, E., C.S. Woolley, M. Frankfurt & B.S. McEwen. 1990. Gonadal steroids regulate dendritic spine density in hippocampal pyramidal cells in adulthood. *J. Neurosci.* **10:** 1286–1291.
3. Woolley, C.S. & B.S. McEwen. 1992. Estradiol mediates fluctuation in hippocampal synapse density during the estrous cycle in the adult rat. *J. Neurosci.* **12:** 2549–2554.
4. Chen, J.R. *et al.* 2009. Gonadal hormones modulate the dendritic spine densities of primary cortical pyramidal neurons in adult female rat. *Cereb. Cortex* **19:** 2719–2727.
5. Dumitriu, D. *et al.* 2008. Rapid estrogen-induced spinogenesis in the ovariectomized rat. Program No. 736.2.2008 Neuroscience Meeting Planner. Society for Neuroscience. Washington, DC.
6. MacLusky, N.J., V.N. Luine, T. Hajszan & C. Leranth. 2005. The 17alpha and 17beta isomers of estradiol both induce rapid spine synapse formation in the CA1 hippocampal subfield of ovariectomized female rats. *Endocrinology* **146:** 287–293.
7. Rudick, C.N. & C.S. Woolley. 2003. Selective estrogen receptor modulators regulate phasic activation of hippocampal CA1 pyramidal cells by estrogen. *Endocrinology* **144:** 179–187.
8. Sandstrom, N.J. & C.L. Williams. 2004. Spatial memory retention is enhanced by acute and continuous estradiol replacement. *Horm. Behav.* **45:** 128–135.
9. Luine, V.N., L.F. Jacome & N.J. Maclusky. 2003. Rapid enhancement of visual and place memory by estrogens in rats. *Endocrinology* **144:** 2836–2844.
10. Teyler, T.J., R.M. Vardaris, D. Lewis & A.B. Rawitch. 1980. Gonadal steroids: effects on excitability of hippocampal pyramidal cells. *Science* **209:** 1017–1018.
11. Brinton, R.D. 1993. 17β-Estradiol induction of filopodial growth in cultured hippocampal neurons within minutes of exposure. *Mol. Cell. Neurosci.* **4:** 36–46.
12. Towle, A.C. & P.Y. Sze. 1983. Steroid binding to synaptic plasma membrane: differential binding of glucocorticoids and gonadal steroids. *J. Steroid Biochem.* **18:** 135–143.
13. Milner, T.A. *et al.* 2001. Ultrastructural evidence that hippocampal alpha estrogen receptors are located at extranuclear sites. *J. Comp. Neurol.* **429:** 355–371.
14. Milner, T.A. *et al.* 2005. Ultrastructural localization of estrogen receptor beta immunoreactivity in the rat hippocampal formation. *J. Comp. Neurol.* **491:** 81–95.
15. Toran-Allerand, C.D. 2004. Estrogen and the brain: beyond ER-alpha and ER-beta. *Exp. Gerontol.* **39:** 1579–1586.
16. Maggiolini, M. & D. Picard. 2010. The unfolding stories of GPR30, a new membrane-bound estrogen receptor. *J. Endocrinol.* **204:** 105–114.
17. Waters, E.M. *et al.* 2009. Estrogen receptor alpha and beta specific agonists regulate expression of synaptic proteins in rat hippocampus. *Brain Res.* **1290:** 1–11.
18. Spencer, J.L. *et al.* 2008. Uncovering the mechanisms of estrogen effects on hippocampal function. *Front. Neuroendocrinol.* **29:** 219–237.
19. Akama, K.T. & B.S. McEwen. 2003. Estrogen stimulates postsynaptic density-95 rapid protein synthesis via the Akt/protein kinase B pathway. *J. Neurosci.* **23:** 2333–2339.
20. Arber, S. *et al.* 1998. Regulation of actin dynamics through phosphorylation of cofilin by LIM-kinase. *Nature* **393:** 805–809.
21. Yildirim, M. *et al.* 2008. Estrogen and aging affect synaptic distribution of phosphorylated LIM kinase (pLIMK) in CA1 region of female rat hippocampus. *Neuroscience* **152:** 360–370.
22. Adams, M.M., R.A. Shah, W.G. Janssen & J.H. Morrison. 2001. Different modes of hippocampal plasticity in response to estrogen in young and aged female rats. *Proc. Natl. Acad. Sci. USA* **98:** 8071–8076.
23. Adams, M.M. *et al.* 2004. Estrogen modulates synaptic N-methyl-D-aspartate receptor subunit distribution in the aged hippocampus. *J. Comp. Neurol.* **474:** 419–426.

24. Foy, M.R., M. Baudry, J.G. Foy & R.F. Thompson. 2008. 17beta-estradiol modifies stress-induced and age-related changes in hippocampal synaptic plasticity. *Behav. Neurosci.* **122:** 301–309.
25. Siegelbaum, S.A. & E.R. Kandel. 1991. Learning-related synaptic plasticity: LTP and LTD. *Curr. Opin. Neurobiol.* **1:** 113–120.
26. Bi, R. *et al.* 2000. The tyrosine kinase and mitogen-activated protein kinase pathways mediate multiple effects of estrogen in hippocampus. *Proc. Natl. Acad. Sci. USA* **97:** 3602–3607.
27. Bi, R., M.R. Foy, R.F. Thompson & M. Baudry. 2003. Effects of estrogen, age, and calpain on MAP kinase and NMDA receptors in female rat brain. *Neurobiol. Aging* **24:** 977–983.
28. Adams, M.M. *et al.* 2002. Estrogen and aging affect the subcellular distribution of estrogen receptor-alpha in the hippocampus of female rats. *J. Neurosci.* **22:** 3608–3614.
29. Savonenko, A.V. & A.L. Markowska. 2003. The cognitive effects of ovariectomy and estrogen replacement are modulated by aging. *Neuroscience* **119:** 821–830.
30. Sherwin, B.B. & J.F. Henry. 2008. Brain aging modulates the neuroprotective effects of estrogen on selective aspects of cognition in women: a critical review. *Front. Neuroendocrinol.* **29:** 88–113.
31. Morrison, J.H., R.D. Brinton, P.J. Schmidt & A.C. Gore. 2006. Estrogen, menopause, and the aging brain: how basic neuroscience can inform hormone therapy in women. *J. Neurosci.* **26:** 10332–10348.
32. Brinton, R.D. 2009. Estrogen-induced plasticity from cells to circuits: predictions for cognitive function. *Trends Pharmacol. Sci.* **30:** 212–222.
33. Lu, K.H., B.R. Hopper, T.M. Vargo & S.S. Yen. 1979. Chronological changes in sex steroid, gonadotropin and prolactin secretions in aging female rats displaying different reproductive states. *Biol. Reprod.* **21:** 193–203.
34. Gilardi, K.V. *et al.* 1997. Characterization of the onset of menopause in the rhesus macaque. *Biol. Reprod.* **57:** 335–340.
35. Rapp, P.R. & D.G. Amaral. 1989. Evidence for task-dependent memory dysfunction in the aged monkey. *J. Neurosci.* **9:** 3568–3576.
36. Keenan, P.A., W.H. Ezzat, K. Ginsburg & G.J. Moore. 2001. Prefrontal cortex as the site of estrogen's effect on cognition. *Psychoneuroendocrinology* **26:** 577–590.
37. Rapp, P.R., J.H. Morrison & J.A. Roberts. 2003. Cyclic estrogen replacement improves cognitive function in aged ovariectomized rhesus monkeys. *J. Neurosci.* **23:** 5708–5714.
38. Hao, J. *et al.* 2007. Interactive effects of age and estrogen on cognition and pyramidal neurons in monkey prefrontal cortex. *Proc. Natl. Acad. Sci. USA* **104:** 11465–11470.
39. Goldman-Rakic, P.S. 1988. Topography of cognition: parallel distributed networks in primate association cortex. *Annu. Rev. Neurosci.* **11:** 137–156.
40. Kasai, H. *et al.* 2003. Structure-stability-function relationships of dendritic spines. *Trends Neurosci.* **26:** 360–368.
41. Bourne, J. & K.M. Harris. 2007. Do thin spines learn to be mushroom spines that remember? *Curr. Opin. Neurobiol.* **17:** 381–386.
42. Pan, F. & W.B. Gan. 2008. Two-photon imaging of dendritic spine development in the mouse cortex. *Dev. Neurobiol.* **68:** 771–778.
43. Matsuzaki, M., N. Honkura, G.C. Ellis-Davies & H. Kasai. 2004. Structural basis of long-term potentiation in single dendritic spines. *Nature* **429:** 761–766.
44. Wang, A.C.-J. *et al.* 2009. Synaptic Estrogen Receptor: Levels in prefrontal cortex predict cognitive performance in female Rhesus Monkey. Program No. 569.7.2009 Neuroscience Meeting Planner. Society for Neuroscience. Chicago, IL.
45. Gibbs, R.B. 2000. Long-term treatment with estrogen and progesterone enhances acquisition of a spatial memory task by ovariectomized aged rats. *Neurobiol. Aging.* **21:** 107–116.
46. McLaughlin, K.J., H. Bimonte-Nelson, J.L. Neisewander & C.D. Conrad. 2008. Assessment of estradiol influence on spatial tasks and hippocampal CA1 spines: evidence that the duration of hormone deprivation after ovariectomy compromises 17beta-estradiol effectiveness in altering CA1 spines. *Horm. Behav.* **54:** 386–395.
47. Hojo, Y. *et al.* 2004. Adult male rat hippocampus synthesizes estradiol from pregnenolone by cytochromes P45017alpha and P450 aromatase localized in neurons. *Proc. Natl. Acad. Sci. USA* **101:** 865–870.
48. Toran-Allerand, C.D., A.A. Tinnikov, R.J. Singh & I.S. Nethrapalli. 2005. 17alpha-estradiol: a brain-active estrogen? *Endocrinology* **146:** 3843–3850.
49. Leranth, C., O. Petnehazy & N.J. MacLusky. 2003. Gonadal hormones affect spine synaptic density in the CA1 hippocampal subfield of male rats. *J. Neurosci.* **23:** 1588–1592.
50. Leranth, C., M. Shanabrough & D.E. Redmond Jr. 2002. Gonadal hormones are responsible for maintaining the integrity of spine synapses in the CA1 hippocampal subfield of female nonhuman primates. *J. Comp. Neurol.* **447:** 34–42.
51. Hajszan, T. *et al.* 2007. Effects of androgens and estradiol on spine synapse formation in the prefrontal cortex of normal and testicular feminization mutant male rats. *Endocrinology* **148:** 1963–1967.

ANNALS OF THE NEW YORK ACADEMY OF SCIENCES
Issue: *Reproductive Aging*

The hypothalamic median eminence and its role in reproductive aging

Weiling Yin[1] and Andrea C. Gore[1,2,3]

[1]Division of Pharmacology and Toxicology, College of Pharmacy, The University of Texas at Austin, Austin, Texas. [2]Institute for Cellular and Molecular Biology, The University of Texas at Austin, Austin, Texas. [3]Institute for Neuroscience, The University of Texas at Austin, Austin, Texas

Address for correspondence: Andrea C. Gore, Ph.D., Division of Pharmacology and Toxicology, The University of Texas at Austin, 1 University Station, A1915 Austin, Texas 78712. andrea.gore@mail.utexas.edu

The median eminence at the base of the hypothalamus serves as an interface between the neural and peripheral endocrine systems. It releases hypothalamic-releasing hormones into the portal capillary bed for transport to the anterior pituitary, which provides further signals to target endocrine systems. Of specific relevance to reproduction, a group of about 1000 neurons in mammals release the gonadotropin-releasing hormone (GnRH) peptide from neuroterminals in the median eminence. During the life cycle, there are dramatic changes in reproductive demands, and we focus this review on how GnRH terminals in the median eminence change during reproductive senescence. We discuss morphological and functional properties of the median eminence, and how relationships among GnRH terminals and their microenvironment of nerve terminals, glial cells, and the portal capillary vasculature determine the ability of GnRH peptide to be secreted and to reach its target in the anterior pituitary gland.

Keywords: median eminence; hypothalamus; GnRH; glia; tanycyte; reproductive aging

Overview of the median eminence

The median eminence is the structure at the base of the hypothalamus where hypothalamic-releasing and hypothalamic-inhibiting hormones converge onto the portal capillary system that vascularizes the anterior pituitary gland. This region acts as a key interface between the neural and endocrine systems involved in hypothalamic-adenohypophysial regulation of reproduction, stress, lactation, growth, and thyroid and metabolic systems. The median eminence is a very unusual neural structure: although it contains nerve terminals and glial cells, it is virtually devoid of synapses, and it has structural properties that distinguish it from other brain regions. In this paper, we provide information about the molecular, anatomical, and physiological features of the median eminence, and their age-related changes. Our focus will be on the hypothalamic system controlling reproduction through the release of the gonadotropin-releasing hormone (GnRH) peptide from nerve terminals in the median eminence. The hypothalamic–pituitary–gonadal (HPG) axis of females includes GnRH cells and terminals in the hypothalamus; the pituitary gonadotropes that produce the gonadotropins, luteinizing hormone (LH) and follicle-stimulating hormone (FSH); and the ovary, which produces steroid (particularly estradiol and progesterone) and protein hormones (Fig. 1). All of the HPG levels must function normally for reproduction to occur, and during aging, each of these levels undergoes changes that may contribute to reproductive dysfunction and ultimately failure. Here we will discuss age-related changes in the hypothalamic median eminence in general, and in GnRH terminals in particular, and provide speculation about how this may influence reproductive senescence in females.

The median eminence is one of eight circumventricular organs—regions surrounding the cerebral ventricles—in the central nervous system (CNS).[1] As such, these brain regions contain porous

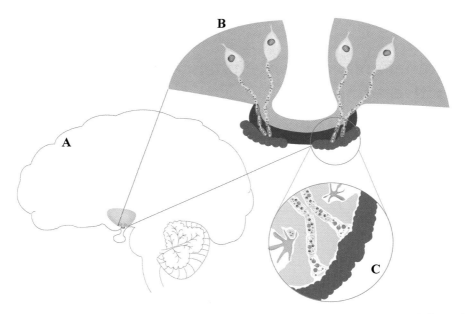

Figure 1. Schematic representation of the hypothalamic–pituitary axis components involved in the control of reproduction. The locations of the base of the hypothalamus (*pink*) and median eminence (*dark red*) are shown in reference to the brain as a whole (panel A). Although the brain is shown sagitally, the median eminence is shown coronally. At the next layer of magnification (Panel B), four GnRH cell bodies are depicted in yellow, with their nerve terminals projecting through the median eminence (*dark red*) to the portal capillaries (*light red*). Panel C is a higher magnification of the GnRH axons terminating at the portal capillaries. Tanycytes (*green*) are interspersed between GnRH and other terminals, and microglia (*dark yellow*) are shown. The blue circles in the axons and terminals in Panels B and C represent large and small secretory vesicles. As the organization of the median eminence in this schematic is linear and the portal capillary vasculature basal lamina border is regular, this drawing depicts the situation in young rats. In aged rats (not shown) these relationships become disorganized.

blood–brain barriers, postulated to enable the circumventricular organs to sense and respond to chemical signals conveyed by the circulatory system within these limited regions in the brain. Two circumventricular organs, the median eminence and the organum vasculosum of the lamina terminalis, are located within the hypothalamus and are involved in neuroendocrine regulation. The median eminence in particular also sends neuronal signals to regulate the peripheral endocrine organs through the hypophyseal portal system (Fig. 1).[2]

The median eminence is one of three portal systems in the human body (the hepatic and renal portal systems being the other two), and as the brain's only portal system, it has unique structural and functional properties. Unlike the general capillary system that drains the blood into the heart directly through a vein, a portal capillary system drains blood into another capillary system through veins. The portal system in the median eminence provides a way for the hypothalamus to communicate with the peripheral endocrine system by both sending and receiving signals through the portal vasculature. It also provides a selective barrier of three layers: the fenestrated endothelial processes, the basal lamina, and glial end feet.[3,4]

Another unique feature of the median eminence is the presence of specialized glial cells, tanycytes, which form a bridge between the portal system and the third ventricle.[3] Tanycytes provide an interconnection between the bloodstream and other parts of the brain through the cerebrospinal fluid (CSF).[5] Thus, the median eminence should be considered to be the communication center from the brain to the rest of the body, via endocrine signals; the route for the body to communicate to the brain, via the leaky fenestrated capillary system; and a route of communication among other brain regions, via the CSF.

Finally, the median eminence, while a neural structure at the base of the hypothalamus, is largely devoid of neural perikarya, dendrites, and synaptic contacts.[6,7] Despite this unusual organization, the nerve terminals, glial cells and the portal vasculature of the median eminence are able to communicate with each other, as evidenced by the presence of many different types of neurotransmitter receptors in this region (reviewed in Ref. 8). The lack of synapses and somatic neural bodies suggests that cell-to-cell communication in the median eminence is probably mainly through volume transmission.[9]

Although the median eminence clearly exhibits important and unique features and roles, it is a surprisingly understudied brain region, a gap that needs to be filled for better appreciating the mechanisms for neuroendocrine control during the life cycle.

The ultrastructure of the median eminence

The ultrastructure of the median eminence has been carefully studied by light, transmission, and scanning electron microscopy. Four distinctive zones can be discerned from dorsal to ventral, as shown in Figure 2: the third ventricular zone, myelinated axon zone, neural profile zone, and capillary zone.[7,10] Under the electron microscope, numerous protrusions (blebs) are distributed on the floor of the third ventricle showing different functions of this area compared to the lateral portion of the third ventricle, which are covered by cilia. In the ventricular zone, ependymal cells and tanycytes line the walls of the third ventricle.[11] In the myelinated axon zone, myelin-covered oxytocin and vasopressin neural axons travel through the median eminence to the posterior pituitary. The neural profile zone contains unmyelinated neuronal processes targeting the portal vasculature in the basal median eminence—these are the hypothalamic-releasing and -inhibiting hormone neural processes. Finally, the capillary zone is characterized by the presence of large neuroterminals, extensive extracellular space and fenestrated capillaries. The extended long processes of tanycytes are seen in all four zones, with tanycytic endfeet in contact with the basal lamina of the fenestrated portal capillary.

The axons of hypothalamic-releasing hormone neurons, such as GnRH, exhibit punctate immunolabeling. For example, GnRH terminals are described as "varicosities" or "puncta"[8,10] because of the discontinuous appearance of the axons that project to the median eminence. Punctae along the neuronal profiles contain abundant large secretory vesicles and small vesicles that are probably on transit to terminals at the portal capillary system (Fig. 2).

GnRH in the median eminence

The median eminence contains neuroterminals that release the hypothalamic-adenohypophysial releasing/inhibiting hormones, among which is GnRH. Although neurons that synthesize these hormones have terminals in the median eminence, the synthesis of these hypothalamic-adenohypophysial hormones occurs in neural cell bodies found in different subdivisions of the hypothalamus. In the case of GnRH, cell bodies are distributed in the preoptic area and medial basal hypothalamus, with some variability across species. The synthesis of GnRH begins with the transcription of its gene to an RNA primary transcript in the nucleus. Following further processing, the mature mRNA is translocated to the cytoplasm. Translation of mRNA occurs in the endoplasmic reticulum and Golgi apparatus, followed by budding of large secretory vesicles, which contain the precursor peptide. Prohormone peptide splicing continues by enzymes in the large secretory vesicles and results in the production of the decapeptide GnRH, which is transported from perikarya to axon to terminal, and stored and released from secretory vesicles in the neuroterminals into the portal vasculature bed in the median eminence[12] (Fig. 1).

Thus, the median eminence is the region where *release* of GnRH and other hypothalamic neuropeptides is controlled, as synthesis occurs elsewhere in the hypothalamus. The microenvironment of the median eminence, including hypothalamic neuroterminals, the surrounding glial cells, and the extracellular fluid that surrounds these structures and the portal vasculature, enables communication among these elements. Considering that the release of GnRH occurs in discrete pulses on the order of approximately hourly intervals in mammals, the coordination of pulses likely depends upon physical and chemical interactions in the median eminence. Indeed, studies showing that explanted median eminence has the capability of releasing GnRH in pulses suggest that there is an intrinsic ability of this region to coordinate this process.[13,14]

GnRH release and reproductive aging

In young adults, the GnRH peptide is released at a frequency of approximately 30–60 min depending upon the species and the study.[13,15,16] This pulsatile pattern of release is critical to the development and

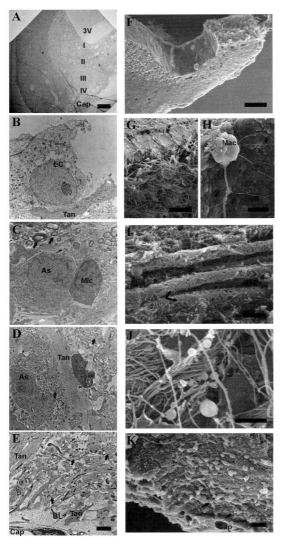

Figure 2. Electron micrographs of the median eminence are shown, with transmission electron microscopy (TEM, A–E) and scanning electron microscopy (SEM, F–K). (A) Half of the median eminence is shown in a low magnification TEM section, oriented with a slight angle to show the lateral median eminence structures from the ventral third ventricle (3V) to the dorsal portal capillary (Cap). Distinctive cellular structures of the four zones (I, II, III, and IV) from the 3rd ventricle to the portal capillary are shown in higher magnification in Panels B, C, D, and E, respectively. The corresponding SEM micrograph (F) shows the median eminence on a 100 μm coronal section. Tissue is oriented at a slight angle to show the floor of the third ventricle and the cryo-fractured lateral median eminence in the upper-right corner. Cellular structures from the third ventricle to the portal capillary in the lateral median eminence are shown in Panels G–K. Tanycyte cell bodies (Tan) and ependymal cells (EC) are shown by TEM (B) and SEM (G). In G, ependymal cells line the lateral walls of the lateral third ventricle. In H, the floor of the third ventricle is covered by small and large bleb like structures (*arrow*). A Kolmer cell (also called an epiplexus macrophage) is shown with long projections extending from its body (Mac). Panels C and I show the myelinated axon zone by TEM and SEM, respectively. In C, an astrocyte (As) and microglia (Mic) cell are indicated. Myelinated axons (*arrows*) are indicated in C and I. Panels D and J show the neural profile zone by TEM and SEM, respectively. In D, neural profiles (*arrow*), astrocyte (As), and tanycyte (Tan) processes are seen. In J, numerous neural profile fibers with varicosities (puncta) are seen. Panels E and K show the pericapillary zone of the median eminence. In E, neuroterminals (*arrow*) containing large dense-core vesicles are in close proximity to tanycyte processes (Tan). Tanycyte endfeet, basal lamina (BL) and capillary epithelial cell form a selective barrier in the portal system. In K, neuroterminals (*arrow*) on the surface of the cryo-fracture are shown to close to the portal capillaries (Cap). Scale bar A = 100 μm, B–E = 2 μm, F = 100 μm, G = 2 μm, H = 20 μm, I–K = 2 μm. Abbreviations: As = astrocyte, EC = ependymal cell, BL = basal lamina, Cap = portal capillaries, I = third ventricular zone, II = myelinated axon zone, III = neural profile zone, IV = portal capillary zone, Mac = macrophage (Kolmer cell), Mic = microglia, Tan = tanycyte, 3V = third ventricle. Scanning EM micrographs courtesy of John Mendenhall.

maintenance of reproductive capacity. In females of those species with spontaneous ovulatory cycles, such as primates and rodents, the amplitude and frequency of GnRH pulses changes across reproductive cycles to control folliculogenesis and ovulation. Although it is now well accepted that GnRH is the primary regulatory hormone of the HPG reproductive axis in adulthood, the role for GnRH in reproductive senescence is less clear. Some of the lack of clarity is due to species differences. In rodents, reproductive failure occurs in mid-life, a period during which the ovaries continue to have the capacity to produce follicles and ovulate (reviewed in Refs. 17 and 18). Thus, reproductive senescence in rodents is likely attributable to a neuroendocrine change. However, it is probably not due to changes in the numbers or morphology of GnRH perikarya, which vary relatively little with age.[19,20] Rather, we believe these differences in function are due to a combination of factors that regulate GnRH cell bodies and dendrites (e.g., neurotransmitters), and potential changes at the level of GnRH terminals in the median eminence, as the latter is the most immediate site for release of the decapeptide. These possibilities are not mutually exclusive. CNS inputs to GnRH perikarya undergo documented changes with aging—for example, glutamatergic stimulation of GnRH neurons decreases with age, and GABAergic inhibition increases with age[21] (reviewed in Ref. 22). The summation of these and other CNS regulatory inputs to GnRH is manifested in rodents as a decline in pulsatile release of GnRH, together with a loss of the surge mode of GnRH/LH release that precedes ovulation. At the level of nerve terminals, there may also be differences in age-related regulatory factors, as the median eminence is rich in neurotransmitters and receptors, but we are not aware of published studies addressing this specific issue.

The diminution of hypothalamic drive with age also relates to the potential decline in both negative and positive feedback regulation of GnRH release by steroid hormones.[23] However, experiments on feedback to GnRH neurons have focused at the level of GnRH perikarya—not the nerve terminals in the median eminence. An important future area of research is to determine whether membrane hormone receptors, such as GPR30, may be expressed on GnRH cells, and if so, whether the GnRH terminals express this receptor. This would enable feedback from estradiol to act directly and rapidly upon GnRH release. It was recently reported using a fetal primate GnRH cell culture that GPR30 is coexpressed in GnRH perikarya.[24] Our laboratory has ongoing experiments designed to evaluate and quantify the coexpression of GPR30 on GnRH terminals in the median eminence of aging rats using double-label immunofluorescence and confocal microscopy. Our preliminary data show that (1) GnRH terminals coexpress GPR30; (2) overall, there is a significant age-related increase in this coexpression; and (3) estradiol upregulates GPR30 coexpression in GnRH terminals in young but not aging rats (M. Noel, S. Ng, K. Resendiz, A.C. Gore, unpublished). These data remain to be confirmed, but they provide tantalizing evidence for a direct site of regulation of estradiol on GnRH release at the level of the median eminence, and for an age-related loss of responsiveness.

As already mentioned, there are species differences in the aging of the hypothalamic GnRH system. Studies in women and nonhuman primates show that the proximal event for menopause is the loss of ovarian follicles through a process involving accelerated follicular atresia. However, neuroendocrine markers do change in primates, and in some cases, precede the period of exponential follicular atresia in a manner that has some parallels to rodents. For example, prior to or during the perimenopause, an erosion of positive feedback effects of estradiol,[25] an elevation of FSH,[26] a decrease in serum anti-Mullerian hormone,[26] and other physiological changes have been reported (reviewed in Ref. 23). In addition, a loss of negative feedback during the perimenopause may not initially be observed in perimenopausal women, but it develops gradually with years postmenopause, suggesting an eventual hypothalamic–pituitary dysregulation.[27,28]

It is extremely difficult to measure GnRH directly from nerve terminals in the median eminence, due to an inability to detect GnRH in peripheral serum, together with the obvious limitations of measuring the release of a hormone from the brain. This has deterred the ability to measure how GnRH release changes with aging. To our knowledge, there are two studies that directly measured GnRH neurosecretion with aging, one in rats and one in nonhuman primates. In the rat, Rubin and Bridges performed push–pull perfusion on GnRH release from the medial basal hypothalamus of ovariectomized rats given steroid hormones to induce positive feedback.[29] They showed a significant decline in GnRH release of the older rats, a finding that mirrored previous work on the diminution of gonadotropin release with aging in rodent species.

The other study that directly measured GnRH release also used push–pull perfusion of the median eminence in young adult and perimenopausal (ovarian intact) rhesus monkeys.[16] We demonstrated a robust increase in pulsatile GnRH release. Mean GnRH release was significantly increased, and although GnRH pulse amplitude was elevated, this latter effect was not significant (probably due to a small sample size, as it was very difficult to obtain very aged female rhesus monkeys). Nevertheless, we observed in three out of four of our perimenopausal monkeys some very elevated GnRH pulses that exceeded even preovulatory GnRH levels.[16] This report of an age-related increase in GnRH has parallels to studies showing increases in LH in rhesus monkeys and women,[30,31] and increases in gonadotropin free-alpha subunit in postmenopausal women.[31] Although there are differences in exactly which parameters of GnRH release increase with age (i.e., mean levels, pulse amplitude, pulse frequency), these are relatively minor compared to the consensus for an age-related upregulation of these neuroendocrine

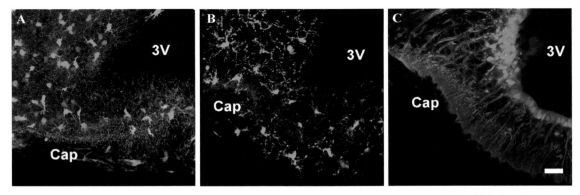

Figure 3. Confocal fluorescence images are shown for transgenic mice with green fluorescent protein (GFP) labeling of astrocytes (A), microglia (B), and tanycytes (C). In all three panels, tissues were subjected to fluorescence immunohistochemistry with a GnRH mouse monoclonal primary antibody (HU11b[34]), detected with an antimouse secondary antibody linked to the fluorophore Texas red. Immunoreactive GnRH neuroterminals (red) are seen in close proximity to all three glial cell types. For astrocytes (A) and microglia (B), these cells are scattered fairly evenly through the median eminence. Tanycyte cell bodies (C) are aligned along the 3rd ventricle, and they extend long processes to the base of the brain at the portal capillary system. Scale bar = 100 μm, Cap = portal capillary region, 3V = third ventricle.

hormones. We interpret these data to mean that a loss of steroid hormone feedback with menopause in primates initially results in an upregulation of GnRH, and subsequently gonadotropin, secretion. However, with greater time postmenopause, the negative feedback effect on GnRH/LH release diminishes,[27] bringing the primate model back in parallel with the rodent model.

Glial cells in the median eminence

Thus far, this paper has focused on the hypothalamic GnRH terminals in the median eminence. However, glial cells constitute a significant part of the median eminence, and they interact extensively with adenohypophysial neurons and the portal capillary system.[10,32] It is important to consider the expression of glia in the median eminence, and further, to understand how they change with aging in their intrinsic properties and in their interrelationships with GnRH terminals.

The median eminence expresses at least three glial cell types—astrocytes, microglia, and tanycytes—as illustrated in Figure 3. For these micrographs, we used hypothalamic tissues from transgenic mice, kindly provided by Dr. Wesley Thompson's laboratory (University of Texas at Austin; protocols were approved by the Institutional Animal Care and Use Committee).[33] These transgenic mice had the green fluorescent protein (GFP) marker driven by promoters for different glial cell type (astrocyte:

S100; microglia: S100; and tanycyte: nestin). Notably, when S100 was used to drive expression of GFP, the glial cells differentiated into either astrocytes or microglia, which could be distinguished by other specific astrocyte and microglia markers and which had very different and easily distinguished differentiated morphological properties; see Ref. 33 for details). Tissues from each transgenic mouse line were immunolabeled for GnRH using the mouse monoclonal antibody, HU11b.[34] The distribution of the three glial cell types, and their close relationship with the GnRH neuroterminals, can be clearly seen in the median eminence (Fig. 3). GnRH axons and terminals are in close proximity to astrocytes, tanycytes, and microglia cells. The microglia and astrocyte signals are evenly distributed in the lateral portion of the median eminence. In the central portion of the median eminence, astrocytes show long extended processes in the myelinated axon zone with some processes extending to the portal capillary zone. Interestingly, nestin-GFP signal was observed in the tanycytes, similar to earlier reports of nestin immunohistochemistry in human hypothalamus.[35] Nestin is an intermediate filament protein that had originally been identified as a marker of neuroepithelial stem/progenitor cells.[36] We observe nestin-GFP signal in the mouse hypothalamus with the same pattern as that described for tanycytes and ependymal cells[10] (Fig. 3). These transgenic mice models provide us with excellent resolution of the

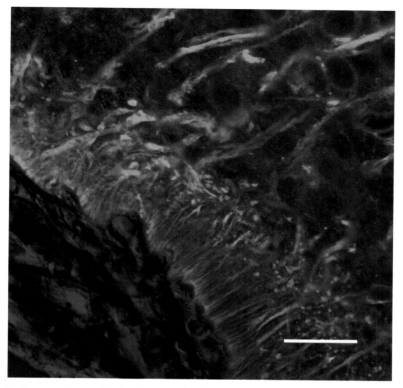

Figure 4. Confocal microscopic image showing the network of GnRH neuroterminals, tanycytes, and their relationship to the portal capillary vasculature of a representative young male nestin-green fluorescent protein (GFP) mouse (a marker for tanycytes). Tanycytic endfeet (seen in the green GFP channel) separate GnRH processes (seen as punctate red immunofluorescence, labeled by Texas red) from the portal capillary bed (DIC signal, pseudocolored in blue). Scale bar = 20 μm.

interaction between GnRH neuroterminals and glia (Fig. 4).

Glial cells and the aging median eminence

Tanycytes are believed to be involved in the transport and release of hormones in the median eminence.[37] It has been proposed that the regulation of GnRH terminals may result from ensheathment by tanycytes,[38] which may allow GnRH terminals access to the portal vasculature or to other neuronal processes. Tanycytes appear to play a role in the remodeling of the aged basal hypothalamus in male and female rats through phagocytotic actions on degenerating neurons.[39,40] Light and electron microscopic studies show that tanycytes undergo morphological modifications with the increased presence of lipid droplets in the cytoplasm of aging male and female rats compared to their younger counterparts.[39] However, relatively little is known about how glial cells in the median eminence undergo age-related changes, and even less about the more specific relationships between GnRH terminals and their glial environment with aging.

We have recently undertaken studies in our laboratory to determine effects of aging on the organization of the median eminence. Using ultramicrotome sections of the median eminence of young, middle-aged and old female rats, we observed significant age-related changes in the structure of the median eminence, and more specifically, substantial changes to the morphology and anatomy of tanycytes.[41] In young female rats, tanycytic cell bodies are lined up along the third ventricle and extend a long process across the median eminence to the portal capillary vasculature (e.g., Figs. 2 and 5). With aging, tanycyte processes become thicker and disorganized in the pericapillary zone, with a loss of perpendicular orientation (Fig. 5). We also noted that the relationships among tanycytes, neural terminals, and the basal lamina of the portal capillary system undergo a progressive disorganization.[41,42] In young rats, the basal laminar boundary is easy to discern, but with aging, the boundary between the neuroterminals of the pericapillary zone and the portal capillaries becomes convoluted (Fig. 5). These structural changes to the organization of the median eminence are likely to alter the function of the selective barrier of the portal system, and alter the access of neuropeptides, such as GnRH, to the portal blood vessels. As the neural–glial relationship is considered to be important for the maintenance of neuroendocrine function,[11,43] and it undergoes age-related changes, we speculate that there is an alteration in the ability of GnRH (and other) terminals to release their neuropeptide effectively. Further research is necessary to ascertain how these processes occur.

Links among GnRH terminals and glia in the aging median eminence

The microenvironment of the median eminence maintains the ability of GnRH neuroterminals to release the decapeptide. Our laboratory performed cryo-embedding immunogold electron microscopy for GnRH using ultrathin sections collected from the caudal median eminence of aging rats. Using this technique, we were able to identify GnRH immunopositive terminals and compare their ultrastructural differences, as well as to study the changes in their surrounding microenvironment with aging.[41] We quantified several properties of the GnRH immunopositive terminals including the size of terminals, the density of secretory vesicles (large dense-core vesicles) within the terminals, the density of immunogold-labeled GnRH peptide, and the area fraction of mitochondria in our rats.[42] Our ultrastructural data suggest that the major change to GnRH immunopositive terminals with age is an increased density of large dense-core secretory vesicles in the terminals. We believe this may reflect a buildup of vesicles due to the loss of ability to release the peptide, consistent with reports of decreased GnRH/LH release with age.[29]

Previously, GnRH terminal and glial interactions have been investigated in young rats, with differences detected in the distance between GnRH terminals and the basal lamina, the ensheathment of GnRH terminals by tanycytes, and the proximity of GnRH terminals to portal capillaries in animals of differing developmental age and hormonal status.[44]

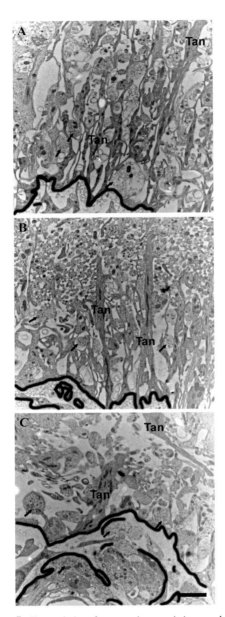

Figure 5. Transmission electron microscopic images showing the organization of the pericapillary zone of the lateral median eminence from representative young (A), middle-aged (B), and old (C) ovariectomized rats. In young rats, tanycyte processes (Tan) are long and narrow, and organized linearly from top (toward the third ventricle) to bottom (toward the portal capillaries, delineated with the thick black line). With aging, tanycyte processes become larger, wider, and disorganized, with a distinct loss of perpendicular orientation. In addition, the boundary between the neuroterminals (arrow) of the pericapillary zone and the portal capillaries (outlined in thick black) becomes very convoluted with age. Scale bar = 2 μm.

Additional transmission electron microscopy studies from our laboratory using postembedding tissues immunolabeled for GnRH provide ultrastructural evidence that the ensheathment of GnRH neuroterminal by glia decreased in old female rats indicating a decreased GnRH–glia interaction during the aging process.[41,42]

Conclusions

A greater understanding of the structure of the median eminence is necessary to fully appreciate the regulation of the release of neurohormones, such as GnRH. The morphology of nerve terminals and glia, particularly tanycytes, undergoes dramatic age-related changes. There are also changes in subcellular properties of GnRH terminals, such as decreased tanycytic ensheathment and an increase in the density of secretory vesicles with aging. Further, the linear organization of tanycytes in the median eminence becomes lost in the aging median eminence. These findings are interpreted to mean that GnRH release in aging rats is impeded by structural organizational changes. By studying ultrastructural properties of the median eminence, and relating them to reproductive status, we are increasing our understanding of the potential role of the hypothalamus in reproductive senescence in female mammals.

Acknowledgments

We thank John Mendenhall (Institute of Cellular and Molecular Biology microscopy facility, University of Texas at Austin) for providing the scanning electron microscopy images shown in Figure 2. We are grateful to Dr. Wesley Thompson and Dr. Yi Zuo for providing glia-GFP mice tissues to study the hypothalamus. Sharon Kim provided expert assistance in the preparation of mouse GFP tissues. Work described in this report was funded by the NIH (AG16765, AG028051 to ACG).

Conflicts of interest

The authors declare no conflicts of interest.

References

1. Johnson, A. 1993. Sensory circumventricular organs and brain homeostatic pathways. *FASEB J.* **7:** 678–686.
2. Green, J. & G. Harris. 1949. Observation of the hypophysioportal vessels of the living rat. *J. Physiol.* **108:** 359–361.
3. Rodriguez, E.M. *et al.* 2005. Hypothalamic tanycytes: a key component of brain-endocrine interaction. *Int. Rev. Cytol.* **247:** 89–164.
4. Jennes, L. & W.E. Stumpf. 1986. Gonadotropin-releasing hormone immunoreactive neurons with access to fenestrated capillaries in mouse brain. *Neuroscience* **18:** 403–416.
5. Kozlowski, G.P. & P.W. Coates. 1985. Ependymoneuronal specializations between LHRH fibers and cells of the cerebroventricular system. *Cell Tissue Res.* **242:** 301–311.
6. Durrant, A. & T. Plant. 1999. A study of the gonadotropin releasing hormone neuronal network in the median eminence of the rhesus monkey (*Macaca mulatta*) using a postembedding immunolabelling procedure. *J. Neuroendocrinol.* **11:** 813–821.
7. Yin, W. *et al.* 2007. Novel localization of NMDA receptors within neuroendocrine gonadotroping-releasing hormone terminals. *Exp. Biol. Med.* **232:** 662–673.
8. Yin, W. & A.C. Gore. 2006. Neuroendocrine control of reproduction aging: roles of GnRH neurons. *Reproduction* **131:** 403–414.
9. Agnati, L.F., M. Zoli, I. Strömberg & K. Fuxe. 1995. Intercellular communication in the brain: wiring versus volume transmission. *Neuroscience* **69:** 711–726.
10. Ojeda, S., A. Lomzicai & U. Sandau. 2008. Glial-gonadotrophin hormone (GnRH) neurone interactions in the median eminence and the control of GnRH secretion. *J. Neuroendocrinol.* **20:** 732–742.
11. Wittkowski, W. 1998. Tanycytes and pituicytes: morphological and functional aspects of neuroglial interaction. *Microsc. Res. Tech.* **41:** 29–42.
12. King, J., S.A. Tobet, F.L. Snavely & A.A. Arimura. 1982. LHRH immunopositive cells and their projections to the median eminence and organum vasculosum of the lamina terminalis. *J. Comp. Neurol.* **209:** 287–300.
13. Terasawa, E. 2001. Luteinizing hormone-releasing hormone (LHRH) neurons: mechanism of pulsatile LHRH release. *Vitamin Horm.* **63:** 91–129.
14. Purnelle, G., A. Gérard, V. Czajkowski & J.P. Bourguignon. 1997. Pulsatile secretion of gonadotropin-releasing hormone by rat hypothalamic explants without cell bodies of GnRH neurons. *Neuroendocrinology* **66:** 305–312.
15. Scarbrough, K. & P.M. Wise. 1990. Age-related changes in pulsatile luteinizing hormone release precede the transition to estrous acyclicity and depend upon estrous cycle history. *Endocrinology* **126:** 884–890.
16. Gore, A.C., B. Windsor-Engnell & E. Terasawa. 2004. Menopausal increases in pulsatile gonadotropin-releasing hormone release in a nonhuman primate (*Macaca mulatta*). *Endocrinology* **145:** 4653–4659.
17. Maffucci, J.A. & A.C. Gore. 2006. Age-related changes in hormone and their receptors in animal models of female reproductive senescence. In *Handbook of Models for Human Aging*. P.M. Conn, Ed.: 533–552. Academic Press, Inc. Oxford, UK.
18. Finch, C.E., L.S. Felicio, C.V. Mobbs & J.F. Nelson. 1984. Ovarian and steroidal influences on neuroendocrine aging processes in female rodents. *Endocrine Rev.* **5:** 467–497.
19. Bestetti, G.E. *et al.* 1991. Functional and morphological changes in the hypothalamopituitary-gonadal axis of aged female rats. *Biol. Reprod.* **45:** 221–228.
20. Hoffman, G.E. & J.J. Sladek. 1980. Age-related changes in dopamine, LHRH and somatostatin in the rat hypothalamus. *Neurobiol. Aging* **1:** 27–37.

21. Neal-Perry, G.S., G.D. Zeevalk, J. Shu & A.M. Etgen. 2008. Restoration of the luteinizing hormone surge in middle-aged female rats by altering the balance of GABA and glutamate transmission in the medial preoptic area. *Biol. Reprod.* **79:** 878–888.
22. Maffucci, J.A. & A.C. Gore. 2009. Hypothalamic neural systems controlling the female reproductive life cycle: gonadotropin-releasing hormone, GABA, and glutamate. *Int. Rev. Cell Mol. Biol.* **274:** 69–127.
23. Wise, P.M. et al. 2002. Neuroendocrine modulation and repercussions of female reproductive aging. *Recent Prog. Horm. Res.* **57:** 235–256.
24. Noel, S.D. et al. 2009. Involvement of G protein-coupled receptor 30 (GPR30) in rapid action of estrogen in primate LHRH neurons. *Mol. Endocrinol.* **23:** 349–359.
25. Weiss, G. et al. 2004. Menopause and hypothalamic-pituitary sensitivity to estrogen. *JAMA* **292:** 2991–2996.
26. Downs, J.L. & H.F. Urbanski. 2006. Neuroendocrine changes in the aging reproductive axis of female rhesus macaques (*Macaca mulatta*). *Biol. Reprod.* **75:** 539–546.
27. Rossmanith, W.G. 1995. Gonadotropin secretion during aging in women: review article. *Exp. Gerontol.* **30:** 369–381.
28. Gill, S., H.B. Lavoie, Y. Bo-Abbas & J.E. Hall. 2002. Negative feedback effects of gonadal steroids are preserved with aging in postmenopausal women. *J. Clin. Endocrinol. Metab.* **87:** 2297–2302.
29. Rubin, B.S. & R.S. Bridges. 1989. Alterations in luteinizing hormone-releasing hormone release from the mediobasal hypothalamus of ovariectomized, steroid-primed middle-aged rats as measured by push-pull perfusion. *Neuroendocrinology* **49:** 225–232.
30. Woller, M.J. et al. 2002. Aging-related changes in release of growth hormone and luteinizing hormone in female rhesus monkeys. *J. Clin. Endocrinol. Metab.* **87:** 5160–5167.
31. Gill, S., J.L. Sharpless, K. Rado & J.E. Hall. 2002. Evidence that GnRH decreases with gonadal steroid feedback but increases with age in postmenopausal women. *J. Clin. Endocrinol. Metab.* **87:** 2290–2296.
32. Garcia-Segura, L. & M. McCarthy. 2004. Minireview: role of glia in neuroendocrine function. *Endocrinology* **145:** 1082–1086.
33. Zuo, Y. et al. 2004. Fluorescent proteins expressed in mouse transgenic lines mark subsets of glia, neurons, macrophages, and dendritic cells for vital examination. *J. Neurosci.* **24:** 10999–11009.
34. Urbanski, H.F. 1991. Monoclonal antibodies to luteinizing hormone-releasing hormone: production, characterization, and immunocytochemical application. *Biol. Reprod.* **44:** 681–686.
35. Baroncini, M. et al. 2007. Morphological evidence for direct interaction between gonadotrophin-releasing hormone neurones and astroglial cells in the human hypothalamus. *J. Neuroendocrinol.* **19:** 691–702.
36. Zimmerman, L. et al. 1994. Independent regulatory elements in the nestin gene direct transgene expression to neural stem cells or muscle precursors. *Neuron* **12:** 11–24.
37. Akmayev, I. & O. Fidelina. 1981. Tanycytes and their relation to the hypophyseal gonadotrophic function. *Brain Res.* **210:** 253.
38. Flament-Durand, J. & J. Brion. 1985. Tanycytes: morphology and functions: a review. *Int. Rev. Cytol.* **96:** 121–155.
39. Brawer, J. & R. Walsh. 1982. Response of tanycytes to aging in the median eminence of the rat. *Am. J. Anat.* **163:** 247–256.
40. Zoli, M. et al. 1995. Age-related alterations in tanycytes of the mediobasal hypothalamus of the male rat. *Neurobiol. Aging* **16:** 77–83.
41. Yin, W., J.M. Mendenhall, M. Monita & A.C. Gore. 2009. Three-dimensional properties of GnRH neuroterminals in the median eminence of young and old rats. *J. Comp. Neurol.* **20:** 284–295.
42. Yin, W., D. Wu, M.L. Noel & A.C. Gore. 2009. GnRH neuroterminals and their microenvironment in the median eminence: effects of aging and estradiol treatment. *Endocrinology* **150:** 5498–5508.
43. Rodriguez, E. et al. 2005. Hypothalamic tanycytes: a key component of brain-endocrine interaction. *Int. Rev. Cytol.* **247:** 89–164.
44. King, J.C. & B.S. Rubin. 1994. Dynamic changes in LHRH neurovascular terminals with various endocrine conditions in adults. *Horm. Behav.* **28:** 349–356.

Animal models of reproductive aging: what can they tell us?

Steven N. Austad

University of Texas Health Science Center San Antonio, San Antonio, Texas

Address for correspondence: Steven N. Austad, Barshop Institute for Longevity and Aging Studies, Department of Cellular and Structural Biology, 15355 Lambda Drive, STCBM 3.100.07, San Antonio, Texas 78245. austad@uthscsa.edu

This commentary explores the relationship between what can be learned about reproductive senescence from studies in the laboratory compared with what can be learned from studies in the field. Laboratory studies allow researchers to isolate and analyze detailed cellular and molecular mechanisms of reproductive senescence, however drawing evolutionary inferences from captive studies can be misleading. The ideal study would combine field and laboratory observations and experiments. As with most other biological phenomena, understanding the nature of genetic and environmental interactions is central to understanding of reproductive aging.

Keywords: reproductive aging; natural selection

Natural selection molds animals to maximize successful reproduction. Consequently, one might expect reproductive senescence to be rare in nature, because if survival to age x is likely, then other things being equal, natural selection should favor cellular, molecular, and physiological mechanisms that preserve reproduction performance to at least that age. In the long-term field studies of baboons described in the chapter by Altmann et al.,[1] this expectation looks to be approximately true—at least for females. From early adulthood until age 18, there appears to be no change in female birth rate. Eighteen years is well beyond the 12 year life expectancy of adult females in this population.[2] Even among the rare females older than 18 years (roughly 3% of the population), some continue to reproduce at high rates, although others cease reproducing altogether (Altmann et al., Fig. 1).

The logic discussed earlier does not apply in a straightforward manner to males. The reason is that at least among polygynous species, such as virtually all mammals, males compete with one another for access to females. Males successful in this competition can obtain very high reproductive rates. Notice in the Altmann et al. chapter (Fig. 1) that the most successful male breeder has roughly three times the offspring production of the average female. But competition can be physically costly. Male reproduction peaks early (at about 9 years of age) and declines relatively abruptly. This could be due to senescence of reproductive system itself, although that seems an unlikely possibility. Testosterone as measured in feces declines with age, but still appears high enough for males to be producing viable sperm until near the end of life. The observed male reproductive decline is more likely the consequence of a deteriorating ability to compete for mates, owing to nonreproductive aspects of physiological senescence, such as loss of strength, speed, judgment, or even the accumulation of irreparable injuries, such as to joints. A very similar age-related pattern of sex differences in reproductive decline is seen in a wild population of the polygynous red deer.[3]

It is worth considering how researchers studying baboon reproductive senescence in a captive population might reach different conclusions from those found in this field study. Captive animals are likely to be better fed, have fewer energy demands, and be subject to little or no reproductive competition compared with their wild counterparts. As a consequence, female baboons from captive colonies usually reach reproductive maturity about 1 year earlier than females from the wild.[2] For males, the lack of reproductive competition would likely make their observed reproductive rate decline more slowly compared with males in the field, reflecting purely the physiological status of the reproductive system rather than male–male competitive ability. Similar differences between field and captive conditions may obtain for other parameters potentially affected by male competitive ability. For instance, body mass index, like reproduction, declines more rapidly with age in males than females in nature. This may reflect physiological senescence but it may also reflect lack of access to the best foraging sites because of male–male competition. It would be interesting to know whether the same pattern obtains in captive populations where such competition is lacking.

The main point I am making is that captive studies are capable of revealing aspects of reproductive senescence that may be masked under field conditions, but that evolutionary inferences drawn from captive studies can be misleading and should be approached with caution.

The value of mechanistic, highly controlled laboratory investigations of reproductive senescence in genetically tractable species is revealed nicely by the chapters by Tatar[4] and Kenyon.[5] In both the laboratory roundworm (*Caenorhabditis elegans*) and fruit fly (*Drosophila melanogaster*), reproductive senescence is easily measured by an age-related decrease in the rate of egg laying, decreasing egg viability, and ultimately cessation of egg-laying altogether. Tatar[4] notes that in the fruit fly, peak reproduction in females occurs 5–10 days after eclosion and thereafter declines. The decline in egg-laying seems to be due both to decreasing proliferation of germline stem cells and to increasing apoptosis among cells of the protective egg chamber in which oocytes mature. The rate of reproductive senescence may be affected by changes in molecular signaling by TGF-β orthologs. Moreover, overexpression of two recently identified genes of unknown function increases late-life fecundity without reducing egg-laying in early life. These studies represent a very refined and steadily advancing understanding of mechanisms underlying female reproductive senescence in flies.

In *C. elegans*, it is well known that under laboratory conditions self-fertilizing hermaphrodites reach peak egg laying 2–4 days after reaching adulthood and that egg-laying decreases rapidly thereafter, ceasing completely by about 7 days of age.[6] Some, but not all, of this decrease in egg laying is due to sperm depletion. This can be determined because if hermaphrodites, instead of self-fertilizing, mate with a male thus obtaining a much greater than usual sperm supply, they lay more eggs than normal. However, even with this superabundant sperm supply, egg production still declines and ceases, although it takes a few days longer than usually. Some longevity mutants—and there are hundreds of them in *C. elegans*—also extend reproductive longevity.[7]

Perhaps the most intriguing relationship between reproduction and longevity in the worm has little to do with reproductive aging itself, but with how the germline influences longevity. As discovered by Kenyon a decade ago, if one destroys the germline itself either during larval life or in adulthood, worms live longer. However, if one destroys the entire gonad which includes the germline plus the somatic gonad, then lifespan is not affected.[8] The mammalian equivalent would be observing a lifespan extension by destroying eggs but no such extension when destroying the entire ovary. For this effect to obtain, Kenyon[5] tells us in her chapter that two genes, the worm insulin receptor (*daf-2*) and a steroid hormone receptor (*daf-12*), are required. However, their activity is required only in the worm intestine, not in the rest of the body.

As interesting as this tinkering under the hood of these model invertebrates is, and as potentially informative about other species—possibly even humans, the evolutionary significance of these findings remains obscure because we know so little about aging, reproduction, or longevity of flies or worms in nature. Even what we know in the laboratory is confined to a few standard environmental conditions.

Carey and Molleman's[9] chapter emphasizes how important the environment is to understanding the biology, including the pattern and timing of reproductive senescence, of any organism. Their chapter focuses on two fly species—the Mexican fruit fly or medfly (*Ceratitis capitata*) and the Mexican fruit fly or mexfly (*Anastrepha ludens*). These authors note that the type of food consumed by fly larvae has an enormous impact on adult reproduction and longevity. Specifically, medfly larva that are fed on plums live more than twice as long, and lay more than twice as many eggs over their lifetime, as flies fed on bananas (both fruits are normal foods of wild medflies), even though their adult diets are identical. The impact of early life events on survival and reproductive senescence are not unprecedented: both the prenatal and early postnatal environments are known to influence longevity and the timing of reproductive senescence in mice.[10,11]

Nutritional experiments with mexflies also show major nutritional effects on both survival and reproduction, and illustrate the principle of life history tradeoffs. Diets that maximize longevity differ from those that maximize reproduction. A fascinating contrast between the medfly and common laboratory flies and worms is that continuous calorie restriction of adult medflies fails to extend life (and indeed shortens it at extreme levels of restriction) as it does in the laboratory species.

The experiments done with the medfly and mexfly make clear that environmental influences on senescence—both in reproduction and survival—are ubiquitous and that a full understanding of any species' biology requires information on its performance under a wide range of conditions including, if possible, its life in the wild.

For instance, it is arguable that *C. elegans* is the best described and defined animal species we know *in the laboratory*. However, even the most rudimentary knowledge of its biology in the field is lacking. We do not know what eats it or how much it eats, although we know it does not eat *E. coli*—its standard laboratory diet. We also know that it lives some 40% longer on bacteria that it might eat in nature (*B. subtilis*) than on *E. coli* and that it lives only one-fifth as long on other potential natural food sources.[12] Furthermore, we do not know its habitat preferences. Although it is often described as a soil dwelling nematode, it is often found on larger soil organisms, such as snails and millipedes.[13] As a consequence of this lack of knowledge, it is difficult to ask evolutionary questions of the worm, which is unfortunate given our exquisite knowledge of its laboratory biology. For instance, is worm reproductive senescence as observed in the laboratory evolutionarily relevant? That is, does reproductive senescence exist in wild populations? Would genetic treatments that extend life and presumably health in the laboratory, also do so in a natural environment?

We do have a possible glimpse of an answer to some of these questions. Although longitudinal studies of *C. elegans* in the wild have never been performed, Van Voorhies *et al.* asked the simple question, how long do worms live under benign laboratory conditions if simply kept in soil as opposed to the agar or liquid media in which they are typically reared?[14] Using natural soil that was either heat-treated to kill all soil organisms or left intact, and using as a food source the same *E. coli* that worms are typically fed in the laboratory, these researchers found that median longevity dropped from 12.0 days on agar plates to 1.5 days on untreated soil and 1.0 days in heat-treated soil. By 6 days of age, fewer than 5% of worms were still alive in either soil type. Perhaps even more surprising than this dramatic fall in survival was the observation than *daf-2* mutants, which in agar live twice as long as wild-type worms, were slightly shorter-lived in heat treated soil.

If these results mimic to any significant degree conditions worms face in nature, then the ramifications are considerable. For one thing, it is theoretically conceivable that worms live no longer than this in nature. Worms reproduce sufficiently quickly that even these shortened lifespans are still more than sufficient to maintain a viable worm population. Second, if these lifespans are representative of the real world, then reproductive senescence and whatever tradeoffs may or may not be observed between early and late reproduction in the laboratory are clearly irrelevant in evolutionary terms. Finally, whether certain genes delay or accelerate senescence may be highly dependent on the environment. Similar results were observed by Mackay's laboratory, which found that in fruit flies chromosomal regions associated with increasing lifespan under one sort of laboratory conditions might be associated with decreasing lifespan under other conditions.[15] Moreover, genes associated with long life in one sex were sometimes associated with short life in the other sex. The impact of environmental contingency on longevity is not unanticipated. Treatments, such as caloric restriction, that may confer long life under benign conditions may shorten life under other conditions, such as during outbreaks of infectious diseases.[16]

The tremendous power of genetic manipulation in model laboratory species has taught us, and will continue to teach us, a tremendous amount. However, we should not let this power distract us from the realization that understanding how different environments interact with genetic variants is equally informative. The real world poses unique challenges both to animals and humans. After all, we are not genetically homogeneous. We do not live under constant disease-free conditions eating a monotonously healthful diet. Understanding reproductive aging and the maintenance of health requires understanding our environment as well as our genes.

Conflicts of interest

The author declares no conflicts of interest.

References

1. Altmann, J. et al. 2010. Life history context of reproductive aging in a wild primate model. *Ann. N.Y. Acad. Sci.* **1204:** 127–138.
2. Bronikowski, A.M. et al. 2002. The aging baboon: comparative demography in a non-human primate. *Proc. Natl. Acad. Sci. USA* **99:** 9591–9595.
3. Nussey, D.H. et al. 2009. Inter- and intrasexual variation in aging patterns across reproductive traits in a wild red deer population. *Am. Nat.* **174:** 342–357.
4. Tatar, M. 2010. Reproductive aging in invertebrate genetic models. *Ann. N.Y. Acad. Sci.* **1204:** 149–155.
5. Kenyon, C. 2010. A pathway that links reproductive status to lifespan in *Caenorhabditis elegans*. *Ann. N.Y. Acad. Sci.* **1204:** 156–162.
6. Chen, J. et al. 2007. A demographic analysis of the fitness cost of extended longevity in *Caenorhabditis elegans*. *J. Gerontol. A Biol. Sci. Med. Sci.* **62:** 126–135.
7. Gems, D. et al. 1998. Two pleiotropic classes of daf-2 mutation affect larval arrest, adult behavior, reproduction and longevity in *Caenorhabditis elegans*. *Genetics* **150:** 129–155.
8. Hsin, H. & C. Kenyon. 1999. Signals from the reproductive system regulate the lifespan of *C. elegans*. *Nature* **399:** 362–366.
9. Carey, J.R. & F. Molleman. 2010. Reproductive aging in tephritid fruit flies. *Ann. N.Y. Acad. Sci.* **1204:** 139–148.
10. Sun, L., A.A. Sadighi Akha, R.A. Miller & J.M. Harper. 2009. Life-span extension in mice by preweaning food restriction and by methionine restriction in middle age. *J. Gerontol. A Biol. Sci. Med. Sci.* **64:** 711–722.
11. vom Saal, F.S. 1989. Sexual differentiation in litter-bearing mammals: influence of sex of adjacent fetuses in utero. *J. Anim. Sci.* **67:** 1824–1840.
12. Garsin, D.A. et al. 2003. Long-lived *C. elegans* daf-2 mutants are resistant to bacterial pathogens. *Science* **300:** 1921.
13. Barrière, A. & M.-A. Félix. 2005. High local genetic diversity and low outcrossing rate in *Caenorhabditis elegans* natural populations. *Curr. Biol.* **15:** 1176–1184.
14. Van Voorhies, W.A., J. Fuchs & S. Thomas. 2005. The longevity of *Caenorhabditis elegans* in soil. *Biol. Lett.* **1:** 247–249.
15. Vieira, C. et al. 2000. Genotype-environment interaction for quantitative trait loci affecting life span in *Drosophila melanogaster*. *Genetics* **154:** 213–227.
16. Ritz, B.W., I. Aktan, S. Nogusa & E.M. Gardner. 2008. Energy restriction impairs natural killer cell function and increases the severity of influenza infection in young adult male C57BL/6 mice. *J. Nutr.* **138:** 2269–2275.

ANNALS OF THE NEW YORK ACADEMY OF SCIENCES
Issue: *Reproductive Aging*

Life history context of reproductive aging in a wild primate model

Jeanne Altmann,[1,2,3] Laurence Gesquiere,[1] Jordi Galbany,[4,5] Patrick O. Onyango,[1] and Susan C. Alberts[2,4]

[1]Department of Ecology & Evolutionary Biology, Princeton University, Princeton, New Jersey. [2]Institute for Primate Research, National Museums of Kenya, Nairobi, Kenya. [3]Department of Veterinary Anatomy and Physiology, University of Nairobi, Nairobi, Kenya. [4]Department of Biology, Duke University, Durham, North Carolina. [5]Secció d'Antropologia, Departament de Biologia Animal, Universitat de Barcelona. Barcelona, Spain

Address for correspondence: Jeanne Altmann, Princeton University, Department of Ecology & Evolutionary Biology, Room 401 Guyot Hall, Princeton, New Jersey 08544. altj@princeton.edu

The pace of reproductive aging has been of considerable interest, especially in regard to the long postreproductive period in modern women. Here we use data for both sexes from a 37-year longitudinal study of a wild baboon population to place reproductive aging within a life history context for this species, a primate relative of humans that evolved in the same savannah habitat as humans did. We examine the patterns and pace of reproductive aging, including birth rates and reproductive hormones for both sexes, and compare reproductive aging to age-related changes in several other traits. Reproductive senescence occurs later in baboon females than males. Delayed senescence in females relative to males is also found in several other traits, such as dominance status and body condition, but not in molar wear or glucocorticoid profiles. Survival, health, and well-being are the product of risk factors in morphological, physiological, and behavioral traits that differ in rate of senescence and in dependence on social or ecological conditions; some will be very sensitive to differences in circumstances and others less so.

Keywords: reproductive aging; baboons; toothwear; body condition; steroid hormones; senescence

Over 30 years ago, Sarah Hrdy introduced what arguably stands as the first published lifespan perspective on female reproductive strategies in wild nonhuman primates.[1–3] Prevailing wisdom at the time was that senescence did not occur in wild populations. In contrast, Hrdy posited that female nonhuman primates experience age-related reproductive decline and a postreproductive stage that presages the long postreproductive period of modern women (see Hawkes and Smith[3a]). Moreover, she suggested that within natural societies of nonhuman primates, such as those of the langur monkeys she studied in Asia, reproductive senescence is associated with fitness benefits.[1–3]

Thus, our study of reproductive aging in wild baboons has roots in Hrdy's perspective on langur reproductive strategies, but relevant data for aging in natural populations were largely unavailable at that time. In fact, patterns of aging in wild animals, not just wild primates, still remain largely undescribed despite their importance to understanding fundamental aspects of life-history evolution, and despite early speculation on their importance in the evolutionary and behavioral ecology of humans and other species.

Using almost four decades of data from the Amboseli Baboon Research Project (ABRP), we are now in a position to evaluate reproductive aging in wild female primates and compare females' reproductive time-course with that of males. In addition, we examine reproductive aging of both sexes in the context of aging patterns in several other traits, addressing the question of whether reproductive aging resembles or stands out from that of other traits. We begin with a general introduction to baboons and their relationships to humans, and then describe the Amboseli baboon population, the ABRP longitudinal life-history research program, and a brief overview of relevant methods.

doi: 10.1111/j.1749-6632.2010.05531.x

Wild baboons as a model for understanding human aging

Baboons (genus *Papio*) are among the best studied of the cercopithecine primates.[4] They are large, diurnal, semi-terrestrial monkeys that are highly selective and flexible foragers. Baboons have a close evolutionary relationship with humans (∼94% sequence similarity genome-wide)[5] and striking behavioral and ecological similarities as well.[6] They have achieved a nearly continental distribution in Africa, occupying habitats ranging from moist evergreen forests to deserts, and from equatorial to temperate regions. Among primates, this positions baboons as second only to humans in geographical and environmental range. Their ability to cope with environmental extremes and with high levels of predictable and unpredictable environmental variability, even during demanding life stages, such as parental care and aging, is similarly shared with humans. Furthermore, baboons show little or no seasonality of reproduction in most habitats. In other words, baboons have both adapted to diverse habitats and, in major aspects of their life histories, including those associated with reproduction, have largely escaped from the seasonal constraints of these diverse habitats. This is a combination of traits that they share with humans but with few other primates. Like many human hunter-gatherer societies, most baboon (sub)species live in stable social groups of 20–100 members, including multiple adults and juveniles of both sexes. Baboons also possess highly differentiated social relationships, and a flexible mating and social system. These traits position baboons as a widespread "generalist" species, like humans.

In the wild, female baboons usually reach menarche early in their fifth year; during adulthood they have highly visible and regular sexual swellings that indicate ovarian cycle phase,[7] and they exhibit skin-color changes on their hindquarters that indicate pregnancy, beginning by the end of the first trimester of their 6-month gestation. Infants are completely dependent on their mothers for almost a year, followed by a second year of increasing independence during which time their mothers usually conceive again. Males experience testicular enlargement and associated onset of sperm production in their sixth or seventh year, but do not become fully adult until they are about 8–10 years old, when they achieve an adult body mass nearly twice that of adult females. Males usually disperse from their group of birth as older subadults or young adults. Females and males do not form permanent mating bonds with each other. Instead, mating takes place in the context of mate-guarding episodes, typically called consortships. Demographic senescence in Amboseli is evident by about 15 years of age.[8] In Amboseli and two other wild populations, the oldest reported individuals were 27-year-old females.[8a] Baboons and humans spend a similar proportion of their lifetime in the several major life stages (infancy, the juvenile period, adolescence, and adulthood), with each stage in baboons—and the total lifespan—lasting about one-third as many years as in humans.

The Amboseli ecosystem, baboon population, and baboon research project

Located at the northwestern base of Mt. Kilimanjaro, the Amboseli-Longido ecosystem is a semi-arid, short-grass savanna in a Pleistocene lakebed. This ecosystem exhibits both seasonal and long-term patterns of environmental change, ones that are typical, in type and magnitude, of the changes that characterized the East African environments in which both humans and baboons evolved.[9,10]

The ABRP, ongoing since 1971, includes detailed longitudinal data on demography, ontogeny, ecology, behavior, parentage, and steroid hormone profiles on known individuals that are part of the robust undisturbed Amboseli baboon population (see www.princeton.edu/∼baboon and publications therein). At any given time, the ABRP monitors approximately 300 individually known animals of all ages. The data reported here were obtained solely on animals that move freely through their environment and subsist entirely on natural food sources found in the wild. We visually recognize all individuals in our five study groups; like humans, baboons have highly distinctive facial and morphological features that make them easily recognizable to experienced observers. All animals in study groups are well habituated to the presence of neutral observers.

Our various methods are described in the next section.

Methods

We employ several distinct research methodologies in collecting individual-based data, including the following for the results presented herein: (1) observational monitoring of demographic and behavioral events for all individuals in the study population, (2) genetic paternity analysis, (3) steroid hormone analysis, and (4) occasional darting to immobilize animals in order to take morphometric measurements and perform other procedures. Data are subsequently stored and accessed through our long-term database, BABASE (www.papio.biology.edu).

Field methods for data collection

Our long-term analyses of data from ABRP depend on continuity and consistency in the quality and intensity of data collection over the years. We achieve this consistency through a number of key resources and strategies, including the presence of long-term observers and a comprehensive set of written protocols for field, lab, and data management, which are publicly available online at the ABRP website (www.princeton.edu/~baboon). Some types of data have been obtained since 1971, e.g., demographic events for some individuals. Other types of data collection began at various times since then. In particular, fecal samples for genetic analysis of paternity are available since 1993 (these samples represent individuals born as early as 1968), fecal samples for hormone metabolite extraction since 2000, and morphological data from 1989–1991 and 2006–2008.

Our observational methods are well documented, and we outline them briefly here. Six days per week, we follow 1–2 social groups of baboons each day for six observation hours per day. Throughout the day, we collect a diverse range of observational data including those of particular relevance for evaluating aging in the traits reported here. We begin each day with a systematic group census to record births, deaths, immigrations, and emigrations. We also record the color of each female's paracallosal skin, and the size and status (turgescent or deturgescent) of her sex skin.[7,11,12] This information allows us to retrospectively determine, for each female on each day of the study, her reproductive condition (pregnant, cycling, lactating) and, if cycling, the day of her cycle relative to ovulation. Throughout the day, while collecting systematic behavioral samples, we collect freshly deposited fecal samples from all individuals; these provide our primary window into the physiology of aging and our primary source of DNA for paternity determination (used for measuring male birth rates).

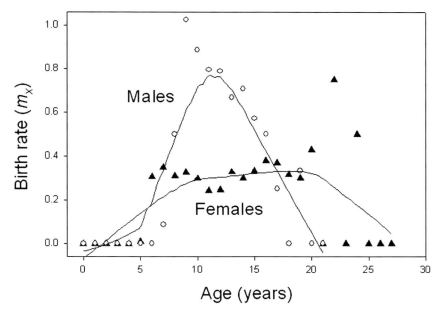

Figure 1. Age-related patterns of birth rates for Amboseli baboons. Female birth rates (filled triangles) are stable until late adulthood. In contrast, male birth rates (open circles; based on genetic analysis) exhibit a strongly age-related pattern, following a similar pattern to dominance rank,[24] with a sharp peak early in adulthood followed by a relatively steep decline. Plotted with lowest curve, window = 0.5.

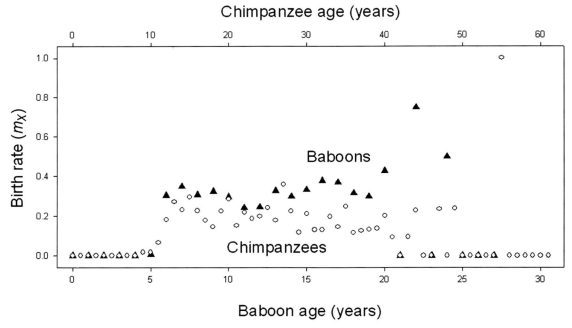

Figure 2. Age-related patterns of birth rates for female Amboseli baboons and for chimpanzees—filled triangles for baboons, open circles for chimpanzees. Baboon data are as in Figure 1. Chimpanzee data are from Ref. 22; note that data for ages 53–61 years in the chimpanzee plot are from a single female. Data are plotted on a scale of two chimpanzee years to one for baboons.

Genetic paternity analysis

Fecal DNA occurs in low quantity and is often degraded. However, extensive work by a number of groups, including ours, has clearly shown that with careful controls the results of fecal DNA analysis are reliable and repeatable.[13–17] We assign paternity based on exclusion and also through the use of the likelihood-based paternity assignment program CERVUS 2.0.[18] With these methods we have unambiguously assigned paternity to approximately 300 individuals born in Amboseli (Alberts et al.[17] and unpublished). Paternity assignments, which are comparable to the known maternities for females, provide the data for evaluating age-related patterns of male birth rates by associating each birth with the age of the infant's father.

Fecal steroid analysis (glucocorticoids and reproductive steroid hormones)

Freshly deposited fecal samples from known individuals are collected in vials prefilled with 95% ethanol to approximate a volumetric ratio of 2:5 feces to ethanol. Samples are stored in Amboseli for less than 2 weeks in a charcoal refrigerator, then are transported first to the University of Nairobi for initial processing and then to Princeton University for hormone extraction and purification followed by assay using an ^{125}I radioimmunoassay.[19–21]

Results and discussion

Age-related changes in reproduction: birth rates and steroid hormones

Baboon reproductive output declines with age, and the patterns of change differ from humans in potentially informative ways. In Amboseli, females continue to produce offspring in their early to mid-20s—equivalent to women in their 60s. In addition, birth rates start to decline only at approximately 18 years of age[8] (Fig. 1, also Ref. 12), an age that is equivalent to the 50s in humans (relying again on the general observation that the various baboon life stages are about one-third as long as the equivalent in humans). The age-related pattern of birth rates in baboon females is essentially the same as that recently reported for wild populations of chimpanzees.[22] Like the baboons, the chimpanzees exhibited declining birth rates only at relatively old ages and complete cessation of births occurred among only the very oldest individuals (Fig. 2).

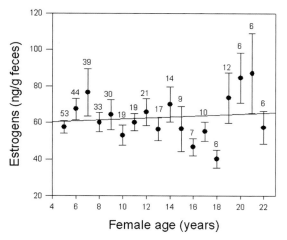

Figure 3. Estrogen concentrations for baboon females are stable into relatively old age, following a similar pattern to that for female birth rates. The plotted value and bars for each year represent the mean and SE across all cycling individuals (pregnant and lactating individuals were excluded). Within any year, each female is represented by only a single data point, which is calculated as the mean of all values obtained for that female during that year of age. The number of individuals contributing to the data for each age appears just above the error bar for that age.

As expected from the stable birth rates across most of adulthood for baboon females, female baboons also maintain relatively high estrogen concentrations into late adulthood (Fig. 3), although high variance characterizes the later age classes, consistent with findings of greater variance in reproductive measures among older women.[23] Sample sizes for individual fecal estrogen (fE) profiles for baboon females do not yet provide adequate statistical power to evaluate longitudinal patterns across adulthood, but the available data indicate individual patterns very similar to the populational one. In addition, we are initiating studies to evaluate age changes in progesterone; a small preliminary populational data set revealed no age-related change in progesterone (fP) concentration or in the fE/fP ratio ($P = 0.909$ and 0.577, respectively).

Male baboons in Amboseli, in contrast to females, experience peak paternity in early adulthood (around 9 years of age), and a steady age-related decline in reproductive output begins immediately thereafter (Fig. 1).[17] The decline in male birthrates occurs as male dominance rank declines because male reproductive output is highly influenced by competition with other males; high-ranking males generally obtain greater access to reproductive females than do low-ranking males.[24]

Adult male baboons also exhibit clear age-related decline in testosterone[21,25] (fT) (Fig. 4A), although testosterone decreases a few years later and the decline proceeds more gradually than the changes in male birth rates. This decline in testosterone occurs when estrogens in cycling females remain high. This population-level pattern could result from a situation in which males with initially high testosterone experience high mortality risk, so that surviving males are disproportionately those with consistently low T, having exhibited little or no decrease in T concentrations during aging. However, initial examination of individual trajectories of T for Amboseli males (Fig. 4B), reveals that declining T concentrations with age are characteristic of 17 of the 19 individual males that we have observed. Declines in T with age are widely documented in men and are often associated with health risks.[26,27] Interestingly, Feldman et al.[26] found steeper declines in their longitudinal study than suggested by cross-sectional trends, perhaps suggesting that men with more steep declines are at higher mortality risk. Future investigations will evaluate predictors and sequelae of individual variability in both peak levels and rate of decline in baboon males.

Preliminary analyses of a small data set (not graphed) indicate that estrogen (fE) concentrations in our population of male baboons increased ($R^2 = 0.396$, $P = 0.016$) and fT to fE ratios decreased with age ($R^2 = 0.558$, $P = 0.002$). High fE in males as in females may be protective against bone loss[28] and cardiovascular disease.[29] However, declines in T-to-E ratio are suggested by some studies of men[30–33] and are thought to have adverse consequences for male longevity.

In summary, in both offspring production and fecal estrogen profiles, female reproductive patterns are age-related only during relatively old ages, and the pattern in offspring production is strikingly similar to that reported for wild chimpanzees. In contrast, male reproductive patterns are highly age-related in both offspring production and fecal testosterone. They exhibit a pattern of rapid rise in early adulthood followed shortly thereafter by steady decline.[21] This decline parallels the well-known pattern of age-related dominance rank in male baboons.

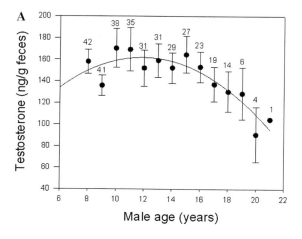

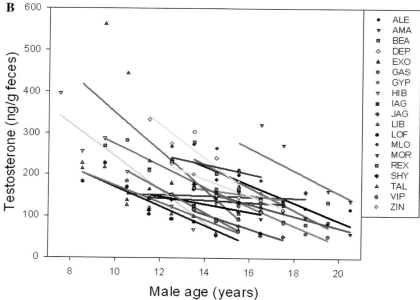

Figure 4. (A) Testosterone concentrations decline with age for baboon males in our sample, starting by mid-adulthood (see Fig. 3 for the contrasting pattern in females). Calculations and symbols as in Figure 3. (B) The population-level decline with age in testosterone concentrations can be explained by decline with age for individual adult male baboons across adulthood (at least 5 years of data for each male). Only 2 of 19 males did not exhibit a pattern of decline.

Age-related patterns of senescence in other morphological and physiological traits

Age-related changes in body mass index (BMI).
In humans, having a low BMI is a known risk factor for mortality, particularly among the elderly,[34] so that the relationship between BMI and mortality risk is U-shaped, with mortality risk increasing at both extremes.[35,36] Studies of western humans have often focused on potential adverse effects of a high BMI (of being overweight or obese) because obesity is so prevalent in these societies. However, the adverse effects of obesity appear to be attenuated in old age, while the adverse effects of being underweight are exacerbated.[36–38]

In Amboseli, healthy baboons that subsist entirely on natural foods exhibit an extraordinarily low level of body fat, 1.9%, and they never approach obesity (although obesity does occur in baboons that visit refuse sites associated with tourist facilities, not the

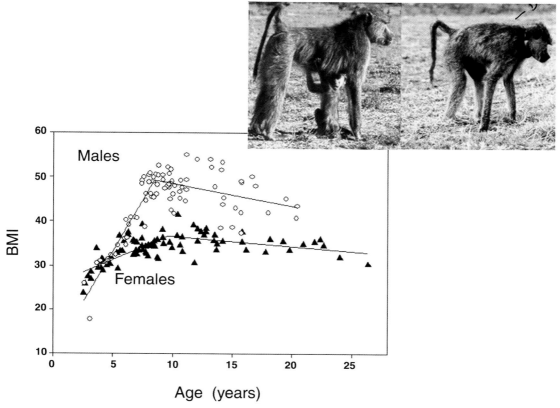

Figure 5. BMI increases for females (filled triangles) during early years of adulthood (until about 10 years of age, approximately 4 years after their first birth) and then declines very gradually during aging. Male (open circles) BMI increases more dramatically until attainment of adulthood (age 8 years) and then begins a more rapid decline than that of female BMI. Plotted lines are for best-fit piecewise regression model, providing a somewhat better fit for males (adj $R^2 = 0.76$) and an equally good fit (adj $R^2 = 0.51$) for females than a quadratic model. Pictured is a female during early adulthood and the same female at age 27, the oldest age yet recorded for wild baboons.

subjects of this study).[39] Not only is obesity absent, we have frequently observed animals that appear to be exceptionally thin, particularly old animals (Fig. 5) or females that are lactating during dry seasons or droughts.

We measured BMI in a subset of the Amboseli baboons that we briefly immobilized through darting. BMI exhibits clear age-related patterns and sex differences during adulthood (Fig. 5). The two sexes do not differ in BMI until the sixth year of life, when females experience their first conception and males enter into the subadult period that is characterized by an adolescent growth spurt not seen in females.[40] This male growth spurt occurs more in weight than in long bones and results in BMI in young adult males that is almost 50% greater, and body mass approximately 100% greater, than that of same-aged females. Male BMI peaks at about the onset of adulthood (8 years of age) and declines steadily throughout adulthood. Females, in contrast, continue to experience gradual increase in BMI during the early years of adulthood, peaking at approximately 10 years of age (4 years after the average age at first birth). Female BMI then declines steadily but at a slower rate than in males, narrowing the sex difference. Very few males live into their 20s,[8] and the limited data for males in the late teens suggest high variability and perhaps differential mortality for those of low BMI. Tooth wear is likely to be a major contributor to changes in BMI and is described below.

Age-related changes in tooth wear. Tooth wear increases with age in many species (including several

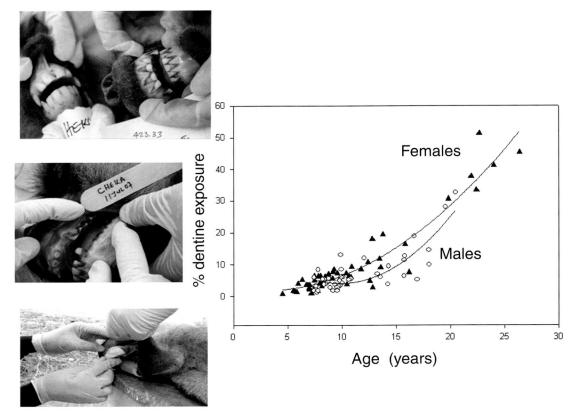

Figure 6. Molar wear (measured as proportion of dentine exposure, PDE) is age-related in both sexes, as seen here for M1—the first molar. M1 wear tends to be as great in females (filled triangles) as in males (open circles) during aging. Quadratic models: adj $R^2 = 0.90$ for females, adj $R^2 = 0.69$ for males. Pictured is a comparison of a young adult female (7 years old) with her elderly mother (21 years old), also a 24-year-old female. Also shown is the molding technique used to obtain casts for measuring PDE.

primates[41–44]). A landmark study in wild lemurs found that increased tooth wear in older mothers was associated with decreased survival of their infants,[43] indicating a strong link between tooth wear and food processing ability. In Amboseli, we have previously documented that periodontal health decreases with age.[45] Further, we have observed dramatic changes in tooth wear with age, using photographs taken during brief immobilizations (Fig. 6).

We recently collected high quality tooth impressions on a subset of baboons that were briefly immobilized through darting. We measured tooth wear as the percentage of dentine exposure on the occlusal surface. Our analysis of tooth wear shows a strong signal of age in both sexes (Fig. 6 for the M1—the first molar).[46,47,47a] Unlike age-related patterns of change in BMI and reproduction, the age-related pattern in molar wear does not exhibit a male-biased aging pattern (Fig. 6). However, qualitative observations suggest that females experience similar aging patterns on their molars as on their other teeth, whereas males experience more major breaks and total loss of these other teeth; this will be the subject of forthcoming evaluation.

Age-related changes in glucocorticoids. Aging-associated hypercortisolism, manifested either in basal levels or in a stress response that is resistant to feedback mechanisms, has been documented in humans and in several studies of nonhuman primates.[48,49] Chronically elevated glucocorticoid levels can jeopardize health and survival[50,51] by decreasing reproductive hormones,

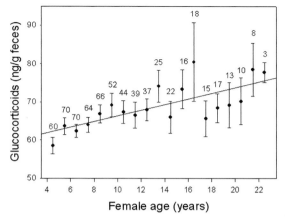

Figure 7. Glucocorticoid concentrations increase gradually with age for females. Calculations and symbols as in Figure 3.

suppressing immunity, promoting atherosclerosis and ulcers, decreasing muscle mass, and impairing growth and tissue repair.[52–54] Although the incidence of cardiovascular diseases and ulcers is unknown for wild baboons, an increased risk of infectious diseases, a decrease in cutaneous wound healing, and loss of muscle mass are all likely to increase mortality risk as well as morbidity in natural populations.[54]

We documented a striking age-related change in adrenocortical axis function in Amboseli baboons, using a rare opportunity in which we collected blood from over 100 animals in 1989 and 1990.[48] Males and females in our oldest age classes experienced both higher basal cortisol levels and increased resistance to downregulation. The single, cross-sectional measurements in that study provided the foundation for our recent ongoing studies of excreted glucocorticoids (fGC). Our initial analyses support our prior finding of an increase throughout adulthood in glucocorticoid concentrations for females (Fig. 7). Males, however, show no simple change with age in fGC in these longer-term analyses, and instead evidence more complex patterns, which are currently under investigation. Thus, glucocorticoids, like molar wear, exhibit age-related changes but not the strong male-biased sex differences in senescence that were evident in BMI, reproductive rates, and reproductive hormones.

Reproductive senescence in the wild: a comparison with other somatic systems and between the sexes

The signature of senescence in wild baboons is clear across diverse morphological, physiological, and behavioral systems, including reproduction of both females and males. However, the timing and pace of senescence differs across systems, in some cases between the sexes, and between baboons and humans, in multiple, informative ways. For example, reproductive senescence for female baboons is evident in birth rates and in estrogen levels only relatively late in life. Moreover, complete cessation of offspring production is experienced only by the very oldest females, those who reach their mid-20s, an age at which body condition as measured by BMI is very low and tooth wear is extensive. The age-related pattern of female baboon birth rates is the same as that in chimpanzee females. It stands in contrast to that observed in women, in which the timing of reproductive senescence far exceeds that of other organ systems, resulting in a long postreproductive period in humans (e.g., in Ref. 55 and sequala).

For baboon males, reproductive decline in both birth rates and testosterone concentrations parallel the early and rapid decline in male dominance status and in body condition (BMI). Thus, a potentially intertwined complex of traits—social, physiological, morphological, and offspring production—exhibit earlier and more rapid decline in males than in females. Recent comparisons of age-related patterns of androgens in men[33] and in several closely related monkey taxa[21] including Amboseli baboons, exhibit interesting predictability across taxa or subsets of human populations that can be related to levels of parental care or mating patterns. These findings suggest important directions for insights into plasticity of senescence patterns in men and nonhuman primate males across a range of taxa, as well as across ecological and social contexts.

Survival, overall health, and well-being reflect senescence and risk factors in many morphological, physiological, and behavioral traits. Each trait follows its own patterns of aging, either independently or in some cases with causative interrelationships. Age-related variability in these traits, in their causative relationships, and in their joint impact, will also exhibit varying degrees of dependence on social or ecological conditions; some will be very sensitive to differences in circumstances and others perhaps less so. The age-related pattern in androgen concentrations among men and among nonhuman primates is one example of a trait that is highly contingent upon social context.[21,33] Another example of condition-dependent aging involves female reproduction in nonhuman primates. Some traits, such as age of menarche and fertility rates, exhibit strong response to differences in food availability; such responses are particularly evident where primates are provisioned by humans, either in captivity or in cases where the nonhuman populations live in close commensalism with humans and their foods. These situations provided most of the examples of postreproductive females first identified by Hrdy,[2] and they constitute the majority of those reported recently across a broad range of primates.[56] In addition, in these situations, we would expect that age-related senescence in body condition and tooth wear will be attenuated, as will demographic senescence. This may result in a greater prevalence of females experiencing a postreproductive period and perhaps, though not necessarily, an extended duration of the postreproductive phase if survival is enhanced more than fertility is (see Hawkes and Smith[3a]).

Understanding the evolutionary history, current diversity, and future potential of aging in humans and our closest relatives will require understanding the plasticities of various age-related traits and the contingencies among them. Nonhuman primate models from a range of ecological conditions will be essential for attaining this understanding, and critical for achieving a breadth of insights into aging.

Acknowledgments

Financial support for ABRP research was provided primarily by NSF IBN-0322613, NSF BSE-0323553, RO3 MH65294, NIA P30AG024361, and the Chicago Zoological Society. We thank the Kenya Wildlife Services, Institute of Primate Research, National Museums of Kenya, and members of the Amboseli-Longido pastoralist communities. Thanks go to the Amboseli field team who contributed to sample and data collection (R.S. Mututua, S. Sayialel, and J.K. Warutere) and to T. Wango for assistance in Nairobi. Thanks to T. Fenn, N. Learn, and L. Maryott for database assistance and to M. Emery-Thompson for the chimpanzee data in Figure 2. This research was approved by the IACUC at Princeton University (protocol #1689, 9 November 2007) and Duke University (#A1830-06-04), and adhered to all the laws and guidelines of Kenya (Kenya Research Permit MOEST 13/001/C351 Vol. II). We are grateful to two anonymous reviewers for helpful comments on an earlier version of the manuscript and to the NIA for supporting the Workshop on Reproductive Aging.

Conflicts of interest

The authors declare no conflicts of interest.

References

1. Hrdy, S.B. 1977. *The Langurs of Abu: Female and Male Strategies of Reproduction*. Harvard University Press. Cambridge, MA.
2. Hrdy, S.B. 1981. "Nepotists" and "altruists": the behavior of old females among macaques and langur monkeys. In *Other Ways of Growing Old: Anthropological Perspectives*. P.T. Amuss & S. Harrell, Eds.: 59–76. Stanford University Press. Stanford.
3. Hrdy, S.B. 1981. *The Woman That Never Evolved*. Harvard University Press. Cambridge.
3a. Hawkes, K. & K.R. Smith. 2010. Do women stop early? Similarities in fertility decline in humans and chimpanzees. *Ann. N.Y. Acad. Sci.* **1204:** 43–53.
4. Swedell, L. & S.R. Leigh. 2006. *Reproduction and Fitness in Baboons*. Springer. New York.
5. Rogers, J. & J.E. Hixson. 1997. Baboons as an animal model for genetic studies of common human disease. *Am. J. Hum. Genet.* **61:** 489–493.
6. Jolly, C.J. 2007. Babooons, mandrills, and mangabeys: afro-papionin socioecology in a phylogenetic perspective. In *Primates in Perspective*. C.J. Campbell, *et al.*, Eds.: 240–251. Oxford University Press. New York.
7. Gesquiere, L.R., E.O. Wango, S.C. Alberts & J. Altmann. 2007. Mechanisms of sexual selection: sexual swellings and estrogen concentrations as fertility indicators and cues for male consort decisions in wild baboons. *Horm. Behav.* **51:** 114–125.
8. Alberts, S.C. & J. Altmann. 2003. Matrix models for primate life history analysis. In *Primate Life Histories and Socioecology*. P.M. Kappeler & M.E. Pereira, Eds.: 66–102. University of Chicago Press. Chicago.
8a. Bronikowski, A. *et al.* 2001. The aging baboon: comparative demography and lifespan heritability in a nonhuman primate model system. *Proc. Natl. Acad. Sci. USA* **99:** 9591–9595.
9. Behrensmeyer, A.K. 2006. Climate change and human evolution. *Science* **311:** 476–478.
10. Potts, R. 1998. Variability selection in hominid evolution. *Evol. Anthropol.* **7:** 81–96.
11. Beehner, J.C. *et al.* 2006. The endocrinology of pregnancy and fetal loss in wild baboons. *Horm. Behav.* **49:** 688–699.
12. Beehner, J.C., D.A. Onderdonk, S.C. Alberts & J. Altmann. 2006. The ecology of conception and pregnancy failure in wild baboons. *Behav. Ecol.* **17:** 741–750.

13. Borries, C. et al. 1999. DNA analyses support the hypothesis that infanticide is adaptive in langur monkeys. *Proc. R. Soc. Lond. B Biol. Sci.* **266**: 901–904.
14. Constable, J., M.V. Ashley, J. Goodall & A.E. Pusey. 2001. Noninvasive paternity assignment in Gombe chimpanzees. *Mol. Ecol.* **10**: 1279–1300.
15. Morin, P.A., K.E. Chambers, C. Boesch & L. Vigilant. 2001. Quantitative polymerase chain reaction analysis of DNA from noninvasive samples for accurate microsatellite genotyping of wild chimpanzees *(Pan troglodytes verus)*. *Mol. Ecol.* **10**: 1835–1844.
16. Vigilant, L., M. Hofreiter, H. Siedel & C. Boesch. 2001. Paternity and relatedness in wild chimpanzee communities. *Proc. Natl. Acad. Sci. USA* **98**: 12890–12895.
17. Alberts, S.C., J. Buchan & J. Altmann. 2006. Sexual selection in wild baboons: from mating opportunities to paternity success. *Anim. Behav.* **72**: 1177–1196.
18. Marshall, T.C., J. Slate, L.E. Kruuk & J.M. Pemberton. 1998. Statistical confidence for likelihood-based paternity inference in natural propulations. *Mol. Ecol.* **7**: 639–655.
19. Khan, M.Z., J. Altmann, S.S. Isani & J. Yu. 2002. A matter of time: evaluating the storage of fecal samples for steroid analysis. *Gen. Comp. Endocrinol.* **128**: 57–64.
20. Lynch, J.W. et al. 2003. Concentrations of four fecal steroids in wild baboons: short-term storage conditions and consequences for data interpretation. *Gen. Comp. Endocrinol.* **132**: 264–271.
21. Beehner, J. et al. 2009. Testosterone related to age and life-history stages in male baboons and geladas. *Horm. Behav.* **56**: 472–480.
22. Thompson, M.E. et al. 2007. Aging and fertility patterns in wild chimpanzees provide insights into the evolution of menopause. *Curr. Biol.* **17**: 2150–2156.
23. Weinstein, M. et al. 2003. Timing of menopause and patterns of menstrual bleeding. *Am. J. Epidemiol.* **158**: 782–791.
24. Alberts, S.C., H.E. Watts & J. Altmann. 2003. Queuing and queue-jumping: long-term patterns of reproductive skew in male savannah baboons, *Papio cynocephalus*. *Anim. Behav.* **65**: 821–840.
25. Sapolsky, R. 1986. Endocrine and behavioral correlates of drought in wild olive baboons (*Papio anubis*). *Am. J. Primatol.* **11**: 217–227.
26. Feldman, H.A. et al. 2002. Age trends in the level of serum testosterone and other hormones in middle-aged men: longitudinal results from the Massachusetts Male Aging Study. *J. Clin. Endocrinol. Metab.* **87**: 589–598.
27. Wu, F.C.W. & A. von Eckardstein. 2003. Androgens and coronary artery disease. *Endocr. Rev.* **24**: 183–217.
28. Kuchuk, N.O. et al. 2007. The association of sex hormone levels with quantitative ultrasound, bone mineral density, bone turnover and osteoporotic fractures in older men and women. *Clin. Endocrinol.* **67**: 295–303.
29. Arnlov, J. et al. 2006. Endogenous sex hormones and cardiovascular disease incidence in men. *Ann. Intern. Med.* **145**: 176–184.
30. Vermeulen, A., J.M. Kaufman & V.A. Giagulli. 1996. Influence of some biological indexes on sex hormone-binding globulin and androgen levels in aging or obese males. *J. Clin. Endocrinol. Metab.* **81**: 1821–1826.
31. Leifke, E. et al. 2000. Age-related changes of serum sex hormones, insulin-like growth factor-1 and sex-hormone binding globulin levels in men: cross-sectional data from a healthy male cohort. *Clin. Endocrinol.* **53**: 689–695.
32. Khosla, S., L.J. Melton, E.J. Atkinson & W.M. O'Fallon. 2001. Relationship of serum sex steroid levels to longitudinal changes in bone density in young versus elderly men. *J. Clin. Endocrinol. Metab.* **86**: 3555–3561.
33. Muller, M. et al. 2003. Endogenous sex hormones in men aged 40-80 years. *Eur. J. Endocrinol.* **149**: 583–589.
34. Wilson, M.M.G. 2001. Bitter-sweet memories: truth and fiction. *J. Gerontol. A Biol. Sci. Med. Sci.* **56**: M196–M199.
35. Harris, T. et al. 1988. Body-mass index and mortality among nonsmoking older persons—the Framingham-Heart-Study. *JAMA* **259**: 1520–1524.
36. Corrada, M.M., C.H. Kawas, F. Mozaffar & A. Paganini-Hill. 2006. Association of body mass index and weight change with all-cause mortality in the elderly. *Am. J. Epidemiol.* **163**: 938–949.
37. Stevens, J. et al. 1998. The effect of age on the association between body-mass index and mortality. *N. Engl. J. Med.* **338**: 1–7.
38. Heiat, A., V. Vaccarino & H.M. Krumholz. 2001. An evidence-based assessment of federal guidelines for overweight and obesity as they apply to elderly persons. *Arch. Intern. Med.* **161**: 1194–1203.
39. Altmann, J. et al. 1993. Body size and fatness of free-living baboons reflect food availability and activity levels. *Am. J. Primatol.* **30**: 149–161.
40. Altmann, J. & S.C. Alberts. 2005. Growth rates in a wild primate population: ecological influences and maternal effects. *Behav. Ecol. Sociobiol.* **57**: 490–501.
41. Morbeck, M.E., A. Galloway & D.R. Sumner. 2002. Getting old at Gombe: skeletal aging in wild-ranging chimpanzees. In *Aging in Nonhuman Primates*. J.M. Erwin & P.R. Hof, Eds.: 48–62. Karger. Basel.
42. Nichols, K.A. & A.L. Zihlman. 2002. Skeletal and dental evidence of aging in captive western lowland gorillas: a preliminary report. In *Aging in Nonhuman Primates*. J.M. Erwin & P.R. Hof, Eds.: 22–31. Karger. Basel.
43. King, S.J. et al. 2005. Dental senescence in a long-lived primate links infant survival to rainfall. *Proc. Natl. Acad. Sci. USA* **102**: 16579–16583.
44. Cuozzo, F.P. & M.L. Sauther. 2006. Severe wear and tooth loss in wild ring-tailed lemurs (*Lemur catta*): a function of feeding ecology, dental structure, and individual life history. *J. Hum. Evol.* **51**: 490–505.
45. Phillips-Conroy, J.E. et al. 1993. Periodontal health in free-ranging baboons of Ethiopia and Kenya. *Am. J. Phys. Anthropol.* **90**: 359–371.
46. Phillips-Conroy, J.E., T. Bergman & C.J. Jolly. 2000. Quantitative assessment of occlusal wear and age estimation in Ethiopian and Tanzanian baboons. In *Old World Monkeys*. P.F. Whitehead & C.J. Jolly, Eds.: 321–340. Cambridge University Press. Cambridge, UK.
47. Kay, R.F. & J.G.H. Cant. 1988. Age assessment using cementum annulus counts and tooth wear in a free-ranging population of *Macaca mulatta*. *Am. J. Primatol.* **15**: 1–15.
47a. Galbany, J., J Altmann, A. Pérez-Pérez & S.C. Alberts. 2010. Age and individual foraging behavior predict tooth wear in

Amboseli baboons. *Amer. J. Phys. Anthro.* (accepted pending revisions).
48. Sapolsky, R.M. & J. Altmann. 1991. Incidence of hypercortisolism and dexamethasone resistance increases with age among wild baboons. *Biol. Psychiatry* **30:** 1008–1016.
49. Goncharova, N.D. & B.A. Lapin. 2002. Effects of aging on hypothalamic-pituitary-adrenal system function in non-human primates. *Mech. Ageing Dev.* **123:** 1191–1201.
50. Sapolsky, R.M. 1994. *Why Zebras Don't Get Ulcers.* W.H. Freeman and Company. New York.
51. Munck, A., P.M. Guyre & N.J. Holbrook. 1984. Physiological functions of glucocorticoids in stress and their relation to pharmacological actions. *Endocr. Rev.* **5:** 25–44.
52. Sapolsky, R.M. 2004. Social status and health in humans and other animals. *Annu. Rev. Anthropol.* **33:** 393–418.
53. Sheffield-Moore, M. & R.J. Urban. 2004. An overview of the endocrinology of skeletal muscle. *Trends Endocrinol. Metab.* **15:** 110–115.
54. Ashcroft, G.S., S.J. Mills & J.J. Ashworth. 2002. Ageing and wound healing. *Biogerontology* **3:** 337–345.
55. Hawkes, K. *et al.* 1998. Grandmothering, menopause, and the evolution of human life histories. *Proc. Natl. Acad. Sci. USA* **95:** 1336–1339.
56. Atsalis, S., S.W. Margulis & P.R. Hof. 2008. *Primate Reproductive Aging: Cross-Taxon Perspectives*. S. Karger AG. Switzerland.

ANNALS OF THE NEW YORK ACADEMY OF SCIENCES
Issue: *Reproductive Aging*

Reproductive aging in tephritid fruit flies

James R. Carey[1,2] and Freerk Molleman[1,3]

[1]Department of Entomology, University of California, Davis, California. [2]Center for the Economics and Demography of Aging, University of California, Berkeley, California. [3]Current address: Institute of Ecology and Earth Sciences, University of Tartu, Vanemuise 46, Tartu, EE-51014, Estonia

Address for correspondence: James R. Carey, Department of Entomology, University of California Davis, Davis, California 95616. jrcarey@ucdavis.edu

The broad objective of this paper is to present an overview and synthesis of selected studies on reproduction and aging in two model tephritid fruit fly species including the Mediterranean fruit fly, *Ceratitis capitata*, and the Mexican fruit fly, *Anastrepha ludens*. We summarize the research findings from empirical studies and modeling investigations involving reproduction in the two tephritid species. At the end we identify and discuss four general principles regarding reproductive aging in tephritids including reciprocity of reproductive and aging costs, qualitative tradeoffs, plasticity of lifespan and reproduction, and life history constraints and determinacy.

Keywords: life tables; lifespan; fecundity; cost of reproduction; senescence; calorie restriction

Introduction

Aging and reproduction in insects are inextricably linked and mutually affecting, the former reducing egg-laying rate at older ages independent of previous egg laying, and the latter increasing the risk of death at each age with increasing reproductive effort. Although the tradeoffs between early reproduction and the risk of death are well known, the reciprocal relationship—i.e., the influence of aging on reproductive output—is less well understood. While age-patterns of reproduction play a role in the evolution of senescence, the *reproductive* system itself can also decrease in performance with age—reproduction may become less intense, of lower quality or cease (postreproductive lifespan/menopause) as an organism ages. Even though insects are important model systems in aging research, reproductive aging has rarely been addressed specifically in this group or is attributed to the depletion of resources.[1] Thus, the main objective of this paper is to present an overview of the relationship between reproduction and aging based on the results of previous research on tephritid fruit flies by Carey et al.[10]

Reproductive aging in tephritids

Both the Mediterranean fruit fly, *Ceratitis capitata*, and the Mexican fruit fly, *Anastrepha ludens*, belong to the dipteran family Tephritidae—a group of about 4000 species referred to as the "true" fruit flies that is distributed throughout most of the world.[2] Members of this group, most of which are roughly the size of a housefly, lay eggs in intact fruit using their sharp ovipositor rather than on decaying fruit as does their tiny, gnat-sized distant relative *Drosophila melanogaster* that is commonly known as the vinegar fly. Although less useful than *D. melanogaster* for research on genetics and development, the medfly and Mexfly are ideal models for demographic research for at least three reasons: (1) because members of these two species are relatively large and robust, they are much less prone to handling injury and can also be easily observed without the aid of a hand lens and microscope; (2) the willingness of females to lay eggs only when presented with an oviposition host enables researchers to manipulate their egg laying independent of dietary manipulations. In contrast, *D. melanogaster* females tend to lay their eggs on their food; and (3) because both of these tephritid species are reared at industrial scales, large numbers of individuals are always available for large-scale demographic studies.

Baseline patterns of reproductive aging

Model tephritids: Medfly and mexfly. The broad goal of the individual-level studies of reproduction

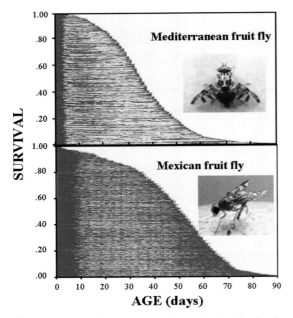

Figure 1. Event history charts of reproduction for female Mediterranean (*top*) and Mexican (*bottom*) fruit flies. *Green* = zero eggs/day; *yellow* = 1–40 and 1–50 eggs/day for medflies and Mexflies, respectively; *red* = >40 and >50 eggs/day for medfly and Mexflies, respectively.[6,23]

in both the medfly[3–5] and the Mexican fruit fly[6] was to gather baseline information on their maturation rates, daily egg-laying patterns, age-specific survival, and lifespan. Survival and reproduction were monitored daily in a total of 1000 individual females of each species given access to *ad libitum* food, the results of which are summarized in Figure 1. Several aspects of this figure merit comment. First, the broad patterns of survival are similar in both cohorts including: (i) gentle decreases in survival for the first 10% of deaths; (ii) steeper declines in survival for the next 80% of deaths; and (iii) long tails for the remaining 10% of deaths. These patterns are manifestations of an underlying mortality schedule in female flies that accelerates at young and middle ages and decelerates at older ages.[7–9] Second, both species of tephritids are relatively long lived with life expectancies at eclosion of the medfly and Mexfly of approximately 36 and 49 days, respectively. Thus, the average female Mexfly lives about a third longer than the average medfly. Third, both tephritid species are capable of laying 30–50 eggs per day during peak reproductive ages and 10–20 eggs per day at many of the older ages. Thus, the observed lifetime egg production in these species is extraordinarily high with fecundity in the medfly ranging from 640 to 1150 eggs/female[10] and in the Mexfly 1400 eggs/female,[6] a fecundity higher than any lifetime rate observed in earlier studies on the medfly.[3,11] Fourth, two broad patterns were evident in egg laying of both species: (i) a weak correlation between longevity and ages of first reproduction and (ii) the retention of egg-laying capabilities at advanced ages.

Main conclusions. The results of these studies of reproduction in the two tephritid fruit fly species: (i) provided important baseline information (i.e., maturation; egg-laying patterns; and postreproduction) for both species; and (ii) revealed that there was no correlation between either the intensity of early and late egg laying or the intensity of early egg laying and longevity.

Medfly biotypes. Because tephritid fruit flies such as the medfly have colonized a number of different temperate and tropical areas throughout the world, between-biotype comparisons of survival and reproduction shed important light on how different environments favor the evolution of different life history traits in these species. Thus, the purpose of the study by Greek entomologist Alexandros Diamantidis and his colleagues[12] was to compare the life history traits of medfly populations obtained from different global regions. In a common garden environment, medfly populations were studied from six global regions including Africa (Kenya), the Pacific (Hawaii), Central America (Guatemala), South America (Brazil), extra-Mediterranean (Portugal), and Mediterranean (Greece). Substantial between-population differences were observed with female life expectancies ranging from 4 weeks in the shortest lived population (Guatemala) to 8 weeks in the longest lived population (Greece) (Table 1). Although Hawaiian and Kenyan females were relatively short-lived, the lifespans of males from these two regions were similar as were those for males of the long-lived populations (Brazil, Portugal, and Greece). Therefore, female cohorts could be classified into either short-lived (Guatemala, Hawaii, and Kenya) or long-lived (Brazil, Portugal, and Greece) biotypes. Although average fecundity rates were similar among populations, substantial differences existed in the age-specific reproduction schedule (Table 1). Short-lived populations (e.g., Guatemala)

Table 1. Longevity (both sexes) and reproduction (females only) in medfly populations originating from six different geographic areas[12]

	Expectation of life (days)		Reproduction (eggs/female)	
	Males	Females	Gross	Net
Guatemala	68.0	48.1	683.5	525.2
Hawaii	106.5	52.1	727.8	569.5
Kenya	115.9	58.3	701.1	655.0
Brazil	122.3	75.7	746.5	545.1
Portugal	107.1	75.6	700.5	549.1
Greece	112.1	72.3	1117.2	631.1

compressed the period of high egg-laying days into a limited "reproductive window" between days 10 and 40. In contrast, long-lived populations (e.g., Greece) spread out high egg-laying activity during their entire oviposition course (i.e., 4+ months). Furthermore, maturation rates as determined by the age at which the first egg of an individual is laid, were highly variable across populations.

Main conclusions. The most important findings with respect to reproduction and aging included: (i) differences among populations in longevity were substantial (e.g., 48 versus 76 days for females); (ii) the gender gap favored males in all population by a wide margin (e.g., nearly 50 days in Kenyan populations); (iii) differences in net reproduction were substantially less relative to differences in longevity; and (iv) all populations experienced clear postreproductive periods in females.

Host effects

The ecology and demography of tephritid fruit flies can not be understood without understanding their relationship with their host fruit, which enters into their life histories in two respects: (i) as their sole source of larval nutrition; and (ii) as the site in which they lay their eggs as adult females. We summarize the results of two studies in this section including one on larval host effects on adult reproduction and survival in the medfly[13] and the other on the effects of host deprivation on the life history traits of medfly females.[14]

Larval host

Inasmuch as the host upon which an adult fly develops as a larvae can have a profound effect on its survival and reproduction, the purpose of a study conducted in Hawaii was to compare the life history traits of medflies reared on a number of different larval hosts.[13] A total of 30 different host species (e.g., plum, peach, mango, and papaya) were seeded with large numbers of medfly eggs, 24 of which produced viable adults in numbers (i.e., 15–60 pairs) that could be used for life table and egg-laying assays. The emerging adults from each host were placed in group cages and monitored for daily egg laying and survival. Large differences were observed in both sex-specific life expectancies and in female age-specific and lifetime reproduction across the 24 host species. Life expectancy was 22 days or greater for cohorts reared from 70% of the hosts. However, the average female fly reared from plum lived nearly twice as long as the average female reared from banana or mammee apple, two hosts that produced the shortest-lived flies. The highest gross and net reproductive rates from plum-reared females were twice the lowest rates that occurred in banana-reared flies as shown in Figure 2. These results reveal that the differences in lifetime reproduction in medfly females reared across these two different hosts were due to a combination of difference in the age-specific levels as well as to the differences in reproductive duration (i.e., lifespans).

Main conclusion. The hosts upon which adult fruit flies develop as larvae can have a profound effect on both longevity (twofold) and lifetime reproduction (2.5-fold). Research is needed on the effects of larval nutrition on adult demographic traits since it would shed important light on the effects of early life conditions on mortality at advanced ages.

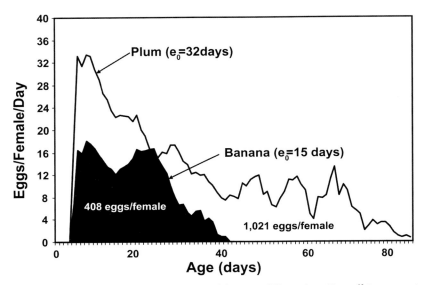

Figure 2. Age-specific reproduction in the medfly females reared from two different larval hosts.[13] Inset contains lifetime reproduction for medflies reared from each host.

Adult host deprivation

Unlike *D. melanogaster* females, which feed on the same rotting fruit as they lay their eggs, adult tephritids lay their eggs in intact host fruit by boring a hole through the host skin using their ovipositor. Therefore the purpose of the study[14] was to quantify the effect of different levels and patterns of host deprivation on survival and reproduction in medfly females that are maintained on a full diet. Female medflies in Hawaii were subjected to four different patterns of host deprivation at each of three levels for a 24-day period. Survival was recorded daily and egg production was recorded on the days in which hosts were present. Results revealed that host deprivation disrupts the reproductive cycle of females and reduces their overall reproductive effort. Reduced reproductive effort at young ages due to host deprivation increases survival (up to twofold) and daily reproduction (up to eightfold) at older ages. Therefore, host deprivation postpones senescence, but nevertheless some reproductive potential is lost. After host deprivation, flies laid large numbers of eggs, and the second day the number was lower than subsequent days (Fig. 3). Medfly egg production is not only affected by a female's chronological age but also by her previous reproductive effort. Any level or pattern of host deprivation increases survival by reducing reproductive effort.

Main conclusion. The results of this study revealed that arresting reproduction in female medflies that are capable of reproducing extends their longevity. Although the same general longevity-extending, nonreproducing outcome is observed in medfly dietary restriction studies, the underlying mechanisms may be different.

Food restriction effects on reproductive aging

As noted in a seminal essay in ecology over two decades ago[15] as well as by many gerontologists concerned with calorie restriction,[16,17] food is the burning question in animal society—i.e., without adequate food individual development is arrested, reproductive rate slows, and mortality risk increases. It follows that any discussion of reproductive aging is incomplete without an understanding of and information on the effects of food on both survival and reproduction.

Dual modes of aging

The goal of the paper describing the phenomenon that was termed "dual modes of aging"[18] was to test the hypothesis that medflies are capable of living to and reproducing at extreme ages if their reproduction is suspended at young ages due to lack of dietary protein. Experiments were conducted using the medfly at the rearing facility in Tapachula,

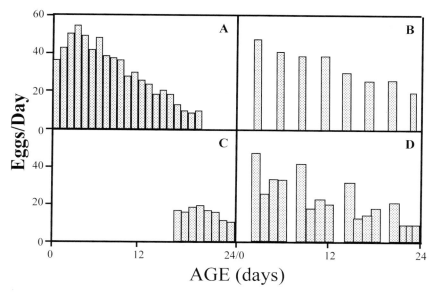

Figure 3. Reproduction in medflies given access to different patterns of oviposition hosts over a 24-day period; (**A**) control (host available all 24 days); (**B**) host available every third day; (**C**) host available days 16–24; (**D**) host available 4 of 5 days.[14] Note the large differences in egg production levels between the control and treatment cohorts for days 21–24.

Mexico. An initial pool of 2500 adults of each sex was maintained in single pair cages with sugar and water. At 30, 60, and 90 days, subgroups of 100 pairs were provided with a full diet (i.e., protein hydrolysate + sugar) *ad libitum* and their reproduction and survival were monitored until the last female died. The experiment revealed three important life expectancy patterns: (i) in all three treatments the remaining life expectancy of flies increased on the days when they were given a full diet after having been maintained on a sugar-only diet; (ii) life expectancy of the full-diet control flies at eclosion (age zero) was similar to remaining life expectancy of the treatment flies at the ages when they were first given a full diet; (iii) remaining life expectancy declines rapidly after medflies are switched to a full diet. One of the most remarkable discoveries was that 4- to 5-month-old medflies were capable of producing moderate numbers of eggs if they had been maintained on a sugar diet for the first 3 months (Fig. 4). A switch to full diet appears to have set the mortality clock back 90 days and to the slowly rising trajectory of the sugar-only cohort rather than the more-rapidly rising trajectories of the other cohorts switched to full diets.

Main conclusions. The finding that female medflies are capable of switching mortality schedules suggests that the reproductive experience of females may include two distinct modes: (i) a waiting mode with low mortality, in which few or even no eggs are produced and (ii) a reproductive mode with prolific egg laying and low initial mortality followed by an acceleration in mortality and a reduction in egg laying. The results of this study also provided evidence for a reproductive cost (lost eggs) of aging; i.e., that some of the eggs that were not laid at young ages due to restricted access to protein source were not recovered at older ages when females were given access to a protein source.

Constant calorie restriction

The purpose of the study summarized here was to investigate medfly longevity and reproduction across a broad spectrum of diet restriction using a protocol similar to those applied to most rodent studies.[19] Age-specific reproduction and age of death were monitored in 1200 individuals of both sexes that were individually maintained on 1-of-12 diets from *ad libitum* to 30% *ad libitum*. Diet was provided in a fixed volume of solution that was fully consumed each day, ensuring control of total nutrient consumption for each fly. Contrary to what would be expected under the "longevity extension through dietary restriction" paradigm,[20] flies

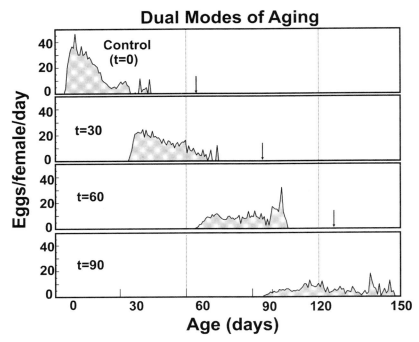

Figure 4. Event history graphs for medflies given access to different calorie restriction treatments.[19] The time t (day) at which flies given access to the full (sugar + protein) diet indicated at left within each panel. Arrows indicate age of death of oldest flies within each treatment.

that were fed less experienced shorter life expectancies. Egg production continuously varied with diet, with average egg production declining more rapidly when diet was restricted to below 70%, and cohorts no longer containing individuals that laid more than 1800 eggs when diet was restricted by more than 50%. Even at the lowest food availability (30%) flies still laid some eggs. Although under food limitation reproduction is partly postponed, a fraction of a females' reproductive potential appears to be completely lost due to aging alone.

Main conclusions. The two main findings from this medfly calorie restriction study in which only calories but not dietary composition were manipulated included: (i) increased longevity was not observed in any of the food restricted cohorts and (ii) reproductive patterns revealed that when medfly females are given the opportunity to reproduce they do so at the expense of increased mortality even under the most harsh dietary conditions.

Nutritional gradients

The goal of this study on the Mexfly[21] was to design an experiment in which both dietary composition and caloric content were altered in the adult food to address the question: Under what nutritional conditions do the longevity-enhancing effects resulting from food restriction either counteract, complement, or reinforce the mortality costs of reproduction? To answer this question a fine-grained dietary restriction study was designed involving 4800 individuals of *A. ludens* in which sex-specific survival and daily reproduction were measured in females in each of 20 different treatments (sugar:yeast ratios) plus four starvation controls. The contour graphs that were constructed from the response surface data reveal two nutritional thresholds: (i) a calorie restriction threshold ($\approx$10%) below, which both female reproduction and survival decline precipitously and (ii) a sugar–yeast (compositional) threshold ($\approx$5%) below which reproduction declines precipitously but where survival decreases gradually. The square contours in the reproductive surface showed that the yeast level sets a threshold in terms of the range of increasing reproduction due to increasing calories, and calories set a threshold in terms of the corresponding domain where yeast increase leads to reproduction increase. The intensity of egg laying is mainly regulated via yeast and not calories. Flies need to have access to high quality foods once they have a minimum of calories needed to sustain minimum longevity without which reproduction would be low.

Main conclusion. The finding that lifespan and reproductive maxima occur at much different nutritional coordinates underscored the well-known tradeoff between reproduction and survival. More importantly the magnitude of the nutritional differences between these maxima suggests that there is no dietary solution to maximize both of these life history parameters; that aging and reproduction are nutritionally inextricably coupled.

Models of reproductive aging

Exponential decline

The goal of the modeling study by statistician Hans Müller and his colleagues[22] was to examine the relationship between reproduction and lifespan in the medfly based on age-specific reproductive data generated in the earlier studies.[3,23] The trajectories of fecundity for individual flies were fitted to data from the peak to day 25 by nonlinear least squares. Egg-laying trajectories at the individual level followed simple exhaustion or decay dynamics and, therefore, the best predictor for subsequent mortality is the rate of decline in egg laying, i.e., the rate at which the egg supply is exhausted, rather than intensity of reproduction. This relationship is presented in Figure 5. Individual egg-laying trajectories rose sharply after egg laying began 5–17 days after emergence, reached a peak and then slowly declined. The rate of decline varied between individuals but one of the key findings was that this rate was approximately constant for each individual. The age trajectory of reproductive decline for each fly was accordingly modeled by the exponential function

$$f(x) = \beta_0 \exp(-\beta_1(x - \theta))$$

where $f(x)$ is the fecundity (eggs/day) of the fly at age x days and θ is the age at peak egg laying. The two parameters β_0 (mean 57.25 ± 16.70), the peak height of the trajectory and β_1, the rate of decline (mean 0.090 ± 0.093), varied considerably from fly to fly. A modest but significant negative correlation between β_0 and β_1 indicated that fecundity tends to decline more slowly for flies with higher peak fecundity. The protracted decline in egg laying after

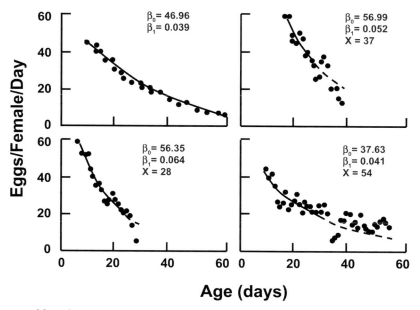

Figure 5. Trajectory of fecundity and mortality in the medfly.[22] The trajectories of fecundity for individual flies are fitted to data from the peak to day 25 (solid line) by nonlinear least squares and predicted thereafter (dashed line). Each panel depicts the reproduction of individual females in which reproductive peak, the rate of decrease (i.e., slope) and age of death are denoted by β_0, β_1, and X, respectively. Note that the greater the slope (β_1), the shorter the individual lifespan for each of the three example (i.e., upper right and two lower panels).

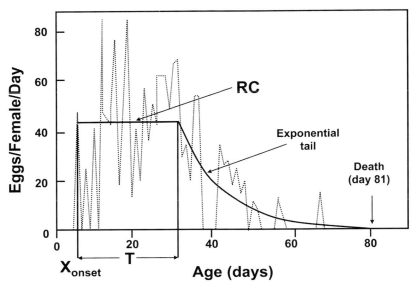

Figure 6. A typical individual fecundity pattern in medfly female.[24] The pattern (solid line) has the following parameters: onset of reproduction $X_{onset} = 4.0$, duration of plateau period, $T = 27.0$, a plateau level, RC = 42.3, and a tail time constant (exponential tail) $\tau_{tail} = 10.6$ in the exponent, $t = x - T - X_{onset}$. Death occurs at day 81.

the initial sharp rise was reasonably well predicted by the exponential model.

Main conclusions. The findings point to a fundamental link between reproductive dynamics and survival. This link is carried by the dynamics of reproduction and not by the absolute magnitude of reproduction as measured by the number of eggs produced.

A three-stage adult life model

The goal of the paper by Novoseltsev et al.[24] was to use individual-level reproductive data derived from four different medfly studies to confirm that (i) individual fecundity in female flies is endowed as a flat pattern with a steady-state period of constant rate of egg laying; (ii) an individual female has three stages in her adult life: maturation, maturity, and senescence; (iii) an individual fecundity pattern has no maximum; and (iv) two natural causes of death exist in flies including premature and senescence-caused. A model developed in a previous study was brought to bear on the four medfly populations that consisted of five parameters including X_{onset} denoting the age of first reproduction, T denoting the duration of the egg-laying plateau, RC denoting the plateau level, τ_{tail} denoting the exponential tail, and LS denoting lifespan. The model, depicted in Figure 6, sheds light on medfly aging and senescence in several respects, the most notable of which is use in comparing differences in the five parameter values across four different medfly populations (Table 2). One of the most remarkable results contained in this table are the large differences between the Greek flies derived from wild populations and the flies from the other populations—the egg-laying plateau (RC) was much lower and the maturation time (X_{onset}), duration of egg laying (T), and lifespan (LS) all much longer. However, the value of the exponential tail (senescence parameter) for the Greek flies did not stand apart from the value of this parameter in the other populations.

Main conclusions. (i) A practical implication is that the model makes it possible to store individual fecundity data in parametric form and (ii) the model parameters can be used for both inter- and intrapopulation comparisons as revealed in comparisons between Greek medflies and medflies from the other populations.

General principles derived from tephritid studies

The results of the studies of reproduction and aging using the two model tephritid species reveal several important relationships between reproduction and aging that are summarized as follows:

Table 2. Parameter values of representative fecundity patterns[24] where rc, t, x, τ_{tail}, and ls denotes height of egg-laying plateau (eggs/female), duration of plateau, age of maturity (first egg), exponential tail, and lifespan, respectively

Population (n)	RC	T	X	τ_{tail}	LS
Mexico (936)	40.5	14.3	6.5	8.1	37.2
Mexico (97)	42.2	11	5.4	4.9	32.8
Greek (50)	14.6	22.3	19.1	6	65.6
Israel (20)	40	10.4	3.2	9.6	32.6

1. *Reciprocity*. Although a lifespan cost of reproduction has received much attention (i.e., the mortality cost of early reproduction), the results of many of the studies summarized in this paper reveal the importance of a reciprocal cost—the reproductive cost due to aging. A gradual decrease in egg production always occurred. Host deprivation as well as sugar–protein manipulations affected the level of egg output in that the high output days of young ages (in the full diet without host-deprivation trials) are never fully recovered. Therefore, there is an important cost to postponing reproduction, which we call a "reproduction cost of longevity," which represents the reciprocal of the more traditional "mortality cost of reproduction."

2. *Qualitative tradeoffs*. Tradeoffs in life history evolution are often viewed in terms of resource allocation (i.e., the "Y-model" in which resources are allocated to either maintenance or reproduction) with ecological tradeoffs as a separate class. Because there are many cases where such resource allocation tradeoffs do not emerge clearly from data, this resource allocation model has been criticized by endocrinologists who view the actions of signaling pathways as the underlying mechanism of tradeoffs.[25] However, this may turn out to be a mere refinement of the "Y" model with the specific regulatory mechanisms added on and different "Y"s for different resources. To better understand tradeoffs we propose to take a more longitudinal view of the life history trajectories of individuals and recognize that there are qualitative tradeoffs that can not be understood simply as a resource allocation issue or an ecological tradeoff: e.g., a unit of reproduction can not simply be exchanged for a unit of lifespan. There are numerous examples of large qualitative differences in the tradeoffs between longevity and reproduction in the tephritid studies that can not be explained as differences in resource allocation.[10]

3. *Plasticity*. Both the lifespan and pattern of reproduction of fruit flies are highly plastic with enormous intercohort and between-biotype differences in both longevity and reproduction. With manipulations of diet and host deprivation, average longevity can differ fivefold. While it may be obvious that reproductive output can be reduced by providing a poor diet or host deprivation, the temporal pattern of reproduction also can be changed dramatically; e.g., dual modes shifting ages of last reproduction from around 30 to over 150 days. This within-species plasticity enables individuals to complete their life cycle under a wide range of environmental conditions. It also suggests that constraints on life history evolution are weak qualitatively. For example individuals may trade late reproduction for some reduction in total reproductive output, but still be able to reproduce successfully.

4. *Constraints and determinacy*. Previous work has advocated against a determined maximum lifespan. Is there a limit on total reproduction? The reproductive potential of fruit flies appears to be predestined by the early life experience in the host environment. Thus, adults eclose with a reproductive endowment that varies considerably by host species, has unknown limits and can be considered indeterminate. Whether they are able to then reach their potential depends on the availability of nutrients in the adult diet (to yolk eggs), host availability and the total number of life days remaining. Because yeast rich diets are best

for yolking eggs, but not for lifespan, this tradeoff may frequently prevent flies from reaching their reproductive potential. Some particular flies, however, have postreproductive lifespans that are greater than their usual interoviposition intervals and this can be interpreted as having reached their potential.

Acknowledgments

Supported by NIA/NIH Grants P01 AG022500–01 and P01 AG08761–10.

Conflicts of interest

The authors declare no conflicts of interest.

References

1. Boggs, C.L. 2009. Understanding insect life histories and senescence through a resource allocation lens. *Funct. Ecol.* **23:** 27–37.
2. Christenson, L.D. & R.H. Foote. 1960. Biology of fruit flies. *Annu. Rev. Entomol.* **5:** 171–192.
3. Carey, J.R. *et al*. 1998. Relationship of age patterns of fecundity to mortality, longevity, and lifetime reproduction in a large cohort of Mediterranean fruit fly females. *J. Gerontol. A Biol. Sci. Med. Sci.* **53A:** B245-B251.
4. Carey, J.R. *et al*. 1998. A simple graphical technique for displaying individual fertility data and cohort survival: case study of 1000 Mediterranean fruit fly females. *Funct. Ecol.* **12:** 359–363.
5. Carey, J.R. 2001. Insect biodemography. *Annu. Rev. Entomol.* **46:** 79–110.
6. Carey, J.R. *et al*. 2005. Biodemography of a long-lived tephritid: reproduction and longevity in a large cohort of Mexican fruit flies, *Anastrepha ludens*. *Exp. Gerontol.* **40:** 793–800.
7. Carey, J.R., P. Liedo, D. Orozco & J.W. Vaupel. 1992. Slowing of mortality rates at older ages in large medfly cohorts. *Science* **258:** 457–461.
8. Carey, J.R. *et al*. 1995. A male-female longevity paradox in medfly cohorts. *J. Anim. Ecol.* **64:** 107–116.
9. Müller, H.-G. *et al*. 1997. Early mortality surge in protein-deprived females causes reversal of sex differential of life expectancy in Mediterranean fruit flies. *Proc. Natl. Acad. Sci. USA* **94:** 2762–2765.
10. Carey, J.R. 2003. *Longevity: the Biology and Demography of Life Span*. Princeton University Press. Princeton.
11. Carey, J.R. *et al*. 2002. Food pulses increase longevity and induce cyclical egg production in Mediterranean fruit flies. *Funct. Ecol.* **16:** 313–325.
12. Diamantidis, A.D. *et al*. 2009. Life history evolution in a globally-invading tephritid: patterns of survival and reproduction in medflies from six world regions. *Biol. J. Linn. Soc.* **97:** 106–117.
13. Krainacker, D.A., J.R. Carey & R. Vargas. 1987. Effect of larval host on the life history parameters of the Mediterranean fruit fly, *Ceratitis capitata*. *Oecologia* **73:** 583–590.
14. Carey, J.R., D. Krainacker & R. Vargas. 1986. Life history response of Mediterranean fruit fly females to periods of host deprivation. *Entomol. Exp. Appl.* **42:** 159–167.
15. White, T.C.R. 1978. The importance of a relative shortage of food in animal ecology. *Oecologia* **33:** 71–86.
16. Holliday, R. 1989. Food, reproduction and longevity: is the extended lifespan of calorie-restricted animals an evolutionary adaptation? *Bioessays* **10:** 125–127.
17. Masoro, E.J. 2003. Subfield history: caloric restriction, slowing aging, and extending life. *Sci. Aging Knowledge Environ.* re2: 1–7.
18. Carey, J.R. *et al*. 1998. Dual modes of aging in Mediterranean fruit fly females. *Science* **281:** 996–998.
19. Carey, J.R. *et al*. 2002. Life history response of Mediterranean fruit flies to dietary restriction. *Aging Cell* **1:** 140–148.
20. Masoro, E.J. 2006. Caloric restriction and aging: controversial issues. *J. Gerontol. A Biol. Sci. Med. Sci.* **61A:** 14–19.
21. Carey, J.R. *et al*. 2008. Longevity-fertility trade-offs in the tephritid fruit fly, *Anastrepha ludens*, across dietary-restriction gradients. *Aging Cell* **7:** 470–477.
22. Müller, H.G., J.R. Carey, D. Wu & J.W. Vaupel. 2001. Reproductive potential determines longevity of female Mediterranean fruit flies. *Proc. R. Soc. Lond. B Biol. Sci.* **268:** 445–450.
23. Carey, J.R. & P. Liedo. 1999. Mortality dynamics of insects: general principles derived from aging research on the Mediterranean fruit fly (Diptera: Tephritidae). *Am. Entomol.* **45:** 49–55.
24. Novoseltsev, V.N. *et al*. 2004. Systemic mechanisms of individual reproductive life history in female medflies. *Mech. Ageing Dev.* **125:** 77–87.
25. Zera, A.J. & L.G. Harshman. 2002. The physiology of life history trade-offs in animals. *Annu. Rev. Ecol. Syst.* **32:** 95–126.

ANNALS OF THE NEW YORK ACADEMY OF SCIENCES
Issue: *Reproductive Aging*

Reproductive aging in invertebrate genetic models

Marc Tatar

Department of Ecology and Evolutionary Biology, Brown University, Providence, Rhode Island

Address for correspondence: Marc Tatar, Box G-W, Brown University, Providence, Rhode Island 02912. Marc_Tatar@Brown.edu

> The invertebrate genetic systems of *Caenorhabditis elegans* and *Drosophila melanogaster* are emerging models to understand the underlying mechanisms of reproductive aging and the relationship between reproduction and lifespan. Both animals show progressive decline in egg production beginning at early middle age, caused in part by reduction in germline stem cell proliferation as well as in survival of developing eggs. Molecular genetic analysis reveals that insulin and TGF-β signaling are regulators of germline stem cell maintenance and proliferation during aging. Furthermore, the lifespan of both *C. elegans* and *D. melanogaster* appears to be regulated by signaling that depends on the presence of germline stem cells in the adult gonad. These invertebrate models provide powerful tools to dissect conserved causes of reproductive aging.

Keywords: *C. elegans*; *D. melanogaster*; stem cells; reproductive aging; insulin; TGF-β

Reproductive senescence is the process whereby adults have progressively reduced fertility and fecundability with advancing age, beginning in human females sometime in the fourth decade and progressing until menopause approximately two decades later.[1] The way we understand such human reproductive senescence is shaped by the fact that all oocytes are produced during prenatal development. Human females are born with a finite pool of arrested primordial follicles. Reproductive senescence may be inevitable because the size of this pool may only decrease with age, and because the quality of the quiescent eggs diminishes with time. Reproductive senescence, however, is not just a phenomenon of mammals. In the nematode *Caenorhabditis elegans* and the fruit fly *Drosophila melanogaster*, fecundity declines beginning at young to middle ages until females cease reproduction while often remaining viable. Yet, in contrast to mammals, primary oocytes are produced from stem or progenitor cells within the mature adult in both the worm and the fly. Given this difference in timing of primary egg production, are there any common threads between the processes of reproductive senescence of mammals and of these invertebrate models? Perhaps there is—these invertebrates regulate aspects of their life history, including development, reproduction, and lifespan, with transforming growth factor-beta (TGF-β) and insulin/IGF hormones. Animals may have common, ancestral signaling networks to regulate the propensity for reproductive decline, and to control the relationship between reproduction and lifespan.

Nematode reproduction and its senescence

The nematode *C. elegans* develops from a zygote to a mature adult through four larval stages. Adults are typically hermaphroditic; males can develop but are rare. Cells that will produce the somatic gonad and the germ lineage of hermaphrodites first arise at early embryogenesis.[2] Cells of the Z1/Z4 lineages go on to produce the somatic tissues of ovary and testes, whereas cells of the Z2/Z3 lineage produce germline progenitors. Germline progenitor cells make a finite number of sperm during the fourth larval stage of development. At the transition to adulthood, these germline progenitor cells switch to become the continually dividing source for oocytes.

C. elegans biologists work with standard inbred stocks. A common wild-type strain is N2. Eggs from this strain hatch and develop to adults in about 3 days, unless they pass into an alternative resistant, nonfeeding dauer state that lasts up to many weeks and then transition back to larval stage 4 before molting to an adult. Adults from both developmental pathways begin to reproduce immediately, fertilizing newly generated eggs with their own sperm. Egg laying peaks in adults at ages 2–3 days and thereafter declines rapidly. Self-fertilizing hermaphrodites lay about 250–300 fertilized eggs by age 4–5 days old. Unfertilized eggs are also oviposited, with a peak at about day 2 but continuing to age 8 or 9 days.[3]

These observed patterns of reproductive decline may arise simply because hermaphrodites are limited by their fixed store of sperm, much as human female reproductive span may be limited in the number of primary follicles. For *C. elegans*, this hypothesis can be addressed experimentally. Mated hermaphrodites show that reproduction is not sperm limited.[4] In these adults, total reproduction increases to ∼430 progeny but egg laying still begins to decline at about adult day 2 and ends at age 8–9 days. A similar pattern occurs in mutants of the F-box-containing protein encoded by *fog-2*, which are self-sterile and must mate to lay their eggs. Importantly, *fog-2* mutant adults mated at late ages produced few early progeny but still displayed a typical pattern of reproductive decline. Not only is reproductive senescence independent of sperm availability, it is not accelerated by early reproduction or caused simply by the metabolic demands of reproduction.

Besides reproductive aging, *C. elegans* is a key model to study genetic factors regulating lifespan. Although these studies largely focus on longevity assurance, some data address the interaction between genes conferring longevity and their impact on reproduction. Insulin/IGF signaling, in particular, has been dissected as a regulator of longevity. Mutants of the insulin receptor encoded by *daf-2* increase longevity about twofold.[5] In self-fertilizing hermaphrodites, several but not all *daf-2* alleles produce some progeny at late ages and delay age-associated change in adult gonad morphology.[6,7] The effect of *daf-2* on the reproductive schedule of mated hermaphrodites has been described only recently.[4] Although the *daf-2(e1370)* allele retards reproductive aging, other alleles of *daf-2* extend longevity without affecting the pattern of reproductive decline. These data suggest that the age distribution of reproductive effort and the timing of reproductive senescence are not inexorably linked to lifespan. For the most part, genes extending the lifespan of *C. elegans* tend to do so by increasing postreproductive life expectancy.[4]

Perhaps the most interesting feature of *C. elegans* reproduction in the context of aging is the way in which it can modulate lifespan. The original study showing extended longevity in mutants of *daf-2* included an interesting control—sterile worms.[5] To rule out an explanation where *daf-2* increased lifespan simply because it reduced fecundity, a lifespan trial with wild-type animals was conducted with worms sterilized by laser ablation of the gonad. Survival in the sterile worms was indistinguishable from that of the fertile worms. Thus, the increase in longevity caused by *daf-2* could not be simply due to some effect on reproduction. But although useful for how we interpret *daf-2*, these data opened a paradox that was largely overlooked at the time. Contrary to expectations based on decades of empirical data in many animals,[8] there was no apparent tradeoff between reproduction and lifespan; eliminating the "cost of reproduction" did not increase longevity.

Fortunately, this paradox was not lost upon the authors. In a subsequent study, worms were made sterile by laser ablation of the gonad and germline precursor cells.[8a] Ablation of Z1 and Z4 caused loss of the full gonad and these sterile worms had normal lifespan. Ablation of Z2 and Z3, however, only eliminated germline precursors, leaving adults with empty somatic gonads. Remarkably, these adults were long-lived. Thus, sterility does not modulate aging but loss of the germline increases lifespan. To interpret these data, Hsin and Kenyon[8a] suggest that an active germline produces a proaging endocrine signal whereas the somatic gonad produces a longevity-assurance signal. Loss of both signals results in no net change in lifespan but loss of the negative germline signal alone permits the longevity assurance signal from the somatic gonad to increase lifespan.

We might understand reproductive senescence, including the process in mammals, if we can identify the somatic and germline longevity regulatory signals proposed to occur in *C. elegans*. The relevant literature is still very young. Initially, Hsin and Kenyon[8a] established that the forkhead transcription factor *daf-16* was required for germline ablation to increase lifespan. DAF-16 is modulated by signals from DAF-2, the *C. elegans* insulin-like receptor. When DAF-2 is activated by insulin-like peptides, a signal is transduced through the cell to keep DAF-16 in the cytoplasm, thereby preventing this transcription factor from inducing genes that slow aging. Thus, whereas *daf-2* mutation is argued to extend lifespan independently of its effects on the germline, the DAF-16 transcription factor must be involved in the way the germline modulates lifespan.

A second endocrine system that integrates reproduction and aging involves *daf-12*. DAF-12 is a nuclear hormone receptor that binds dafachronic acid, a bile acid-like steroid.[9] *daf-12* was originally identified by its mutant effects on dauer formation and was subsequently described to influence metabolism, developmental timing, and longevity.[10] In unfavorable environments, the insulin/IGF and TGF-β signals are weak, and unliganded DAF-12 interacts with a corepressor to promote transcriptional programs favoring dauer and longevity.[10] DAF-12 also plays a role in the way the germline modulates longevity because the extended lifespan of germline deficient adults requires wild-type *daf-12*.[11]

Despite the power of these insights, neither insulin/IGF signaling nor *daf-12* yet describes how germ cells communicate with the soma to control aging, or reciprocally how reproductive senescence contributes to the somatic aging process. But at this early stage we might suggest a place to look. As noted earlier, nematodes possess TGF-β signaling pathways. Although typically studied in the contexts of dauer formation and growth, TGF-β was recently shown to be a regulator of *C. elegans* adult lifespan.[11] This is important because the TGF-β superfamily includes the reproductively important mammalian hormones activin and inhibin.

The TGF-β pathway of *C. elegans* signals through two branches, the dauer branch and the so-called small/male abdomen branch.[12] In the dauer branch, a TGF-β like hormone encoded by *daf-7* activates a TGF Type I/Type II receptor complex encoded by *daf-1* and *daf-4*.[13] This receptor then phosphorylates SMADs (homologs of both the drosophila protein, mothers against decapentaplegic (MAD) and the *C. elegans* protein SMA), which are proteins that repress TGF-β activated transcription factors that otherwise promote dauer formation. The small/male abdomen branch uses a different hormone as its ligand, the Dpp/BMP-like (DBL-1). Through a receptor with units encoded by *daf-4* and *sma-6*, DBL-1 leads to phosphorylation of a different set of Smads that help induce genes essential for cell growth and the development of male traits. Early studies of *C. elegans* TGF-β found no specific effect on lifespan even though TGF-β controls dauer, and this dauer regulation and lifespan are linked through shared insulin signaling.[6] Reappraisal by Shaw *et al.* solved this problem.[11] TGF-β mutants cause embryos to prematurely hatch within the mother so that the ability of some mutants to slow aging was obscured by the high age-independent matricide. To solve this problem Shaw chemically inhibited egg development. In this condition, TGF-β pathway mutants of the dauer formation branch increase longevity up to twofold, and in an insulin-signal dependent manner.

It is too early to yet understand how these *C. elegans* genes slow aging or whether they also affect reproductive senescence but parallels to the mammalian TGF-β family suggest that solving these questions is just a matter of time. *daf-4* is homologous to the mammalian activin-type IIB receptor (ActRIIB).[14] ActRIIB mutant mice have reduced follicle-stimulating hormone (FSH), and perhaps as a consequence, delayed fertility, follicular atresia and reduced number of corpora lutea.[15] In mammals, activin binds to ActRII, which then makes a functional complex with receptor Type I. Activin ligands themselves are dimers made of activin β-subunits. Inhibins are also dimers, made from the inhibin α-subunit with one activin β-subunit. Inhibins interact with the ActRIIB but do not recruit receptor Type I, thus competitively inhibiting the potential effects of activin-dependent signaling.[16] Inhibin-B, in particular, is associated with human reproductive aging because it is produced by activated follicles.[17,18] As the number of recruited follicles declines with age, the pituitary sees less inhibin-B. Activin is now unopposed, and FSH is produced at high levels characteristic of perimenopause and menopause. If *C. elegans* modulates their ovarian axis with TGF-β feedback signaling in a way similar to mammals, we might predict there will be a decline in some inhibin-like subunit with age and a correlated increase in gonadotropic hormone, perhaps one of the worm's 38 insulin-like peptides.[19]

Drosophila reproduction and its senescence

Both male and female *D. melanogaster* produce gametes as adults from populations of stem cells within their gonads. Here we focus on female reproductive aging where output is easy to quantify in terms of egg production. Because male reproductive success is difficult to measure, much less is known about male reproductive aging although recent elegant work has documented changes in fertility, sperm production, and stem-cell functions with age.[20–22]

Females eclose as adults with a pair of ovaries that already produce early-stage egg chambers.[23] Each ovary consists of ∼10–20 ovarioles. At the terminal end of each ovariole is the germarium that contains a stem cell niche for germline stem cells (GSCs).[24] Egg chambers are formed in the germarium and progressively develop as they transit down the ovariole until they are fertilized at the posterior end of the common oviduct and then deposited. A typical germarium contains two or three GSCs. As these divide, one daughter cell remains as a GSC whereas the other differentiates into a cystoblast that subsequently undergoes four mitotic divisions to produce an interconnected set of 15 nurse cells with one oocyte. Follicle cells derived from somatic stem cells of the germarium encapsulate the germline-cyst cells to form an egg chamber.

Age-specific egg production is qualitatively similar among laboratory strains of *D. melanogaster*, although patterns vary somewhat in quantitative detail.[25–28] Egg production for females maintained on a normal diet is very low in newly eclosed females, peaks at about age 5–10 days where 20–50 eggs are laid per day, and declines thereafter until about 30–50 days old. There are few data on postreproductive life expectancy; unpublished observations from our group found old females to live about 2–3 days after last egg although some environmental manipulations could dramatically increase this duration. Reproductive senescence is also apparent at the level of egg quality.[25,26] Nearly all eggs hatch when laid by young females, but this rate progressively declines to about 50% in females 30 days old. Likewise, egg morphological defects increase with female age. Besides changes in external markers of reproduction, there is a progressive shift in mitochondria DNA composition among eggs from aged females that carried two mitochondria DNA length variants within their GSCs.[29]

Several mechanisms may contribute to reproductive senescence in female Drosophila.[30] At a morphological scale, old females do not lay fewer eggs because they have reduced number of ovarioles.[26] However, within each ovariole the number of GSCs is reduced with age (from ∼ 2.2 to 2.6 at <1 week to ∼1.7 to 2.0 at >5 weeks) but these differences are too small to account for the substantial drop in egg production.[24,26,31,32] In contrast, the rate of GSC proliferation is markedly reduced with age.[26,31] In ovarioles with two GSCs, independent methods to measure the rate of GSC divisions are reported to show 1.7-fold reduction from 3–25 days and 0.50-fold reduction thereafter. In the earlier period, these rates may account for the observed change in egg number but not so after 25 days old where egg production falls more than sixfold. Finally, apoptosis of aging egg chambers are a major factor contributing to reproductive senescence. Apoptosis increases in the egg chambers of older females such that at age 40 days more than half of the eggs are marked for degredation.[26]

The decrease in egg survival in aged Drosophila may parallel processes that occur in follicles of aging humans where cumulative damage from reactive oxygen molecules may impair the viability of developing antral follicles.[33] With Drosophila it is possible to experimentally explore this idea.[31] Overexpression of antioxidant super oxide dismutase (SOD) in GSCs or in the GSC niche increased the number of GSCs in late age females. This study did not address, however, whether SOD overexpression rescued the apoptosis of developing egg chambers in old females. In an ancillary observation, expression of heat shock proteins in GSCs reduces fertility,[34,35] and oxidative stress is known to induce the heat shock response.[36] Oxidative stress in aged follicles of mammals and in the newly produced egg chambers of aged Drosophila may contribute to reduced survival and quality of eggs during reproductive senescence.

Molecular analysis with Drosophila can also distinguish between effects on reproductive senescence from within the germline relative to those from without. These studies differentiate the roles played by the GSCs, the ovarian stem cell niche, and the systemic physiology external to the ovary. Developmental studies show that TGF-β from the stem cell niche is required for the early proliferation and maintenance of GSCs.[37–39] TGF-β ligands are secreted from niche cells and are received by the GSCs through their ActRI/ActRII receptor.[31,37] Changes in such TGF-β signaling may contribute to fly reproductive senescence.[31] Measured by downstream reporters within GSCs, TGF-β signaling is reduced in aged females. This decline appears to be functional because the experimental expression of a gene encoding the TGF-β ligand in niche cells starting at young ages increases egg production at late ages. Thus, reproductive senescence to some extent must be caused by declines in the performance of tissues outside of the GSCs, and in particular, by changes in the somatic niche.

Like TGF-β, insulin-like peptide is gonadotropic in the fly.[40–42] Insulin-like peptide stimulates GSC proliferation in response to food in young adults. The primary source of insulin-like peptides in the adult is brain neurosecretory cells.[43] Ablation of insulin-like peptides in the brain of young females retards GSC proliferation.[42] Likewise, inhibiting the insulin receptor of the gonad retards oogenesis. Because old adults eat less as measured by total uptake of carbon and nitrogen to somatic and egg production,[44] oogenesis could be affected by low protein intake in two ways. It may reduce the secretion of insulin-like peptides, and it will reduce the supply of nutrients needed to provision developing eggs. The pattern of insulin-like peptide expression with age has yet to be fully documented, although insulin signaling appears to be essential for germline maintenance in females aged for 2 weeks.[24,26,31,32] Little is also known about the impact of feeding with age on reproduction except that starvation can induce apoptosis and egg resorption in young flies, although oogenesis can be restored by refeeding at advanced ages at the time when previously fed females are already postreproductive.[28,40] Whether these diet-mediated events are caused by the direct limitation of nutrients or through hormonal signals is unknown.

A recent approach to understanding reproductive aging with Drosophila used a mutant screen to find genes that increase late age fecundity.[27] Transposon insertions that conditionally overexpress a unique gene were tested to see if they prevented the age-dependent decline in egg production. Overexpression of two genes increased late fecundity. These genes, named *magu* and *hebe*, also increased adult lifespan by ∼5–30%. Only *magu* has a recognizable mammalian homolog, which is SMOC2, an extracellular matrix factor known to stimulate cell proliferation and angiogenesis.[45] The function of *magu* in Drosophila is unknown, although it often occurs in tissues with stem cells (including the ovary) suggesting as a hypothesis that it may play a role in the maintenance of GSC proliferation.[27]

Finally, as with *C. elegans*, Drosophila is a key tool to understand the relationship between somatic aging, lifespan, and reproduction. Past studies using experimental evolution were readily able to increase late age reproduction by direct selection, often at the same time increasing lifespan and decreasing early reproduction.[8,46,47] Although longevity and late reproduction in this context may be evolutionarily constrained by tradeoffs with early-age fitness traits, many results also suggest that at least in some environments there is no necessary connection between survival and the schedule of reproduction. Genetic manipulation of several genes, such as *indy*, *ecdysone receptor*, or *foxo* (the Drosophila homolog of *C. elegans daf-16*), can all increase lifespan without reducing either late or early reproduction.[48–50] Whether these manipulated genes also delay the rate of decline of reproduction with age has not been carefully examined. Conversely, sterility *per se* does not extend fly lifespan. Mutations of the gene *egalitarian* blocks normal egg chamber development but mutant females are not longer lived than controls[51] and as observed in flies made sterile during development by mutations in the genes germ cell-less and tudor.[52] But again as in *C. elegans*, reducing the number of GSCs in Drosophila is sufficient to extend lifespan.[51] This condition is accompanied by an interesting endocrine response whereby mRNA for insulin-like peptides become highly induced in the fly brain yet insulin signaling in peripheral somatic tissue appears to be very low.

One possible solution to this apparent paradox of hyper-insulin coupled with low insulin response may involve the insulin-like binding protein ImpL2, which is the homolog of mammalian IGFBP7.[53,54] ImpL2 binds to and antagonizes insulin-like peptides,[53,54] and expression of *ImpL2* is greatly increased in the GSC-knockout adults.[51] A potential explanation for the behavior of this system reasons that insulin-like peptides are gonadotropic and are negatively regulated by some ovary-derived signal that is induced by the germline, much as mammalian FSH is modulated by inhibin derived from follicles. Knockout of the fly germline would release this negative feedback, leading to abundant production of insulin-like peptides. However, although these peptides would be appropriate to stimulate the germline they would also be expected to accelerate somatic aging because reduction of insulin signaling has been shown to increase fly lifespan.[55] Production of ImpL2 might block the somatic effects of the insulin-like peptide yet still permit activity at the ovary if the hormone-binding protein complex is processed in a tissue-specific manner. Analysis of this proposed pituitary-gonad-like axis in Drosophila has the potential to model key features of reproductive control that deteriorate with age in mammals.

Conclusions

Despite the differences in the way female reproduction is organized in mammals relative to Drosophila and C. elegans, we are finding potential parallels in their molecular systems of proliferation and maintenance. Current analyses aim to better describe the brain/ovary axis of these invertebrates. It will identify signals from the brain that control germline proliferation and, conversely, signals from the gonad that provide feedback to the gonadotropic signals of the brain. It will identify signals from the gonad that directly or indirectly modulate lifespan. And it will likewise further resolve how reduced germline function and egg chamber survival are caused by changes within the germline itself with age, in the somatic environment of the gonad, and most likely in the interaction of these cell populations. These cellular events are likely to have conserved properties in the process of mammalian ovarian aging.

Conflicts of interest

The author declares no conflicts of interest.

References

1. te Velde, E.R. & P.L. Pearson. 2002. The variability of female reproductive ageing. *Hum. Reprod. Update* **8:** 141–154.
2. Albert Hubbard, E.J. & D. Greenstein. 2000. The *Caenorhabditis elegans* gonad: a test tube for cell and developmental biology. *Dev. Dyn.* **218:** 2–22.
3. Kadandale, P. & A. Singson. 2004. Oocyte production and sperm utilization patterns in semi-fertile strains of *Caenorhabditis elegans*. *BMC Dev. Biol.* **4:** 3.
4. Hughes, S.E., K. Evason, C. Xiong & K. Kornfeld. 2007. Genetic and pharmacological factors that influence reproductive aging in nematodes. *PLoS Genet.* **3:** e25.
5. Kenyon, C. et al. 1993. A *C. elegans* mutant that lives twice as long as wild type. *Nature* **366:** 461–464.
6. Larsen, P.L., P.S. Albert & D.L. Riddle. 1995. Genes that regulate both development and longevity in *Caenorhabditis elegans*. *Genetics* **139:** 1567–1583.
7. Garigan, D. et al. 2002. Genetic analysis of tissue aging in *Caenorhabditis elegans*: a role for heat-shock factor and bacterial proliferation. *Genetics* **161:** 1101–1112.
8. Bell, G. & V. Koufopanou. 1986. The cost of reproduction. In *Oxford Surveys in Evolutionary Biology*, Vol. 3. R. Dawkins & M. Ridley, Eds.: 83–131. Oxford University Press. Oxford.
8a. Hsin, H. & C. Kenyon. 1999. Signals from the reproductive system regulate the lifespan of C-elegans. *Nature* **399:** 362–366.
9. Motola, D.L. et al. 2006. Identification of ligands for DAF-12 that govern dauer formation and reproduction in *C. elegans*. *Cell* **124:** 1209–1223.
10. Magner, D.B. & A. Antebi. 2008. *Caenorhabditis elegans* nuclear receptors: insights into life traits. *Trends Endocrinol. Metab.* **19:** 153–160.
11. Shaw, W.M. et al. 2007. The *C. elegans* TGF-β Dauer pathway regulates longevity via insulin signaling. *Curr. Biol.* **17:** 1635–1645.
12. Savage-Dunn, C. 2005. TGF-β signaling. In *WormBook*, The *C. elegans* Research Community, Eds.: WormBook. doi/10.1895/wormbook.1.22.1. http://www.wormbook.org.
13. Ren, F. et al. 1996. Control of *C. elegans* larval development by neuronal expression of a TGF-β homolog. *Science* **272:** 1389–1391.
14. Estevez, M. et al. 1993. The *daf-4* gene encodes a bone morphogenetic protein receptor controlling *C. elegans* dauer larva development. *Nature* **365:** 644–649.
15. Matzuk, M.M., T.R. Kumar & A. Bradley. 1995. Different phenotypes for mice deficient in either activins or activin receptor type II. *Nature* **374:** 356–360.
16. Lewis, K.A. et al. 2000. Betaglycan binds inhibin and can mediate functional antagonism of activin signaling. *Nature* **404:** 411–414.
17. Klein, N.A. et al. 2004. Age-related analysis of inhibin A, inhibin B, and activin a relative to the intercycle monotropic follicle-stimulating hormone rise in normal ovulatory women. *J. Clin. Endocrinol. Metab.* **89:** 2977–2981.
18. Welt, C.K., D.J. McNicholl, A.E. Taylor & J.E. Hall. 1999. Female reproductive aging is marked by decreased secretion of dimeric inhibin. *J. Clin. Endocrinol. Metab.* **84:** 105–111.
19. Pierce, S.B. et al. 2001. Regulation of DAF-2 receptor signaling by human insulin and *ins-1*, a member of the unusually large and diverse *C. elegans* insulin gene family. *Genes Dev.* **15:** 672–686.
20. Boyle, M., C. Wong, M. Rocha & D.L. Jones. 2007. Decline in self-renewal factors contributes to aging of the stem cell niche in the Drosophila testis. *Cell Stem Cell* **1:** 470–478.

21. Wallenfang, M.R., R. Nayak & S. DiNardo. 2006. Dynamics of the male germline stem cell population during aging of *Drosophila melanogaster*. *Aging Cell* **5:** 297–304.
22. Cheng, J. *et al.* 2008. Centrosome misorientation reduces stem cell division during ageing. *Nature* **456:** 599–604.
23. Spradling, A.C. 1993. Developmental genetics of oogenesis. In *The Development of Drosophila melanogaster*, Vol. 1. M. Bate & A. Martinex-Arias, Eds.: 1–70. Cold Spring Harbor Laboratory Press. Cold Spring Harbor, NY.
24. Xie, T. & A.C. Spradling. 2000. A niche maintaining germ line stem cells in the Drosophila ovary. *Science* **290:** 328–330.
25. David, J., Y. Cohet & P. Fouillet. 1975. The variability between individuals as a measure of senescence: a study of the number of eggs laid and the percentage of hatched eggs in the case of *Drosophila melanogaster*. *Exp. Gerontol.* **10:** 17–25.
26. Zhao, R., Y. Xuan, X. Li & R. Xi. 2008. Age-related changes of germline stem cell activity, niche signaling activity and egg production in Drosophila. *Aging Cell* **7:** 344–354.
27. Li, Y. & J. Tower. 2009. Adult-specific over-expression of the Drosophila genes *magu* and *hebe* increases life span and modulates late-age female fecundity. *Mol. Genet. Genomics* **281:** 147–162.
28. Good, T. & M. Tatar. 2001. Age-specific mortality and reproduction respond to adult dietary restriction in *Drosophila melanogaster*. *J. Insect Phys.* **47:** 1467–1473.
29. Kann, L.M., E.B. Rosenblum & D.M. Rand. 1998. Aging, oocyte turnover, and the evolution of germline heteroplasmy for mitochondrial DNA length variants in *Drosophila melanogaster*. *Proc. Natl. Acad. Sci. USA* **95:** 2372–2377.
30. Waskar, M., Y. Li & J. Tower. 2005. Stem cell aging in the Drosophila ovary. *AGE* **27:** 201–212.
31. Pan, L. *et al.* 2007. Stem cell aging is controlled both intrinsically and extrinsically in the Drosophila ovary. *Cell Stem Cell* **1:** 458–469.
32. Hsu, H.-J. & D. Drummond-Barbosa. 2009. Insulin levels control female germline stem cell maintenance via the niche in Drosophila. *Proc. Natl. Acad. Sci. USA* **106:** 1117–1121.
33. Tatone, C. *et al.* 2008. Cellular and molecular aspects of ovarian follicle ageing. *Hum. Reprod. Update* **14:** 131–142.
34. Wang, Z. & S. Lindquist. 1998. Developmentally regulated nuclear transport of transcription factors in Drosophila embryos enable the heat shock response. *Development* **125:** 4841–4850.
35. Silbermann, R. & M. Tatar. 2000. Reproductive costs of heat shock protein in transgenic *Drosophila melanogaster*. *Evolution* **54:** 2038–2045.
36. Ahn, S.-G. & D.J. Thiele. 2003. Redox regulation of mammalian heat shock factor 1 is essential for *Hsp* gene activation and protection from stress. *Genes Dev.* **17:** 516–528.
37. Xie, T. & A.C. Spradling. 1998. *decapentaplegic* is essential for the maintenance and division of germline stem cells in the Drosophila ovary. *Cell* **94:** 251–260.
38. Song, X. *et al.* 2004. Bmp signals from niche cells directly repress transcription of a differentiation-promoting gene, *bag of marbles*, in germline stem cells in the Drosophila ovary. *Development* **131:** 1353–1364.
39. Chen, D. & D. McKearin. 2003. Dpp signaling silences *bam* transcription directly to establish asymmetric divisions of germline stem cells. *Curr. Biol.* **13:** 1786–1791.
40. Drummond-Barbosa, D. & A.C. Spradling. 2001. Stem cell and their progeny respond to nutritinoal changes during Drosophila oogenesis. *Dev. Biol.* **231:** 265–278.
41. Drummond-Barbosa, D. & A.C. Spradling. 2004. α-Endosulfine, a potential regulator of insulin secretion, is required for adult tissue growth control in Drosophila. *Dev. Biol.* **266:** 310–321.
42. LaFever, L. & D. Drummond-Barbosa. 2005. Direct control of germline stem cell division and cyst growth by neural insulin in Drosophila. *Science* **309:** 1071–1073.
43. Rulifson, E.J., S.K. Kim & R. Nusse. 2002. Ablation of insulin-producing neurons in flies: growth and diabetic phenotypes. *Science* **296:** 1118–1120.
44. O'Brien, D.M., K.J. Min, T. Larsen & M. Tatar. 2008. Use of stable isotopes to examine how dietary restriction extends Drosophila lifespan. *Curr. Biol.* **18:** R155–R156.
45. Vannahme, C. *et al.* 2003. Characterization of SMOC-2, a modular extracellular calcium-binding protein. *Biochem. J.* **373:** 805–814.
46. Rose, M.R. 1984. Laboratory evolution of postponed senescence in *Drosophila melanogaster*. *Evolution* **38:** 1004–1010.
47. Partridge, L., N. Prowse & P. Pignatelli. 1999. Another set of responses and correlated responses to selection on age at reproduction in *Drosophila melanogaster*. *Proc. Biol. Sci.* **266:** 255–261.
48. Rogina, B., R.A. Reenan, S.P. Nilsen & S.L. Helfand. 2000. Extended lifespan conferred by cotransporter gene mutations in Drosophila. *Science* **290:** 2137–2140.
49. Simon, A.F., C. Shih, A. Mack & S. Benzer. 2003. Steroid control of longevity in *Drosophila melanogaster*. *Science* **299:** 1407–1410.
50. Hwangbo, D.S. *et al.* 2004. Drosophila dFOXO controls lifespan and regulates insulin signaling in brain and fat body. *Nature* **429:** 562–566.
51. Flatt, T. *et al.* 2008. Drosophila germ-line modulation of insulin signaling and lifespan. *Proc. Natl. Acad. Sci. USA* **105:** 6368–6373.
52. Barnes, A.I. *et al.* 2006. No extension of lifespan by ablation of germ line in Drosophila. *Proc. Biol. Sci.* **273:** 939–947.
53. Honegger, B. *et al.* 2008. Imp-L2, a putative homolog of vertebrate IGF-binding protein 7, counteracts insulin signaling in Drosophila and is essential for starvation resistance. *J. Biol.* **7:** 10.
54. Arquier, N. *et al.* 2008. Drosophila ALS regulates growth and metabolism through functional interaction with insulin-like peptides. *Cell Metab.* **7:** 333–338.
55. Broughton, S. & L. Partridge. 2009. Insulin/IGF-like signaling, the central nervous system and aging. *Biochem. J.* **418:** 1–12.

ANNALS OF THE NEW YORK ACADEMY OF SCIENCES
Issue: *Reproductive Aging*

A pathway that links reproductive status to lifespan in *Caenorhabditis elegans*

Cynthia Kenyon

Department of Biochemistry and Biophysics, University of California, San Francisco, California

Address for correspondence: Cynthia Kenyon, Mission Bay Genentech Hall MC2200, 600 16th Street, Room S312A, San Francisco, California 94158-2517. cynthia.kenyon@ucsf.edu

In the nematode *Caenorhabditis elegans* and the fruit fly Drosophila, loss of the germline stem cells activates lifespan-extending FOXO-family transcription factors in somatic tissues and extends lifespan, suggesting the existence of an evolutionarily conserved pathway that links reproductive state and aging. Consistent with this idea, reproductive tissues have been shown to influence the lifespans of mice and humans as well. In *C. elegans*, loss of the germ cells activates a pathway that triggers nuclear localization of the FOXO transcription factor DAF-16 in endodermal tissue. DAF-16 then acts in the endoderm to activate downstream lifespan-extending genes. DAF-16 is also required for inhibition of insulin/insulin-like growth factor 1 (IGF-1) signaling to extend lifespan. However, the mechanisms by which inhibition of insulin/IGF-1 signaling and germline loss activate DAF-16/FOXO are distinct. As loss of the germ cells further doubles the already-long lifespan of insulin/IGF-1 pathway mutants, a better understanding of this reproductive longevity pathway could potentially suggest powerful ways to increase healthy lifespan in humans.

Keywords: germ cells; DAF-12; nuclear hormone receptor; aging

The aging process touches everyone's life, and, as a scientific problem, it is fascinating in its own right. What determines the rate at which we age? Aging was once thought to "just happen." We wear out, like old shoes. Yet, the aging process is not completely random and stochastic. Different species can have dramatically different lifespans, and, even within a single animal species, the rate of aging can be influenced by environmental factors, such as the level or quality of nutrients, various stressors, temperature, and sensory cues.[1] The mechanisms by which environmental conditions influence lifespan are now under investigation in many labs, and in many cases, genetic pathways employing classical regulatory proteins, such as kinases and transcription factors, play important roles. Moreover, some mechanisms that influence lifespan have been conserved during evolution. For example, reducing the level of insulin/insulin-like growth factor 1 (IGF-1) hormone signaling extends the lifespan of worms, flies, and mammals.[1,2] The mechanism of this lifespan extension has been explored extensively, particularly in worms,[1,3–6] where inhibition of insulin/IGF-1 signaling has been shown to activate specific transcription factors, including the FOXO-family member DAF-16/FOXO. DAF-16/FOXO, in turn, increases lifespan by up- or downregulating a wide variety of metabolic, cell-protective, chaperone, and anti-immunity genes that appear to act cumulatively to extend lifespan. Mutations that alter the activity of this pathway can extend *Caenorhabditis elegans*' lifespan up to 10-fold.[7] DNA variants in a *FOXO* gene have now been associated with increased longevity in seven human populations across several continents,[1,8–11] indicating that human longevity, too, is likely susceptible to the effects of this transcription factor. In many organisms, reducing insulin/IGF-1 signaling delays the onset of age-related diseases and reduces their severity, suggesting that this pathway couples the normal aging process to age-related disease susceptibility.[6]

The relationship between aging and reproduction is particularly fascinating, given the significance of both processes in the life of an individual and the success of the species. For example, in humans, female menopause may promote longevity by eliminating the chance that an older woman will die during childbirth. Recently, a fascinating relationship between reproduction and aging has been revealed in small organisms; specifically, in the nematode *C. elegans* and the fruit fly Drosophila. In both species, the germ cells; that is, the cells that give rise to sperm and oocytes, influence the aging of the whole animal. If the germline precursor cells are removed in either worms or flies, lifespan is extended by 40–60%.[12,13] These animals not only live longer, they also remain youthful and active longer than normal, suggesting that loss of the germline precursor cells extends lifespan because it slows the animals' rate of aging. Thus, in these animals, the germ cells do not only produce the next generation, they also influence the lifespan of the body in which they reside. This review will describe what we know so far about how this pathway affects the aging process in the animal in which it has been studied most extensively, *C. elegans*.

In *C. elegans*, the germline can be removed by killing the two germline precursor cells (Z2 and Z3; Fig. 1) with a laser microbeam at the time of hatching.[13] Or, the germline can be removed genetically;[14] for example, by shifting animals carrying the temperature-sensitive mutation *glp-1(e2141)* to the nonpermissive temperature, which forces mitotically dividing germline stem cells to exit mitosis and enter meiosis. Lifespan can also be extended when the germline stem cells are forced into meiosis during adulthood.[14] At this time, the animals contain germ cells arrested in meiosis as well as mature sperm and oocytes, and they are producing progeny. This finding, along with others, singles out the germline stem cells (as opposed to sperm, oocytes, or meiotic cells) as being especially important in affecting aging. In addition, this finding indicates that adult tissues are susceptible to the lifespan-extending effects of germline removal. A similar situation exists in flies.[12] How this system might have arisen during evolution is unknown, but one hypothesis is that it might confer a selective advantage by allowing the somatic tissues to "wait" for the germline to mature before aging progresses too extensively.[13] Such a system might help to coordinate the timing of reproduction with aging.

How does germline loss extend lifespan? Because reproduction is an expensive process metabolically, one could imagine that lifespan is extended simply because in the absence of reproduction, more

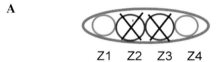
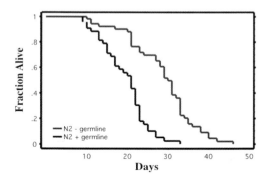

Figure 1. Removing the germline of *C. elegans* extends lifespan. (A) At the time of hatching, the animal's entire reproductive system contains only four cells, so using a laser beam to ablate either the precursors of the germline (Z2 and Z3) or the precursors of all of the reproductive tissues (Z1–Z4) with a laser microbeam is straightforward. Killing Z2 and Z3 in this way extends lifespan ∼60%. (B) The diagram shows the reproductive system of the animal when it is an adult. At this stage, it contains mature sperm (squares) and oocytes (large ovals), and progeny are being produced (not shown). In addition, the germline contains cells arrested in meiosis (blue circles) and proliferating germline stem cells (yellow circles). When the germline stem cells are forced into meiosis during adulthood using a temperature-sensitive mutation that inhibits signaling required for germline stem cell proliferation, then lifespan is increased.

resources can be devoted to cell and tissue maintenance. This idea, that there is a "cost of reproduction," has been put forth by evolutionary biologists, and there are many examples of reciprocal relationships between reproduction and longevity in nature and in the laboratory.[15] In our case, this may be part of the answer, but there seems to be more to it. When the entire reproductive system of *C. elegans* is removed [that is, the cells that give rise to the somatic reproductive tissues (Z1 and Z4) as well as the cells that give rise to the germline (Z2 and Z3)], lifespan is not increased.[13] Animals that lack their entire reproductive systems are also sterile, so these findings argue against models invoking a simple cost of reproduction. Instead, these findings indicate that both the germline and the somatic reproductive tissues play an active role in influencing lifespan. Specifically, the germline and the somatic reproductive tissues exert counterbalancing influences on lifespan, with the germline preventing, and the somatic gonad promoting, lifespan extension.

If active signaling is required for loss of the germ cells to extend lifespan, then what are these signals and how are they communicated to nonreproductive tissues? It seems likely that steroidal hormone signaling plays an important role. Reduction-of-function mutations in the gene *daf-12*, which encodes a nuclear hormone receptor (NHR), or in genes like *daf-9*, which encode proteins that synthesize sterol ligands for DAF-12, prevent loss of the germline from extending lifespan.[13,16,17] (For a summary of genes in this pathway, see Table 1.)

DAF-12/NHR is not the only transcription factor required for loss of the germ cells to extend lifespan. The DAF-16/FOXO transcription factor, described above, is required as well.[13] Interestingly, DAF-16/FOXO appears to act in the intestine of *C. elegans* to extend lifespan when the germline is removed.[18] Under these conditions, DAF-16/FOXO accumulates primarily in intestinal nuclei. This tissue appears to play a central role in this pathway, as expression of DAF-16/FOXO exclusively in the intestine can completely rescue the long lifespan of germline-defective *daf-16(-)* mutants.[19] The intestine of *C. elegans* appears to be the animal's entire endoderm. *C. elegans* does not have a distinct adipose tissue, liver or pancreas; but the intestine stores fat (as does adipose), produces yolk (as does the liver), and produces some important insulin-like peptides, such as INS-7[20] (as does the pancreas). It is

Table 1. Some genes required for loss of the germline to extend the lifespan of *C. elegans*. For references, see text.

Gene	General function	Response to germline loss	Role in insulin/IGF-1 pathway
daf-16	FOXO-family transcription factor	Localizes to intestinal nuclei during adulthood. Acts in the intestine to extend lifespan	Required for insulin/IGF-1-pathway mutants to live long. Localizes to nuclei in many tissues throughout life
daf-12	Nuclear hormone receptor	Is partially required for DAF-16 nuclear localization, but has another, unknown, function in this pathway	Not required for lifespan extension
daf-9	Cytochrome P450 required for DAF-12-ligand biosynthesis	Partially required for DAF-16 nuclear localization	Not required for lifespan extension
tcer-1	Ortholog of the human transcription elongation/splicing factor TCERG1	Intestinal expression increases. Required for the upregulation of some, but not all DAF-16-dependent target genes	Not required for lifespan extension
kri-1	Ortholog of the human disease gene *KRIT1*. Contains protein interaction domains (ankyrin repeats)	Intestinal protein required for DAF-16 nuclear localization and *tcer-1* upregulation upon germline loss	Not required for lifespan extension
K04A8.5	Fat lipase	Upregulated by DAF-16 in the intestine. Could potentially produce a downstream lifespan-extending signal from the intestine to other tissues	Partially required for lifespan extension

possible that DAF-16/FOXO has a conserved lifespan-extending function in the intestine/adipose tissue, as overexpression of the *Drosophila* DAF-16/FOXO ortholog only in adipose tissue extends fly lifespan.[12] Loss of the mouse insulin receptor specifically in adipose tissue also extends lifespan.[21] This intervention would be predicted to activate mouse FOXO proteins in this tissue. How DAF-16 activity in the *C. elegans* intestine extends the lifespan of the entire animal is not known, but a fat lipase called K04A8.5 may be involved.[22] The gene encoding this lipase is upregulated by DAF-16/FOXO in the intestines of germline-defective animals, and its function is required for lifespan extension. It is possible that this lipase is involved in the synthesis of downstream signals from the intestine that influence the lifespan of the other tissues in the animal. The apparently conserved ability of adipose tissue to produce lifespan extending signals in response to FOXO activity makes this issue particularly interesting.

How does loss of the germ cells activate DAF-16/FOXO? One obvious question, given DAF-16/FOXO's central role in the insulin/IGF-1 pathway, is whether loss of the germ cells activates DAF-16/FOXO by inhibiting the activity of the insulin/IGF-1 pathway. Also, is the DAF-12 steroid signaling system involved? Is the somatic gonad required? The answers to these questions are not known in detail; but at least part of the story has begun to emerge (and is summarized in Figs. 2 and 3).

First, the pathway that triggers DAF-16/FOXO nuclear localization is at least partially distinct from the pathway that localizes DAF-16 to nuclei when insulin/IGF-1 signaling is inhibited. In long-lived *daf-2* (insulin/IGF-1-receptor) mutants, DAF-16/FOXO accumulates in most or all somatic cell nuclei throughout life.[23–25] In contrast, when the germline is removed, DAF-16/FOXO does not exhibit nuclear accumulation until adulthood, and then it localizes primarily to nuclei in the intestine. Second, several genes that are required for DAF-16/FOXO nuclear localization and lifespan extension in response to loss of the germ cells are not required for lifespan extension in response to reduced insulin/IGF-1 signaling. *daf-12*/NHR is one such gene[13,26] (discussed further). Another is *kri-1*, the *C. elegans* homolog of the human disease gene *KRIT1*, which encodes an intestinal ankyrin-

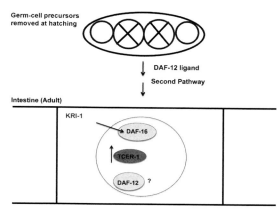

Figure 2. Loss of the germline precursor cells at the time of hatching (circles with Xs) triggers important changes in intestinal cells of the adult. Killing the germ cells at the time of hatching triggers several important changes in the intestine of the adult. The transcription factor DAF-16/FOXO accumulates in nuclei, and the level of the putative transcription-elongation factor TCER-1 rises. Both of these events are completely dependent on *kri-1*. DAF-16 nuclear localization is partially dependent on the DAF-12 steroid signaling pathway, but TCER-1 upregulation is independent of DAF-12. Therefore, there must be a second gonad-to-intestine signaling pathway. The site of action of DAF-12 itself is not known (question mark). It could potentially act in the intestine.

domain containing protein. *kri-1* is required for loss of the germline to mediate DAF-16/FOXO nuclear localization and to extend lifespan.[18] In contrast, *kri-1* is not required for *daf-2* inhibition to extend lifespan.[18] Moreover, apart from DAF-16/FOXO's nuclear accumulation, the requirements for DAF-16/FOXO-dependent gene expression are different between germline-defective animals and insulin/IGF-1-signaling mutants. A putative transcription elongation factor called TCER-1 (a *C. elegans* homolog of human TCERG1) is required for lifespan extension and for the increased expression of many DAF-16-target genes in germline-deficient animals, but it is not required for lifespan extension or *daf-16*-dependent gene regulation in insulin/IGF-1 mutants.[27] It is not clear why TCER-1 should be required for DAF-16 target-gene expression in response to germline ablation, but (for at least some of the same genes) not in response to inhibition of insulin/IGF-1 signaling, and this is an interesting question. Whatever the answer, it is clear that many aspects of the reproductive longevity pathway distinguish it from the insulin/IGF-1 pathway. If loss of the germline extended lifespan simply by inhibiting insulin/IGF-1 signaling, then the

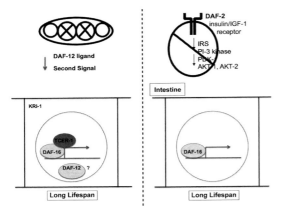

Figure 3. The reproductive and insulin/IGF-1 pathways are distinct. The lifespan extension produced by loss of the germ cells requires KRI-1, DAF-12, and TCER-1, whereas the lifespan extension produced by *daf-2(e1370)* receptor mutations does not. (The question mark following "DAF-12" refers to our uncertainty about the site of action of DAF-12.) Moreover, TCER-1 is required for the increased expression of a set of DAF-16-regulated genes in response to germline loss, but TCER-1 is not required for lifespan extension or for the expression of these DAF-16-regulated genes in insulin/IGF-1 mutants (at least not for the genes that have been examined). However, it is important to note that mutations in the insulin/IGF-1 pathway can affect the operation of the germline pathway (see text), so it is not yet clear whether the two pathways act completely independently of one another. The drawings depict working models: for example, TCER-1 and DAF-16 are hypothesized to interact on individual promoters, but this has not been shown directly.

behavior of DAF-16/FOXO would be expected to be more similar in the two pathways than it actually is. Finally, removing the germlines of animals carrying *daf-2*/insulin/IGF-1-receptor mutations further doubles the already-long lifespans of the animals.[13] This effect is consistent with the idea that these two pathways are not the same.

How is intestinal DAF-16/FOXO informed about the state of the germline? The *daf-12*-dependent steroid signaling pathway appears to play a role. In germline-deficient animals lacking *daf-12*/NHR or the DAF-12-ligand-synthesizing genes *daf-9* or *daf-36*, intestinal DAF-16/FOXO nuclear localization is incomplete.[17,18] Administering the DAF-12 ligand dafachronic acid to germline-deficient *daf-9* mutants (which can not make dafachronic acid) restores full DAF-16 nuclear localization and lifespan extension.[17] This finding suggests that the DAF-12 signaling pathway plays a role in mediating communication between the reproductive system and the intestine.

In addition to promoting DAF-16/FOXO nuclear localization, DAF-12 has another, unknown, function in this pathway.[18] It is possible to force DAF-16 nuclear localization by mutating the AKT-phosphorylation sites on DAF-16 through which insulin/IGF-1 signaling prevents DAF-16 nuclear accumulation in normal intact animals. This constitutively nuclear mutant DAF-16 protein can substitute for wild-type DAF-16 and extend lifespan in germline-deficient animals. However, this lifespan extension is still dependent on DAF-12. This finding indicates that DAF-12 has another function that is essential for lifespan extension, in addition to its role in DAF-16 nuclear localization. The nature of this other function is unknown.

The fact that some DAF-16/FOXO nuclear localization takes place when the germline is removed in *daf-12(-)* or *daf-9(-)* mutants suggests that there is a second pathway that informs the intestine about the status of the germline. This interpretation is consistent with another observation[27]: When the germline is removed, the level of TCER-1 rises in the intestine. This increase requires KRI-1, but it is completely independent of DAF-12/NHR. It will be very interesting to learn the identity of this *daf-12*-independent gonad-to-intestine signaling pathway.

Many aspects of this longevity system remain mysterious. Why is the somatic gonad required for loss of the germline to extend lifespan? The somatic gonad is not required for germline loss to trigger DAF-16 nuclear localization[28] or TCER-1 upregulation,[27] but it is required for germline loss to induce at least some DAF-16-dependent transcription.[28] Perhaps these DAF-16-dependent genes are essential for lifespan extension. Whether the somatic gonad has other functions that do not involve DAF-16 is unknown.

Curiously, in strong *daf-2(-)* mutants, the somatic reproductive tissues are no longer required for loss of the germline to further extend lifespan.[13,28] Why inhibiting insulin/IGF-1 signaling removes the requirement for the somatic gonad is not clear. One could imagine that the insulin pathway is downstream of the somatic gonad; that is, that the somatic gonad extends lifespan in germline-deficient animals by inhibiting insulin/IGF-1 signaling. However, this model does not explain the tissue-specific localization of DAF-16, or why *kri-1* and *tcer-1* are required for germline loss, but not insulin/IGF-1-pathway inhibition, to extend lifespan. It is possible that the longevity requirement normally fulfilled by the somatic gonad can also be fulfilled, possibly in a different way, by inhibition of insulin/IGF-1 signaling.

Another intriguing question is how the germline stem cells "tell" the animal that they are present. If the presence or loss of the germline activates a signaling pathway, what is the initiating event? How is it linked to germline stem cell proliferation?

Finally, it is interesting to ask whether this same reproductive signaling pathway, or a molecular variant, might influence lifespan in higher animals. There are several intriguing parallels with *Drosophila* that suggest a common evolutionary origin. First, as mentioned above, forcing the germline stem cells to exit mitosis and enter meiosis during adulthood extends the lifespan of both adult flies and adult worms.[12,14] Moreover, in flies, as in worms, this treatment activates DAF-16/FOXO activity.[12] In contrast to the situation in worms, when the germline is removed in *Drosophila* during development, lifespan is not extended.[29] The reason for this is unknown, but it is tempting to speculate that lifespan is not extended because in flies (unlike in worms), early loss of the germline prevents the correct development of the somatic reproductive tissues.

Less is known about mammals; however transplanting the ovaries of young mice into old females extends lifespan.[30,31] Thus, signals from reproductive tissues can influence mammalian lifespan. Whether there are any additional similarities between the worm pathway and the pathway triggered by this transplantation in mice is not known. It is interesting to wonder whether the loss of oocytes might extend the lifespan of human females, who live longer than men. This is of course completely unknown, but, interestingly, if the ovaries as well as the germ cells are absent in postmenopausal women, then the rate of all-cause mortality (including age-related diseases) increases.[32] It will be fascinating to watch this interesting new field grow and begin to encompass higher organisms. Whether or not human lifespan is influenced in the same way by reproductive tissues, if we learn how the reproductive systems of smaller animals, like worms and flies, regulate DAF-16/FOXO activity, it may be possible to use this information to intervene downstream of the reproductive pathway and extend healthy lifespan without affecting the germline itself. Consistent with this idea, in *C. elegans,* overexpressing the putative transcription-elongation factor gene *tcer-1* extends the lifespan of worms that have an intact reproductive system and are fully fertile.[27] This lifespan extension correlates with the upregulation of many germline-specific DAF-16 target genes, and it is dependent on DAF-16 activity. The evidence that FOXO proteins influence human lifespan[1,8–11] makes the possibility of harnessing this information to influence human health and longevity seem increasingly possible.

Acknowledgments

I thank the members of my laboratory, and the reviewers, for helpful comments and suggestions.

Conflicts of interest

The author declares no conflicts of interest.

References

1. Kenyon, C.J. 2010. The genetics of aging. *Nature* **464:** 504–512. Review.
2. Bartke, A. 2008. Insulin and aging. *Cell Cycle* **7:** 3338–3343.
3. Kleemann, G.A. & C.T. Murphy. 2009. The endocrine regulation of aging in *Caenorhabditis elegans*. *Mol. Cell. Endocrinol.* **299:** 51–57.
4. Panowski, S.H. & A. Dillin. 2009. Signals of youth: endocrine regulation of aging in *Caenorhabditis elegans*. *Trends Endocrinol. Metab.* **20:** 259–264.
5. Tatar, M., A. Bartke & A. Antebi. 2003. The endocrine regulation of aging by insulin-like signals. *Science* **299:** 1346–1351.
6. Kenyon, C. 2005. The plasticity of aging: insights from long-lived mutants. *Cell* **120:** 449–460.
7. Ayyadevara, S., R. Alla, J.J. Thaden & R.J. Shmookler Reis. 2008. Remarkable longevity and stress resistance of nematode PI3K-null mutants. *Aging Cell* **7:** 13–22.
8. Willcox, B.J. *et al.* 2008. FOXO3A genotype is strongly associated with human longevity. *Proc. Natl. Acad. Sci. USA* **105:** 13987–13992.
9. Anselmi, C.V. *et al.* 2009. Association of the FOXO3A locus with extreme longevity in a southern Italian centenarian study. *Rejuvenation Res.* **12:** 95–104.
10. Pawlikowska, L. *et al.* 2009. Association of common genetic variation in the insulin/IGF1 signaling pathway with human longevity. *Aging Cell* **8:** 460–472.
11. Flachsbart, F. *et al.* 2009. Association of FOXO3A variation with human longevity confirmed in German centenarians. *Proc. Natl. Acad. Sci. USA* **106:** 2700–2705.
12. Flatt, T. *et al.* 2008. Drosophila germ-line modulation of insulin signaling and lifespan. *Proc. Natl. Acad. Sci. USA* **105:** 6368–6373.
13. Hsin, H. & C. Kenyon. 1999. Signals from the reproductive system regulate the lifespan of *C. elegans*. *Nature* **399:** 362–366.

14. Arantes-Oliveira, N., J. Apfeld, A. Dillin & C. Kenyon. 2002. Regulation of life-span by germ-line stem cells in *Caenorhabditis elegans*. *Science* **295:** 502–505.
15. Partridge, L., D. Gems & D.J. Withers. 2005. Sex and death: what is the connection? *Cell* **120:** 461–472.
16. Gerisch, B. *et al*. 2001. A hormonal signaling pathway influencing *C. elegans* metabolism, reproductive development, and life span. *Dev. Cell* **1:** 841–851.
17. Gerisch, B. *et al*. 2007. A bile acid-like steroid modulates *Caenorhabditis elegans* lifespan through nuclear receptor signaling. *Proc. Natl. Acad. Sci. USA* **104:** 5014–5019.
18. Berman, J.R. & C. Kenyon. 2006. Germ-cell loss extends *C. elegans* life span through regulation of DAF-16 by kri-1 and lipophilic-hormone signaling. *Cell* **124:** 1055–1068.
19. Libina, N., J.R. Berman & C. Kenyon. 2003. Tissue-specific activities of the *C. elegans* DAF-16 protein in the regulation of lifespan. *Cell* **115:** 489–502.
20. Murphy, C.T. *et al*. 2003. Genes that act downstream of DAF-16 to influence the lifespan of *Caenorhabditis elegans*. *Nature* **424:** 277–283.
21. Bluher, M., B.B. Kahn & C.R. Kahn. 2003. Extended longevity in mice lacking the insulin receptor in adipose tissue. *Science* **299:** 572–574.
22. Wang, M.C., E.J. O'Rourke & G. Ruvkun. 2008. Fat metabolism links germline stem cells and longevity in *C. elegans*. *Science* **322:** 957–960.
23. Henderson, S.T. & T.E. Johnson. 2001. daf-16 integrates developmental and environmental inputs to mediate aging in the nematode *Caenorhabditis elegans*. *Curr. Biol.* **11:** 1975–1980.
24. Lin, K., H. Hsin, L. Libina & C. Kenyon. 2001. Regulation of the *Caenorhabditis elegans* longevity protein DAF-16 by insulin/IGF-1 and germline signaling. *Nat. Genet.* **28:** 139–145.
25. Lee, R.Y., J. Hench & G. Ruvkun. 2001. Regulation of *C. elegans* DAF-16 and its human ortholog FKHRL1 by the daf-2 insulin-like signaling pathway. *Curr. Biol.* **11:** 1950–1957.
26. Gems, D. *et al*. 1998. Two pleiotropic classes of daf-2 mutation affect larval arrest, adult behavior, reproduction and longevity in *Caenorhabditis elegans*. *Genetics* **150:** 129–155.
27. Ghazi, A. & C. Kenyon. 2009. A transcription elongation factor that links signals from the reproductive system to lifespan extension in *C. elegans*. *PLoS Genet.* **5:** e1000639. Epub 2009 Sep 11. PMID: 19749979.
28. Yamawaki, T.M. *et al*. 2008. Distinct activities of the germline and somatic reproductive tissues in the regulation of *Caenorhabditis elegans* longevity. *Genetics* **178:** 513–526.
29. Barnes, A.I. *et al*. 2006. No extension of lifespan by ablation of germ line in Drosophila. *Proc. Biol. Sci.* **273:** 939–947.
30. Cargill, S.L., J.R. Carey, H.G. Müller & G. Anderson. 2003. Age of ovary determines remaining life expectancy in old ovariectomized mice. *Aging Cell* **2:** 185–190.
31. Mason, J.B., S.L. Cargill, G.B. Anderson & J.R. Carey. 2009. Transplantation of young ovaries to old mice increased life span in transplant recipients. *J. Gerontol. A Biol. Sci. Med. Sci.* **64:** 1207–1211.
32. Parker, W.H. *et al*. 2009. Ovarian conservation at the time of hysterectomy and long-term health outcomes in the nurses' health study. *Obstet. Gynecol.* **113:** 1027–1037.

ANNALS OF THE NEW YORK ACADEMY OF SCIENCES
Issue: *Reproductive Aging*

Health consequences of reproductive aging: a commentary

Siobán D. Harlow

Department of Epidemiology, University of Michigan, Ann Arbor, Michigan

Address for correspondence: Siobán D. Harlow, Department of Epidemiology, University of Michigan, 109 Observatory Street, Ann Arbor, Michigan 48104. harlow@umich.edu

This commentary discusses the intersection of human ovarian and somatic aging. It argues for re-contextualizing estrogen's role in and impact on ovarian aging and, more broadly, on women's health, considering in particular the importance of timing, dose, and the broader endocrine milieu. Distinguishing between current clinical needs and optimizing women's future options, the paper outlines an approach to broadening the research agenda to better understand the role of ovarian aging in supporting the metabolic demands of longevity. Three overarching issues important to consider explicitly as we pursue research on the health correlates of reproductive aging are discussed, including implications of a lifespan approach, population diversity, and selection bias.

Keywords: ovarian aging; somatic aging; women's health

The papers in this section address the intersection of human ovarian and somatic aging and discuss emerging evidence that suggests a need to re-contextualize estrogen's role in and impact on ovarian aging and, more broadly, women's health. As articulated in these papers—timing matters, dose matters, and the broader endocrine milieu matters. It also matters whether our goal is to address current health needs with currently available options, for example, the problem that confronts physicians endeavoring to alleviate a woman's suffering with currently available options, or to optimize the options women may have in 10 or 20 years time.

To achieve the latter goal of optimizing future options, more explicit research efforts are needed:

(1) To develop an integrative understanding of the role of estrogen in physiology and health at different points across the life course;
(2) To place the physiologic impact of estrogen in the context of other critical and bioactive reproductive hormones (e.g., follicle-stimulating hormone [FSH]);
(3) To weigh estrogen's importance relative to other aspects of physiology and in the context of environmental insults.

This commentary reflects on the contributions of three of the workshop presentations published here as well as on comments made in other presentations during this workshop session. I elaborate on three overarching issues important to consider more explicitly as we pursue research on the health correlates of reproductive aging—implications of a lifespan approach, population diversity, and selection bias.

An important question raised in the context of this session was whether an adaptive role of a low estrogen state might be articulated for the postmenopausal period as has been proposed for the period of lactation.[1,2] For example, a biologic advantage has been proposed for hot flashes that benefits the newborn and for activation of the hypothalamic–pituitary–ovarian (HPO) axis to attenuate maternal stress responses and enhance metabolic and immune responses.[1–3] Presenters questioned, however, whether the necessary compensatory mechanisms are established for a low estrogen state in the postmenopause to offset the negative consequences of loss of bone and increases in arterial tone, C-CRP, and LDL-C, etc. This excellent question requires more exploration.

Aging ova and the known exponentiation of risk of chromosomal disruptions in pregnancies after age 40[4,5] may be the most parsimonious biologic explanation for ovarian senescence. Selection factors that enhance longevity may well not have ovarian origins. Alternatively, ovarian senescence may promote healthy aging by reducing the energy demands of repeated pregnancy and continued menstruation.

Workshop presenters argue that a low estrogen state is also adaptive in prolonging birth intervals and reducing fertility during periods of food scarcity. I noted with particular interest graphs reproduced from Emory Thompson et al.[6] comparing fertility and survivorship in various populations and species. These graphs suggested that the !Kung, with the narrowest fertility curve and limited reproduction after age 35, experience a survivorship of 50% at age 60–64 and of 20% at age 70, which is remarkable in a subsistence population. We might wish to consider further what this relationship suggests about environmentally moderated reproductive function and healthy aging.

Although not an expert on the evolutionary biology of aging literature, I would nonetheless ask whether we adequately understand the metabolic demands of longevity or the adaptive immune responses that promote longevity. From an evolutionary perspective we might inquire about the metabolic demands of taking on increased responsibility as a gatherer,[7] how stamina or strength is best preserved, especially in the context of aging physiologic systems, and whether the relative increase in androgens, in fact, promotes survival.

The controversy in women's health regarding whether the postmenopausal low estrogen state should be construed as a deficiency state or an adaptive response to the demands of aging was highlighted by an analogy, and the critique of that analogy, made during the workshop between menopausal women and the witch in Hansel and Gretel. It was suggested that the physique of the old crone may be easily explained when framed through the lens of estrogen deprivation. Reconsidering this analogy, we are obliged to consider what alternative, perhaps environmental, explanations might be provided for this old woman's condition. Is her frailty fully determined by her postmenopausal state or might somatic aging contribute? The story, being set in the context of famine and dire poverty, suggests a potential role for malnutrition while vitamin D deficiency might also play a role given a life spent residing within the deep dark wood. The witch is, after all, a hardy survivor—the children's mother is dead (perhaps a maternal death) and their stepmother too does not survive the end of the story.

Alternatively, it may prove more productive to consider the differences between applying a construct of adaptation versus adjustment. If we can accept a value in limiting the reproductive period and the consequent necessity of a transition in the reproductive hormone milieu, how do we conceptualize a period of adjustment and what do we know about generating the optimal postreproductive physiologic state. The portrayal of aging as a state of menopausal estrogen deprivation can not in the end suffice, because the postmenopausal state is universal for all surviving women, yet decrepitude is not, and a place must be created for incorporating the fact of healthy aging. We have, perhaps, in much of our research failed to properly delineate bothersome symptoms of transition from chronic morbidity. Using the analogy of the stages of grief, we may need to consider more thoughtfully what constitutes normative adjustment versus an unresolved or morbid response.

Clinicians confront the suffering of a specific portion of the postmenopausal population. Hormone therapy (HT) helps ameliorate suffering. Yet the patient represents just one, relatively small, portion of the population distribution. Our scientific task is to consider more rigorously what combination of baseline risk, environmental modifiers, and patterns of reproductive aging are most indicative of risk.

One important lesson of the paradigm shift in the post-Women's Health Initiative (WHI) era[8] is the importance of adopting a life course perspective and of balancing multiple competing risks within populations and individuals. Pinkerton and Stovall[9] consider in careful detail the emerging clinical evidence for a differential impact of HT exposure given age at initiation and underlying disease status, particularly in the context of cardiovascular disease and dementia. The WHI and other data presented in this paper suggest the possibility of critical windows for intervention. Data emerging from the longitudinal cohort studies of midlife women, discussed by Ferrell and Sowers,[10] also suggest that the midlife represents a critical period, setting the stage for women's long-term health across multiple dimensions. Traub and Santoro[11] review the complex interrelationships that occur in the aging process between changes in the HPO axis and the somatotrophic axis. What remains unclear is the extent to which physiologic changes leading to cardiovascular disease, for example, originate in ovarian senescence or simply track with it. Further conceptualization is needed to articulate the specific meaning of a critical window as a period of adjustment, adaptation, or onset of decline. Is it that women must optimize opportunities to safely traverse this period of adjustment, or is it a window after which there is no safe haven? In either case, the unanswered questions include: how early might intervention be warranted and in whom?; what are the critical markers of preexisting abnormality?; and what might the risks of intervention be to women with a preexisting condition?

The epidemiologic literature generally supports an association between earlier age at menopause and mortality (all-cause and cardiovascular disease),[12–16] with some evidence supporting an association also with cardiovascular disease incidence.[17,18] However, these data are not fully consistent regarding the relative importance of premature menopause and surgical menopause in defining these risks. The extent to which risk in the latter subgroup is a consequence of early and/or abrupt ovarian senescence versus a consequence of the underlying health issue that led to the premature/early menopause/hysterectomy is underinvestigated and not resolved. Similarly, our understanding of the role of poor health in advancing age at menopause in general remains inadequate. Yet, evidence continues to build suggesting that poor health (diabetes, depression, heart disease) is associated with, or leads to, earlier age at menopause.[19–23] Furthermore, several factors including body mass index, smoking, and socioeconomic status, are common correlates of menopause, menstrual characteristics across the lifespan, and mortality.[19,20,22,24–26] Addressing these interrelationships will be critical to evaluating the timing of risk onset and the extent to which ovarian senescence does, or does not, play a causal role is defining women's risk status.

Ferrell and Sowers[10] describe the longitudinal epidemiologic studies characterizing reproductive aging in population-based samples and discuss their contribution to our understanding of reproductive aging and ovarian senescence in population-based samples of women. These studies typically capture all or part of the lifespan relevant to the "critical window" hypothesis and provide insights into factors associated with population variability in reproductive aging and, frequently, the dynamic interface between reproductive and somatic aging.

Recent data from these studies document the relatively late occurrence of the decline in estrogen in relation to other endocrine changes characteristic of the reproductive aging process including decline in anti-mullerian hormone (AMH), decline in inhibin and rise in FSH. This fact suggests the potential for more precisely delineating the timing of specific physiologic changes of somatic aging in relation to specific physiologic changes of reproductive aging as well as their relevant endocrine correlates. These authors specifically point to the interdependence of reproductive and somatic aging, stressing the importance of teasing out the impact of aging (i.e., aging of the hypothalamic-pituitary axis) on signals classically defined as those of reproductive aging (e.g., FSH).

Traub and Santoro[11] summarize the emerging evidence of the role of estrogen in supporting the functional integrity of the somatotrophic axis during the menstrual cycle, its association with declines in insulin growth factor with aging and its relationship to cardiovascular risk and bone mineral density. They also summarize the evidence related to aging of the adrenal axis and its complex relationship to ovarian aging and chronic disease. They argue that low DHEAS levels have been associated with a range of health issues including cardiovascular disease, depression, and diabetes that previously have frequently been attributed to menopause and estrogen deficiency.

Of critical importance here is the need to establish:

(1) What role a woman's status at the onset of the menopausal transition plays in defining her risk profile;
(2) At what step in the reproductive aging process decrements in physiologic function are or *are not* initiated or accelerated; and
(3) Whether patterns or rates of change in reproductive function matter to health and well-being at the end of the transition.

Once these questions are answered, we may perhaps have an answer to the question "How important then is estrogen really?"

A lifespan perspective

The critical window hypothesis has focused attention on the importance of midlife to health later in life. Yet, the midlife spans about one-fifth of a long-lived individual's lifespan, and normal menopause itself may occur across a 20-year span from approximately age 40–60. There is also substantial variability in the duration of the menopausal transition, especially the early transition. Longitudinal studies of reproductive age and midlife women have documented that poor risk profiles for a range of conditions, including low bone mineral density, often are established prior to the onset of the menopausal transition.[26–28] We have yet to undertake studies with sufficient numbers of women enrolled in their peak reproductive or true late reproductive stage to provide a clear understanding of how baseline risk status segregates women's experience as they transition through the menopause or this presumed critical window. Thus, it is difficult to know whether the risk we observe at later life stages is dominated by baseline risk, for example, the failure to deposit adequate bone to weather the transition period. Most current studies enroll prevalent cohorts, that is women still menstruating, thus the at-risk women are frequently excluded.[29,30]

Although growing, the literature relating menstrual characteristics across the lifespan to other parameters of health and disease remains somewhat limited. Nonetheless, emerging data suggest that reproductive physiology tracks in particular ways across the life course and that the ovarian/somatic interlinkage also tracks in particular ways across the life course. For example,

(1) timing of menarche is associated with chronic disease risk but is not clearly correlated with peak reproductive menstrual characteristics[31] or with age at menopause;
(2) long-normal menstrual cycles (30–31 days) are associated with higher fecundity[32] and later age at menopause[31];
(3) menstrual cycle length and variability are associated with chronic disease risk[25]; and
(4) timing of menopause is associated with chronic disease risk and mortality.[13,14,16]

Several questions about lifelong trajectories and the interlinkage between reproductive function and health across these trajectories remain, including:

(1) careful specification of the timing of cardiovascular insults and explication of what determines the ability of the cardiovascular system to weather the endocrine shift that is characteristic of menopause;
(2) explication of the health benefits of being postmenopausal, for example in terms of energy conservation, the importance of androgens to maintaining musculoskeletal health and whether there may be an evolutionary advantage to observed shifts in metabolism and to metabolic syndrome;
(3) what the relative health advantage of an osteoporotic versus osteoarthritic profile may be.

Learning from population variability

To date, the primary focus of the longitudinal population cohort studies of midlife women has been to describe the typical or dominant pattern of reproductive aging. However, these studies have the potential to provide important information on common variants from this typical or dominant pattern, as well as on genetic, behavioral, and/or environmental factors that might be associated with the population distribution of variant patterns. Increasing scientific focus on common variants of the reproductive aging process may facilitate understanding of the biologic tradeoffs that accompany these different trajectories and may provide insights into differential patterns of disease risk.

Gorrindo et al.[33] have created an algorithm for typing women's lifetime menstrual histories into five categories: stable, stable but with greater variability, oscillating/erratic cycle lengths with a downward trend in cycle length over time, oscillating/erratic cycle lengths with a no downward trend in cycle length over time, and highly erratic. We are currently working on an approach for modeling cycle variability before and after the menopausal transition that allows for uncertainty in the typing of an individual woman's profile and permits us to relate profile types to covariates in the context of a regression model (Elliott, Huang, and Harlow, submitted for publication). Pinkerton and Stovall[9] recommend increased focus on the subpopulation of women with surgical menopause. Oligomenorrheic women are another subgroup that warrants further research focus. As we learn more about variability in the reproductive aging process, it may be possible to define important subgroups in relationship to the probability of healthy aging and to clarify what the evolutionary advantage might be for each different population subgroup.

Selection and bias

In evaluating studies, whether they involve clinical populations or longitudinal population based samples, increased attention to selection factors and estimating bias is warranted. As noted earlier, most longitudinal cohort studies enroll prevalent cohorts with eligibility defined by presence of menses, excluding (or truncating) women who have already achieved their final menstrual period (FMP) at the time of enrollment. We have shown through simulations that bias can be substantial and standard errors severely underestimated in a naïve analysis which ignores this left truncation and/or left censoring and that this bias increases with increasing age of the prevalent cohort (Cain *et al.*, submitted for publication). Research attempting to define critical thresholds for chronic disease risk may need to pay particular attention to issues of selection and bias if presence of underlying disease is associated with increased likelihood of being postmenopausal at a given age. Issues of generalizability, or lack thereof, may lead to mis-estimation of the association between reproductive and somatic aging. As noted by Ferrell and Sowers[10] most of the cohort studies have captured relatively few hysterectomized women or women who are early or late transitioners.

Additional selection issues can arise when data are analyzed prior to all members of the cohort achieving the FMP or other markers of the menopausal transition. Current estimates of the timing of AMH decline, for example, reflect results from the early completers in the studies from which they are derived. Thus, published estimates of average time to FMP following AMH decline may well change as more women in these studies reach transition and menopause.

In summary, the papers in this section have raised important questions about the interrelationships between ovarian and somatic aging. Although knowledge of these interrelationships has increased in the last decade, a fundamental set of questions remains unanswered, that is to what extent, or in what context, is

(1) ovarian aging *correlated* with somatic aging, that is, it is an early marker of aging and should inform timely interventions;

(2) ovarian aging a *trigger for* somatic aging, consistent with the critical window hypothesis, such that our clinical goal should be to prolong ovarian function; or,

(3) ovarian aging *triggered by* somatic aging or underlying disease such that our clinical goal should be to treat the somatic disease.

Conflicts of interest

The author declares no conflicts of interest.

References

1. Lightman, S.L. *et al.* 2001. Peripartum plasticity within the hypothalamo-pituitary-adrenal axis. *Prog. Brain Res.* **133:** 111–129.
2. Brunton, P.J., J.A. Russell & A.J. Douglas. 2008. Adaptive responses of the maternal hypothalamic-pituitary-adrenal axis during pregnancy and lactation. *J. Neuroendocrinol.* **20:** 764–776.
3. Mantzoros, C.S. 2000. Role of leptin in reproduction. *Ann. N.Y. Acad. Sci.* **900:** 174–183.
4. Pellestor, F., T. Anahory & S. Hamamah. 2005. Effect of maternal age on the frequency of cytogenetic abnormalities in human oocytes. *Cytogenet. Genome Res.* **111:** 206–212.
5. Hook, E.B. 1990. Chromosome abnormalities in older women by maternal age: evaluation of regression-derived rates in chorionic villus biopsy specimens. *Am. J. Med. Genet.* **35:** 184–187.
6. Emory Thompson, M. *et al.* 2007. Aging and fertility patterns in wild chimpanzees provide insights into the evolution of menopause. *Curr. Biol.* **17:** 2150–2156.
7. Kaplan, H. *et al.* 2010. Learning, menopause, and the human adaptive complex. *Ann. N.Y. Acad. Sci.* **1204:** 30–42.
8. Writing Group for the Women's Health Initiative Investigators. 2002. Risks and benefits of estrogen plus progestin in healthy post-menopausal women. *JAMA* **288:** 321–333.
9. Pinkerton, J.V. & D.W. Stovall. 2010. Reproductive aging, menopause, and health outcomes. *Ann. N.Y. Acad. Sci.* **1204:** 169–178.
10. Ferrell, R.J. & M.F. Sowers. 2010. Longitudinal epidemiologic studies of female reproductive aging. *Ann. N.Y. Acad. Sci.* **1204:** 188–197.
11. Traub, M.L. & N. Santoro. 2010. Reproductive aging and its consequences for general health. *Ann. N.Y. Acad. Sci.* **1204:** 179–187.
12. Snowdon, D.A. *et al.* 1989. Is early natural menopause a biologic marker of health and aging? *Am. J. Public Health* **79:** 709–714.
13. Jacobsen, B.K., I. Heuch & G. Kvåle. 2003. Age at natural menopause and all-cause mortality: a 37-year follow-up of 19,731 Norwegian women. *Am. J. Epidemiol.* **157:** 923–929.
14. Cooper, G.S. *et al.* 2000. Age at menopause and childbearing patterns in relation to mortality. *Am. J. Epidemiol.* **151:** 620–623.

15. Brett, K.M. & G.S. Cooper. 2003. Associations with menopause and menopausal transition in a nationally representative US sample. *Maturitas* **45:** 89–97.
16. Mondul, A.M., C. Rodriguez, E.J. Jacobs & E.E. Calle. 2005. Age at natural menopause and cause-specific mortality. *Am. J. Epidemiol.* **162:** 1089–1097.
17. Gordon, T., W.B. Kannel, M.C. Hjortland & P.M. McNamara. 1978. Menopause and coronary heart disease. *Ann. Intern. Med.* **89:** 157–161.
18. Colditz, G.A. *et al.* 1987. Menopause and the risk of coronary heart disease in women. *N. Engl. J. Med.* **316:** 1105–1110.
19. Bromberger, J.T. *et al.* 1997. Prospective study of the determinants of age at menopause. *Am. J. Epidemiol.* **145:** 124–133.
20. Gold, E.B. *et al.* 2001. Factors associated with age at natural menopause in a multiethnic sample of midlife women. *Am. J. Epidemiol.* **153:** 865–874.
21. Harlow, B.L., D.W. Cramer & K.M. Annis. 1995. Association of medically treated depression and age at natural menopause. *Am. J. Epidemiol.* **141:** 1170–1176.
22. Rödström, K. *et al.* 2003. Evidence for a secular trend in menopausal age: a population study of women in Gothenburg. *Menopause* **10:** 538–543.
23. Dorman, J.S. *et al.* 2001. Menopause in type 1 diabetic women: is it premature? *Diabetes* **50:** 1857–1862.
24. Willett, W. *et al.* 1983. Cigarette smoking, relative weight, and menopause. *Am. J. Epidemiol.* **117:** 651–658.
25. Harlow, S.D. & S.A. Ephross. 1995. Epidemiology of menstruation and its relevance to women's health. *Epidemiol. Rev.* **17:** 265–286.
26. Sowers, M.F, B. Shapiro, M.A. Gilbraith & M. Jannausch. 1990. Health and hormonal characteristics of premenopausal women with lower bone mass. *Calcif. Tissue Int.* **47:** 130–135.
27. MacMahon, B. *et al.* 1982. Age at menarche, probability of ovulation and breast cancer risk. *Int. J. Cancer* **29:** 13–16.
28. Bainbridge, K.E. *et al.* 2002. Natural history of bone loss over 6 years among premenopausal and early post-menopausal women. *Am. J. Epidemiol.* **156:** 410–417.
29. Dennerstein, L. *et al.* 2000. A prospective population-based study of menopausal symptoms. *Obstet. Gynecol.* **96:** 351–358.
30. Sowers, M.F. *et al.* 2000. Design, survey sampling and recruitment methods of SWAN: a multi-center, multi-ethnic, community-based cohort study of women and the menopausal transition. In *Menopause: Biology and Pathobiology*. Vol. 32. J. Wren, R.A. Lobo, J. Kelsey & R. Marcus, Eds.: 175–188. Academic Press. San Diego.
31. Lisabeth, L., S.D. Harlow & B. Qaqish. 2004. Marginal models for mean and variance: a new approach to modelling menstrual variation across the menopausal transition. *J. Clin. Epidemiol.* **57:** 484–496.
32. Small, C.M. *et al.* 2006. Menstrual cycle characteristics: associations with fertility and spontaneous abortion. *Epidemiology* **17:** 52–60.
33. Gorrindo, T. *et al.* 2007. Lifelong menstrual histories are typically erratic and trending: a taxonomy. *Menopause* **14:** 74–88.

Reproductive aging, menopause, and health outcomes

JoAnn V. Pinkerton[1] and Dale W. Stovall[2]

[1]Department of Obstetrics and Gynecology, Divisions of Midlife, University of Virginia Health System, Charlottesville, Virginia. [2]Reproductive Endocrinology, University of Virginia Health System, Charlottesville, Virginia

Address for correspondence: JoAnn V. Pinkerton, M.D., Box 801104, University of Virginia Health System, Charlottesville, Virginia 22908. jvp9u@virginia.edu

Changes in ovarian hormone production may affect numerous health outcomes including vasomotor symptoms, cardiovascular disease (CVD), osteoporosis, cognition, depression, mood disorders, sexual function, and vaginal atrophy. We will compare age-related changes to those associated with reproductive aging and menopause and the effects of estrogen therapy on selected health outcomes. Hormone therapy (HT) reduces frequency and severity of hot flashes, prevents bone loss and osteoporotic fractures, and relieves vaginal atrophy. Nonhormone therapy trials with antidepressants or gabapentin for hot flash relief are promising. To date, clinical trial data are insufficient to recommend the use of HT for prevention or treatment of CVD, mood disorders, cognition, or sleep disorders. For some disease states, such as CVD and cognition, a "critical time window" has been proposed but not proven, such that estrogen use early in the menopause transition may be beneficial while estrogen use later in life would lead to increased health risks.

Keywords: reproductive aging; menopause; osteoporosis; cognition; cardiovascular disease; vaginal atrophy

Introduction

Sorting out the effects of age from hormone changes associated with menopause is difficult because hormone events do not happen in isolation, although much has been learned regarding different hormone receptors and various organ systems. In this article, we will discuss reproductive aging and menopause with a focus on health outcomes, specifically looking at vasomotor symptoms (VMS), cardiovascular disease (CVD), osteoporosis, cognition, depression, mood disorders, sexual function, and vaginal atrophy. Data are presented on effects of hormone therapy (HT) including the proposed "critical time window" for CVD and cognition where proponents hypothesize that estrogen use early in the menopause transition may have beneficial health effects while estrogen later in life leads to increased health risks.

Defining menopause and stages of menopause

The perimenopause is a physiologic transition characterized by depletion of functional ovarian follicles with mean onset between 45.5 and 47.5 years. The average age of menopause is 51.8 years with variability by race and ethnicity.[1] As the ovarian follicular pool declines, inhibin B and anti-mullerian hormone (AMH) levels fall and follicle-stimulating hormone (FSH) levels rise, reflecting a loss of inhibin restraint with low AMH levels being most predictive of final menstrual period (FMP).[2] Elevated FSH is believed to maintain estradiol levels and sustain ovulation during the early menopausal transition. FSH levels fluctuate during the menopausal transition,[3] making it difficult to use a single FSH measurement to identify menopausal stage until after menopause has occurred. Estradiol levels decline in the months leading up to the FMP. The proportion of ovulatory cycles declines over the 6 years prior to the FMP from over 60% to approximately 10% in the last few months.[3]

The Straw Criteria developed at the Stages of Reproductive Aging Workshop (STRAW) is often used to classify stages of menopausal transition (Fig. 1).[4] In the early transition (stage-2) changes occur in the normal menstrual cycle of >7 days. By late transition (stage-1), two or more menstrual cycles have been missed with at least one intermenstrual

Table 1. Gaps in knowledge

1. Specific neurobiological pathways associated with VMS including gonadal steroids and neurotransmitters.
2. The etiology and long-term health effects of persistent vasomotor effects on bone density, vasculature, cardiovascular risk factors, and cognition.
3. Interaction of HT with age at natural or surgical menopause in light of the postulated critical window of estrogen therapy (ET) on CVD and cognition.
4. Effect of different progestogens on cardiovascular risk.
5. Effect of gonadal steroids on cognition for women with natural or surgical menopause.
6. Effect of ET on decreasing risks of bilateral oophorectomy before age 50.
7. Long-term consequences on cognition of pre- or postmenopausal bilateral oophorectomy.
8. Effect of estrogen agonists/antagonists with and without estrogen on cognition and CVD.
9. Efficacy and safety of long-term testosterone therapy on cardiovascular risk, sexuality, breast cancer, bone loss, and cognition.
10. If diminished ovarian reserve predicts lower BMD at menopause, how can this be determined and/or prevented?
11. In humans, the direct effects of FSH, inhibin B, or AMH on bone.
12. How does one select the timing and most effective therapy for osteoporotic fracture prevention?
13. The effects of antigonadotropin therapy on prevention of bone loss.
14. The minimal effective dose of vaginal estrogen to prevent or reverse genitourinary atrophy.
15. For hypoestrogenic women at risk for urinary tract infection (older, or women on aromatase inhibitors), the effect of topical estrogen on urethral atrophy.
16. Degree of systemic absorption of vaginal estrogen after atrophy improves and risk of stimulation of breast or endometrium.
17. In women utilizing aromatase inhibitors, how much systemic absorption of vaginal estrogen occurs and do the systemic levels decrease over time as the genitourinary atrophy is reversed?

---

interval of 60 days or more. The menopausal transition ends after 12 months of otherwise unexplained amenorrhea, followed by the postmenopause. Early postmenopause is defined as 5 years from the FMP with late postmenopausal years extending beyond those first 5 years.[4]

Menopausal symptoms

Significant differences have been found for menopausal symptoms across racial and ethnic groups.[5] These symptoms, which include hot flashes and night sweats (VMS), disturbances in mood, cognition, sleep, vaginal dryness, pain with intercourse or loss of libido, may affect up to 80% of women during the menopause transition and significantly affect quality-of-life.

The pathophysiology of hot flashes is not well understood. Hot flashes are generally mild and transient, although variable in frequency, severity, duration, and persistence. Hot flashes are associated with peripheral venous dilation with increased skin temperature and blood flow within the first few seconds of the flash. Sweating and skin conductance increase, triggered by small elevations in core body temperature.[6] While core temperature elevations occur in both symptomatic and asymptomatic women, the thermoneutral zone is narrowed in symptomatic women[7] by 0.4° C and is virtually absent (0° C) in those with severe symptoms. Estrogen levels are similar in symptomatic and asymptomatic women, however, estrogen withdrawal is associated with hot flashes and estrogen therapy (ET) reduces hot flashes.[8] The health significance of persistent postmenopausal VMS is unknown.

Elevated sympathetic activation and neurotransmitter levels of serotonin and norepinephrine associated with fluctuations in circulating estrogens may contribute to hot flushes.[6,7] Placebo controlled clinical trials for VMS treatment commonly demonstrate a placebo effect as high as 50%.[8] Nonhormone therapies may reduce hot flashes through effects on central neuromodulators.

Stages	-5	-4	-3	-2	-1	0	+1	+2
Terminology:	Reproductive			Menopausal Transition			Postmenopause	
	Early	Peak	Late	Early	Late*		Early*	Late
				Perimenopause				
Duration of Stage:	Variable			Variable			1 yr	4 yrs	Until Demise
Menstrual Cycles	Variable to regular	Regular		Variable cycle length (> 7 days different from normal)	≥ 2 skipped cycles and an interval of amenorrhea (≥ 60 days)	†Amenor 12 mos	None	
			Length decreases ~2 days					
Endocrine	Normal FSH		↑ FSH	↑ FSH			↑ FSH	

*Stages most likely to be characterized by vasomotor symptoms; †amenor = amenorrhea
↑ = Elevated

Figure 1. Stages of reproductive aging from the STRAW. Reproduced with permission from Ref. 4.

Hormone therapy

HT is currently the only FDA approved treatment of hot flashes with improvement in 75–80% of women.[8] The Women's Health Initiative (WHI) revealed that after 5 years of treatment with estrogen and progesterone (E+P) there is a slight increased risk for breast cancer, CVD, stroke and venous thromboembolic events (VTEs), and a decreased incidence of fractures and colon cancer.[9] The estrogen alone (E alone) arm showed increased risk of heart events, stroke, VTEs, and dementia with a decreased risk of breast cancer at 6.7 years.[10] The risk of breast cancer appeared to be associated with the duration and possibly type of exposure to estrogen and progestins; some studies have suggested less risk with progesterone.[11] In 2007, a WHI reanalysis by Rossouw et al.[12] including only those women under 60 and within 10 years of menopause found a decrease in mortality in women between ages 50 and 59. Women who began E alone within 10 years of menopause or before the age of 60 years achieved a cardioprotective effect. HT is recommended for the relief of VMS after risk and benefit analysis and with periodic reviews of the decision to continue therapy.[13] Extended use of E alone or E+P may be appropriate[13] after informed discussion for persistent VMS after failing an attempt to withdraw from HT, or for women with moderate to severe VMS at high risk for osteoporotic fracture or for whom alternative therapies are not appropriate. Lower than standard doses of estrogen and progestin have been found[14] to relieve VMS and vaginal atrophy, protect the endometrium, and prevent early postmenopausal bone loss[15] with less bleeding or breast tenderness.[16] Attempts over time should be made to lower the dose, discontinue HT, or introduce alternate therapies.

No persistent increased cardiovascular risks were found following discontinuation of E+P arm in the WHI trial (mean 2.4 years of follow-up).[17] However, a 12% higher global risk index was found with an increased risk of fatal and nonfatal malignancies (more lung cancers in E+P group) and a loss of fracture protection. The increased risk of breast cancer associated with the use of E+P declined markedly soon after discontinuation of combined HT.[18]

Nonhormone therapies

Promising nonhormone therapies for VMS evaluated in small randomized clinical trials (RCT) include paced respiration/relaxation, hypnosis, yoga, and acupuncture. Neither homeopathic remedies, magnet therapy, nor foot reflexology out-performed placebo for relief of VMS. Limited data from RCTs suggest that dong quai, evening primrose oil, ginseng extract, Kava Kava, and vitamin E are either ineffective or not clinically significant in relieving VMS. Possibly efficacious remedies include phytoestrogens, black cohosh, and newer herbal preparations. RCTs evaluating neuroactive agents including selective serotonin reuptake inhibitors (fluoxetine, paroxetine, citalopram, mirtazapine), serotonin norepinephrine reuptake inhibitor (venlafaxine, desvenlafaxine), and gabapentin demonstrate that increased serotonergic activity within the central nervous system reduces hot flashes greater than placebo. Although none of these treatments approaches the effectiveness of ET, they may be an alternative for women unable or unwilling to use HT for the management of hot flashes.[19]

Reproductive aging, menopause, and CVD

Women have a lower incidence of CVD compared to men until about 10 years after menopause, when the risk approaches that of men. This is thought to be due to the beneficial effects of premenopausal estrogen on the arterial and cardiac endothelium and musculature.[20] With menopause and declining estrogen levels, metabolic changes occur that increase risk of CVD, including changes in lipid profiles.[20] Elevated triglycerides seen after menopause are an independent CVD risk factor for women.[21] Following menopause, insulin sensitivity declines and a "metabolic syndrome"[22] may develop with an increase in android fat deposition in the central abdominal region and a relative decrease in the gynoid fat deposition around the hips and thighs with associated insulin resistance, hyperinsulinemia, and elevated serum free fatty acids.

HT and CVD

Women experiencing premature menopause have a two- to threefold increase in coronary heart disease (CHD) risk. Estrogens exert atherosclerosis protective effects via effects on lipids, nitric oxide, and direct effects on vasculature.[23] Over 40 observational studies of HT and CHD suggested that the average relative risk for CHD was 40–50% lower among current or previous users of HT compared to never users ($P < 0.001$).[24] However, large RCTs, such as the Heart and Estrogen/Progestin Replacement Study (HERS) trial, showed that women with CHD experienced an increased risk for cardiac events if they took HT. A 2005 Cochrane review, which included WHI and HERS,[25] showed no benefit in either secondary or primary prevention of CVD events.

The timing of initiation of HT may play a role in the development of CVD. A study in macaque monkeys found the protective effect of ET was lost if started more than 2 years after menopause. In women approximately 10–20 years past menopause, the benefits of estrogen on the vasculature appear to be lost as older vessels have reduced vascular responsiveness and lower expression of estrogen receptors.[26] Once complex atherosclerotic plaques are present, the addition of estrogen has been shown to increase matrix metalloproteinases expression, which leads to plaque instability and rupture, thereby increasing the risk of heart attack and stroke.[26]

In 2007, the WHI reanalysis of E alone[12] revealed a significant reduction was found in the risk of heart attack, coronary bypass, or stent for those aged 50–59 at initiation of the study, with no benefit for those aged 60–69 or 70–79. In the E+P arm, there was a trend to reduction of CHD risk for those aged 50–59 whereas harm was seen for those aged 70–79. The "critical time window" hypothesis suggests that estrogen may be beneficial in younger menopausal women but not in older with a "window of opportunity" if HT is initiated within 5–6 years of menopause.[21] Following closure of the WHI trial in 2002, women aged 50–59 years who continued ET were found to have markedly reduced calcification scores compared with women taking placebo (83.1 versus 123.1; $P = 0.02$), indicating that estrogen begun early and continued for at least 8 years had a beneficial effect on coronary vessels with lower levels of coronary artery calcium by computer tomography scanning.[27] In 2009, Prentice's et al.[28] reanalysis of the WHI did not support the "critical window" hypothesis as similar risks for all ages were found for VTE, stroke, and hip fracture, with elevations in breast cancer, total cancer, and global index.

Effects of bilateral oophorectomy on morbidity and mortality

In 2009, Parker et al.[29] reported after 24 years of follow-up from the Nurses Health Study that bilateral oophorectomy at hysterectomy for benign disease without ET, compared to ovarian conservation, was associated with a decreased risk of breast and ovarian cancer but an increased risk of all-cause mortality, fatal and nonfatal CHD, and lung cancer. Among 13,305 women with ovarian conservation, 34 (0.26%) died from ovarian cancer. Prophylactic oophorectomy did not improve survival at any age; there were fewer breast, ovarian, and total cancers but overall risk of cancer death was greater among those with oophorectomy.[29] Parker et al. found, in a subset analysis of never users of ET, that bilateral oophorectomy < age 50 was associated with an increased risk of all-cause mortality, fatal and nonfatal CHD, and stroke with no difference in total cancer risk. On the basis of an approximate 35-year lifespan after surgery, Parker *et al.* calculated one additional death for every nine oophorectomies performed.[29] Women with hysterectomy for benign disease at average risk of ovarian cancer were calculated to benefit from ovarian conservation until at least age 65.[29]

Reproductive aging, menopause, and osteoporosis

The lifetime probability of an osteoporotic-associated fracture in women at the vertebral body, distal radius, or hip is approximately 30%.[30] However, it remains difficult to identify those women at highest risk of fracture to determine when and how to target treatments. Beginning before menopause and for first few years after menopause, a phase of accelerated bone loss occurs resulting in the loss of up to 5% of trabecular bone per year, and 2–3% of cortical bone per year compared to normal loss of 1–3% per year.[30] This accelerated loss in bone density is primarily the result of an increase in the number and activity of the osteoclastic cells, which absorb bone, and are inhibited by estrogen.[30] Later, osteoclastic activity declines and the rate of bone loss slows to 1–2% or less per year.[30] Premenopausal women with declining ovarian reserve and those with VMS have been shown to have increased bone turnover and bone loss; this has been seen in Caucasians, African–American, and Asian women.[31] Postmenopausal women with osteopenia accounted for half of the fractures observed in the National Osteoporosis Risk Assessment (NORA) study.[32] Modifiable risk factors for increased bone loss include estrogen deficiency, low vitamin D levels, low calcium intake, sedentary lifestyle, cigarette or alcohol use, and medications associated with bone loss.[33]

HT and bone

Estrogen acts directly on estrogen receptors in bone, reducing bone turnover and bone loss. The WHI E+P and the E alone studies confirmed a fracture-preventing role of estrogens with a 34% reduction in hip fractures among the 8506 women receiving estrogen and progestin,[9] a 35% relative risk reduction (RRR) in vertebral fractures, a 29% RRR in wrist fractures, and a 24% RRR in total fractures.[34] In a systematic review of RCT,[35] ET for an average of 6.2 years reduced incident fractures by 52% (95% CI 18–64%). Lower than standard doses prevent bone loss and bone turnover with milder effects on bone density if given with adequate calcium and vitamin D, although fracture efficacy data are not available.[15] Discontinuation of HT leads to rapid bone loss of 3–6% during the first year and loss of fracture protection.[17]

Reproductive aging, menopause, and cognition

A decline in memory and cognitive function is considered a consequence of aging. Cognitive performance normally remains stable over many years, with only slight declines in short-term memory and reaction times.[36] Memory disturbance is one of the most reliable indicators of progression from age-associated changes in cognition to dementia, such as Alzheimer's disease (AD).

Although subjective memory complaints are common around the time of natural menopause, possibly related to sleep deprivation or stress, it is difficult to document objective memory deficits. Henderson and Sherwin[37,38] found that the natural menopausal transition was not accompanied by substantial changes in cognitive ability; however, surgical menopause was accompanied by cognitive impairment particularly verbal episodic memory. In contrast, Vearncombe and Pachana[39] found that surgical menopause was associated with specific deficits in visual and verbal memory and verbal fluency domains in small RCTs, but larger RCTs generally found no effect of surgical menopause on cognitive functioning. Rocca *et al.*[40] reported from the Mayo cohort Study of Oophorectomy and Aging, that bilateral oophorectomy prior to natural menopause was associated with an increased risk of Parkinsonism, cognitive impairment, dementia, and depression and anxiety. For those who were younger at the time of surgery or who discontinued estrogen prior to age 50, there was an increased risk of dementia.[41]

HT and cognition

Loss of ovarian hormones after menopause has been hypothesized to contribute to cognitive aging and dementia. Studies on neurons[42] have shown that estrogen initiated while the neurons are still healthy, prior to insults, such as free radicals, β-amyloid, or ischemia, helps preserve neuron function and increases neurons with generation of ATP (energy) in mitochondria of neurons. However, if diseased or abnormal neurons are given estrogen, estrogen increases neuronal degeneration.

In regards to premature surgical menopause (bilateral oophorectomy), evidence from short-term RCTs suggested that HT might benefit verbal memory. Meta-analyses of observational studies on HT[43] implied up to a 44% decrease of AD risk in HT users compared with nonusers with a lower risk of AD with longer duration of HT and a delay in the onset of AD among hormone users who go on to develop dementia. However, in the WHI Memory Study (WHIMS) women 65 or older who took either E alone or E+P for 5 years had an increased risk of developing dementia with an additional 23 cases of dementia per 10,000 women per year.[44] Brain magnetic resonance imaging scans obtained in a subset of 1403 WHIMS women aged 71–89 years[45] showed that conjugated equine estrogen with or without medroxyprogesterone acetate was associated with greater brain atrophy among women 65 years and older with adverse effects most evident in women experiencing cognitive deficits before initiating HT.

The positive observational data demonstrating a reduction in the risk of AD in women who began estrogen or HT at the time of the menopause support a "critical window" hypothesis[38,46] that estrogen begun later in menopause appears detrimental, whereas early initiation of estrogen may reduce dementia risk. However, design issues complicate the observational studies and supporting RCT evidence for the "critical window" in humans is currently lacking.

Reproductive aging, menopause, and mood disorders

Longitudinal studies suggest that hormonal changes related to the menopause transition may play a role in the onset of depression.[47] Risk factors for developing depression in menopause include prior episodes of depression, stress, body mass index, lack of live births, history of postpartum depression, family history of depression, history of sexual abuse, and presence of hot flashes.[48,49] Women with ongoing depressive symptoms despite being on antidepressants were most likely to enter menopause early (RR = 2.7).[48] In the presence of VMS, menopausal women have been found to have lower sleep efficiency, more sleep complaints, higher insomnia, and rates of depression.[50]

Estrogen therapy

ET has not been shown to prevent or treat depression in asymptomatic (without hot flashes) menopausal women. Data from the HERS trial suggested that for symptomatic postmenopausal women, HT was associated with lower depressive symptoms.[51] Two small RCT demonstrated that transdermal estradiol had antidepressant efficacy.[52] Compelling evidence is lacking regarding the antidepressant efficacy of HT alone. Beneficial effects of HT may be more likely to occur in women with concurrent menopausal and depressive symptoms.

Reproductive aging: sexual functioning and dysfunction

Sexual function is affected by both age and menopause.[53] Decreased interest (hypoactive desire), responsiveness, frequency, or change in subjective feelings for partner are common complaints of menopausal women. Vaginal dryness and pain with intercourse increase over the menopausal transition[54] with self-reported sexual dysfunction increasing from 42% in the early menopausal transition to 88% after 8 years. Dennerstein et al.[54] found a decline in sexual function with age and duration of relationship, with natural menopause creating an additional negative effect linked to declining levels of estradiol but not to declining androgens. Cross-sectional and longitudinal studies have suggested that bilateral oophorectomy has a greater effect on sexual functioning with lower sexual desire and less coital pleasure than hysterectomy alone felt to be due to combined loss of estrogen and testosterone.[55,56]

Hypoactive sexual desire disorder (HSDD) includes an absence of sexual thoughts and fantasies, decreased attraction to others, and rare initiation of sexual activity.[57] In the Women's International Study of Health and Sexuality (WISHeS) study, HSDD was associated with significant psychological, emotional, and interpersonal distress; decreased sexual and partner satisfaction; and decreased physical and mental health status.[58]

Estrogen

The interplay between hormonal factors and sexual function at menopause is complex. Oral estrogens improved sexual desire and arousal, despite a negative impact on available testosterone that results from the induction of sex hormone-binding globulin production, which lowers circulating levels of free testosterone up to 42%.[59] Estradiol given transdermally to postmenopausal women up to 24 months significantly improved satisfaction scores for sexual activity, sexual fantasies, degree of sexual enjoyment, vaginal lubrication, and frequency of sexual activity.[60] Local vaginal estrogen improves urogenital atrophy, increases blood flow, and sexual response.[61]

Androgens

Serum androgen levels decline steeply in the early reproductive years prior to menopause leading to circulating testosterone and free testosterone levels nearly 50% lower in postmenopausal women.[56] The postmenopausal ovary appears to continue to produce testosterone.[56] Sexual function improvements[62] or reductions[54] do not appear linearly related to serum testosterone levels. Measuring serum testosterone levels has not been found to be of clinical value in sexual disorders.[63]

Surgical menopause (bilateral oophorectomy) pre- or postmenopausally is associated with a rapid decline (up to 50%) in testosterone production.[56] RCTs have demonstrated the efficacy of transdermal testosterone 300 mcg/day for treatment of HSDD in both surgically and naturally menopausal women[62,64,65] with increases in sexual desire and in frequency of satisfying sexual activity.[64] Side effects of testosterone therapy include clitoromegaly, reversible hirsutism and acne, and irreversible deepening of the voice. Safety concerns include unclear effects on CVD and potential effects on the breast and ovary.

Vaginal atrophy

Aging and progressive declines in estrogen after menopause lead to symptoms of vaginal dryness in up to 50% of postmenopausal women, with dyspareunia, postcoital bleeding, vaginismus, and vulvar pruritus.[61] Decreases in blood flow lead to a paler, less well-nourished epithelium with decreased collagen and elastic fibers and eventually introital stenosis. Over time, the vaginal epithelium becomes thinner and more fragile with development of superficial vulvovaginal fissures and vaginal epithelial petecheia, increases in vaginal pH from 3.8 to 4.5 to 6.0–7.0, vulnerability to opportunistic organisms, and eventually a rigid and contracted vagina.[66]

Vaginal health is sustained with higher circulating levels of androgens and in women who are frequently sexually active, the latter thought to be due to increased blood flow with sexual arousal, continued mechanical distention, and possibly benefits from the seminal fluid itself. Regular use of vaginal moisturizers along with water soluble vaginal lubricants are effective at decreasing friction during sexual activity and improve dyspareunia and vaginal dryness up to 60%.[67] The moisturizers adhere to the vaginal epithelium and retain moisture, but can not reverse atrophic changes and are not as effective as vaginal estrogens.

Estrogen therapy

Atrophic vaginitis is most effectively treated with the use of exogenous estrogen. A Cochrane review[68] ($N = 2129$) found that all formulations of local estrogen, cream, ring, or low-dose tablets, were more effective than placebo or nonhormone gels in relieving symptoms of dryness, pruritus, and burning, and in improving scores for vaginal moisture, vaginal pH, vaginal elasticity, and fluid volume. Systemic low doses of estrogen, CE 0.3 mg daily and transdermal doses of 12.5 mcg daily, showed improvement in vaginal maturation index and resolved symptoms. Local vaginal ET has been shown to reduce the frequency of urinary tract infections in menopausal women.[69] Improvement in vulvovaginal health is noted within a few weeks of beginning therapy, although longer therapy is needed to maintain significant improvement.

Concern about endometrium and breast

Low-dose vaginal estrogen preparations[70] appear to have a safe endometrial profile, but vary in their ability to stimulate the endometrium.[71] In the Cochrane review, no cases of endometrial cancer were reported, with rare cases of endometrial hyperplasia.[69] Although the risk of endometrial stimulation with vaginal administration of unopposed estrogen in women with an intact uterus appears to be minimal, low doses of vaginal estrogen could lead to endometrial proliferation.[72] There are currently no evidence-based recommendations for endometrial monitoring or progestin dosing for women using unopposed low-dose vaginal ET. Reports are conflicting regarding breast safety. Rosenberg et al.[73]

found no association between local low potency ET and either ductal or lobular breast cancer; however, Kendall et al.[74] found that the vaginal estradiol tablet significantly increased systemic estradiol levels, which raises concern in regards to this effect on estradiol suppression achieved by aromatase inhibitors in women with breast cancer.

Summary

Improved biological understanding is needed to understand health risks associated with reproductive aging. HT has been shown to relieve VMS, prevent bone loss and fractures and relieve vaginal atrophy, however its efficacy for prevention or treatment of CVD, mood, cognition, or sleep disorders has not been proven. A "critical time window" of beneficial effects of estrogen use early in the menopause transition has been postulated but not proven. Research is needed to better understand the etiology of hot flashes and the usefulness of nonhormone treatments as well as the efficacy, safety and timing of estrogen, progestogen, and androgen therapies for postmenopausal health. Understanding the neurobiological pathways of ovarian failure may lead to safer therapies.

Conflicts of interest

The authors declare no conflicts of interest.

References

1. Randolph, J.F. et al. 2004. Change in estradiol and follicle-stimulating hormone across the early menopausal transition: effects of ethnicity and age. *J. Clin. Endocrinol. Metab.* **89:** 1555–1561.
2. Sowers, M.R. et al. 2008. Anti-mullerian hormone and inhibin B in the definition of ovarian aging and the menopause transition. *J. Clin. Endocrinol. Metab.* **93:** 3478–3483.
3. Burger, H.G., G.E. Hale, L. Dennerstein & D.M. Robertson. 2008. Cycle and hormone changes during perimenopause: the key role of ovarian function. *Menopause* **15:** 603–612.
4. Soules, M.R. et al. 2001. Executive summary: Stages of Reproductive Aging Workshop (STRAW) Park City, Utah, July, 2001. *Menopause* **8:** 402–407.
5. Gold, E.B. et al. 2006. Longitudinal analysis of the association between vasomotor symptoms and race/ethnicity across the menopausal transition: study of women's health across the nation. *Am. J. Public Health* **96:** 1226–1235.
6. Freedman, R.R. 2005. Pathophysiology and treatment of menopausal hot flashes. *Semin. Reprod. Med.* **23:** 117–125.
7. Freedman, R.R. & M. Subramanian. 2005. Effects of symptomatic status and the menstrual cycle on hot flash-related thermoregulatory parameters. *Menopause* **12:** 156–159.
8. Maclennan, A.H., J.L. Broadbent, S. Lester & V. Moore. 2004. Oral oestrogen and combined oestrogen/progestogen therapy versus placebo for hot flushes.[update of Cochrane Database Syst Rev. 2001;(1):CD002978; PMID: 11279791]. [Review] [113 refs]. *Cochrane Database Syst. Rev.* **4:** 002978.
9. Rossouw, J.E. et al. 2002. Risks and benefits of estrogen plus progestin in healthy postmenopausal women: principal results From the Women's Health Initiative randomized controlled trial. *JAMA* **288:** 321–333.
10. Anderson, G.L. et al. 2004. Effects of conjugated equine estrogen in postmenopausal women with hysterectomy: the Women's Health Initiative randomized controlled trial. [see comment]. *JAMA* **291:** 1701–1712.
11. Fournier, A., F. Berrino & F. Clavel-Chapelon. 2008. Unequal risks for breast cancer associated with different hormone replacement therapies: results from the E3N cohort study. *Breast Cancer Res. Treat.* **107:** 103–111.
12. Rossouw, J.E. et al. 2007. Postmenopausal hormone therapy and risk of cardiovascular disease by age and years since menopause. *JAMA* **297:** 1465–1477.
13. Utian, W.H. et al. 2008. Estrogen and progestogen use in postmenopausal women: July 2008 position statement of The North American Menopause Society. *Menopause* **15:** 584–602.
14. Utian, W.H. et al. 2001. Relief of vasomotor symptoms and vaginal atrophy with lower doses of conjugated equine estrogens and medroxyprogesterone acetate. *Fertil. Steril.* **75:** 1065–1079.
15. Lindsay, R., J.C. Gallagher, M. Kleerekoper & J.H. Pickar. 2002. Effect of lower doses of conjugated equine estrogens with and without medroxyprogesterone acetate on bone in early postmenopausal women. *JAMA* **287:** 2668–2676.
16. Ettinger, B. 2005. Vasomotor symptom relief versus unwanted effects: role of estrogen dosage. *Am. J. Med.* **118:** 74S–78S.
17. Heiss, G. et al. 2008. Health risks and benefits 3 years after stopping randomized treatment with estrogen and progestin. *JAMA* **299:** 1036–1045.
18. Chlebowski, R.T. et al. 2009. Breast cancer after use of estrogen plus progestin in postmenopausal women. *N. Engl. J. Med.* **360:** 573–587.
19. Loprinzi, C.L. et al. 2009. Newer antidepressants and gabapentin for hot flashes: an individual patient pooled analysis. *J. Clin. Oncol.* **17:** 2831–2837.
20. Mosca, L. 2000. The role of hormone replacement therapy in the prevention of postmenopausal heart disease. *Arch. Intern. Med.* **160:** 2263–2272.
21. Stevenson, J.C. 2009. HRT and cardiovascular disease. *Best Pract. Res. Clin. Obstet. Gynaecol.* **23:** 109–120.
22. Spencer, C.P., I.F. Godsland & J.C. Stevenson. 1997. Is there a menopausal metabolic syndrome? *Gynecol. Endocrinol.* **11:** 341–355.
23. Mendelsohn, M.E. & R.H. Karas. 2005. Molecular and cellular basis of cardiovascular gender differences. *Science* **308:** 1583–1587.
24. Grodstein, F., J.E. Manson & M.J. Stampfer. 2006. Hormone therapy and coronary heart disease: the role of time since menopause and age at hormone initiation. *J. Womens Health (Larchmt)* **15:** 35–44.

25. Gabriel, S.R. et al. 2005. Hormone replacement therapy for preventing cardiovascular disease in post-menopausal women. *Cochrane Database Syst. Rev.* (2): CD002229.
26. Clarkson, T.B. & S.E. Appt. 2005. Controversies about HRT–lessons from monkey models. *Maturitas* **51:** 64–74.
27. Manson, J.E. et al. 2007. Estrogen therapy and coronary-artery calcification. *N. Engl. J. Med.* **356:** 2591–2602.
28. Prentice, R.L. et al. 2009. Benefits and risks of postmenopausal hormone therapy when it is initiated soon after menopause. *Am. J. Epidemiol.* **1:** 12–23.
29. Parker, W.H. et al. 2009. Ovarian conservation at the time of hysterectomy and long-term health outcomes in the Nurses' Health Study. *Obstet. Gynecol.* **113:** 1027–1037.
30. Gallagher, J.C. 2007. Effect of early menopause on bone mineral density and fractures. *Menopause* **14:** 567–571.
31. Popat, V.B. et al. 2009. Bone mineral density in estrogen deficient young women. *J. Clin. Endocrinol. Metab.* **7:** 2277–2283.
32. Siris, E.S. et al. 2004. Predictive value of low BMD for 1-year fracture outcomes is similar for postmenopausal women ages 50-64 and 65 and older: results from the National Osteoporosis Risk Assessment (NORA). *J. Bone Miner. Res.* **19:** 1215–1220.
33. Kanis, J.A., H. Johansson, A. Oden & E.V. McCloskey. 2009. Assessment of fracture risk. *Eur. J. Radiol.* **71:** 392–397.
34. Cauley, J.A. et al. 2003. Effects of estrogen plus progestin on risk of fracture and bone mineral density: the Women's Health Initiative randomized trial. [see comment]. *JAMA* **290:** 1729–1738.
35. Farquhar, C. et al. 2009. Long term hormone therapy for perimenopausal and postmenopausal women. *Cochrane Database Syst. Rev.* (2): CD004143.
36. Rubin, E.H. et al. 1998. A prospective study of cognitive function and onset of dementia in cognitively healthy elders. *Arch. Neurol.* **55:** 395–401.
37. Henderson, V.W. & B.B. Sherwin. 2007. Surgical versus natural menopause: cognitive issues. *Menopause* **14:** 572–579.
38. Henderson, V.W. 2009. Aging, estrogens, and episodic memory in women. *Cogn. Behav. Neurol.* **22:** 205–214.
39. Vearncombe, K.J. & N.A. Pachana. 2009. Is cognitive functioning detrimentally affected after early, induced menopause? *Menopause* **16:** 188–198.
40. Rocca, W.A. et al. 2008. Long-term risk of depressive and anxiety symptoms after early bilateral oophorectomy. *Menopause* **15:** 1050–1059.
41. Rivera, C.M., B.R. Grossardt, D.J. Rhodes & W.A. Rocca. 2009. Increased mortality for neurological and mental diseases following early bilateral oophorectomy. *Neuroepidemiology* **33:** 32–40.
42. Brinton, R.D. 2005. Investigative models for determining hormone therapy-induced outcomes in brain: evidence in support of a healthy cell bias of estrogen action. *Ann. N.Y. Acad. Sci.* **1052:** 57–74.
43. LeBlanc, E.S., J. Janowsky, B.K.S. Chan & H.D. Nelson. 2001. Hormone replacement therapy and cognition: systematic review and meta-analysis. *JAMA* **285:** 1489–1499.
44. Shumaker, S.A. et al. 2003. Estrogen plus progestin and the incidence of dementia and mild cognitive impairment in postmenopausal women: the Women's Health Initiative Memory Study: a randomized controlled trial. *JAMA* **289:** 2651–2662.
45. Resnick, S.M. et al. 2009. Postmenopausal hormone therapy and regional brain volumes: the WHIMS-MRI Study. *Neurology* **72:** 135–142.
46. Sherwin, B.B. 2009. Estrogen therapy: is time of initiation critical for neuroprotection? *Nat. Rev. Endocrinol.* **5:** 620–627.
47. Harsh, V., S. Meltzer-Brody, D.R. Rubinow & P.J. Schmidt. 2009. Reproductive aging, sex steroids, and mood disorders. *Harv. Rev. Psychiatry* **17:** 87–102.
48. Clayton, A. & C. Guico-Pabia. 2008. Recognition of depression among women presenting with menopausal symptoms. *Menopause* **15:** 758–767.
49. Bromberger, J.T. et al. 2007. Depressive symptoms during the menopausal transition: the Study of Women's Health Across the Nation (SWAN). *J. Affect. Disord.* **103:** 267–272.
50. Eichling, P.S. & J. Sahni. 2005. Menopause related sleep disorders. *J. Clin. Sleep Med.* **1:** 291–300.
51. Hlatky, M.A. et al. 2002. Quality-of-life and depressive symptoms in postmenopausal women after receiving hormone therapy: results from the Heart and Estrogen/Progestin Replacement Study (HERS) trial. *JAMA* **287:** 591–597.
52. Schmidt, P.J. 2005. Mood, depression, and reproductive hormones in the menopausal transition. *Am. J. Med.* **118**(Suppl. 12B): 54–58.
53. Basson, R. 2005. Women's sexual dysfunction: revised and expanded definitions. *CMAJ* **172:** 1327–1333.
54. Dennerstein, L., P. Lehert, H. Burger & J. Guthrie. 2005. Sexuality. *Am. J. Med.* **118**(Suppl. 12B): 59–63.
55. Dennerstein, L., P. Koochaki, I. Barton & A. Graziottin. 2006. Hypoactive sexual desire disorder in menopausal women: a survey of Western European women. *J. Sex. Med.* **3:** 212–222.
56. Davison, S.L. et al. 2005. Androgen levels in adult females: changes with age, menopause, and oophorectomy. *J. Clin. Endocrinol. Metab.* **90:** 3847–3853.
57. Basson, R. et al. 2004. Revised definitions of women's sexual dysfunction. *J. Sex. Med.* **1:** 40–48.
58. Leiblum, S.R. et al. 2006. Hypoactive sexual desire disorder in postmenopausal women: US results from the Women's International Study of Health and Sexuality (WISHeS). *Menopause* **13:** 46–56.
59. Casson, P.R. et al. 1997. Effect of postmenopausal estrogen replacement on circulating androgens. *Obstet. Gynecol.* **90:** 995–998.
60. Nathorst-Boos, J. et al. 1993. Is sexual life influenced by transdermal estrogen therapy? A double blind placebo controlled study in postmenopausal women. *Acta Obstet. Gynecol. Scand.* **72:** 656–660.
61. Lara, L.A. et al. 2009. The effects of hypoestrogenism on the vaginal wall: interference with the normal sexual response. *J. Sex. Med.* **6:** 30–39.
62. Shifren, J.L. et al. 2006. Testosterone patch for the treatment of hypoactive sexual desire disorder in naturally menopausal women: results from the INTIMATE NM1 Study. *Menopause* **13:** 770–779.

63. Davis, S.R., S.L. Davison, S. Donath & R.J. Bell. 2005. Circulating androgen levels and self-reported sexual function in women. *JAMA* **294:** 91–96.
64. Buster, J.E. *et al*. 2005. Testosterone patch for low sexual desire in surgically menopausal women: a randomized trial. *Obstet. Gynecol.* **105:** 944–952.
65. Davis, S.R. *et al*. 2006. Efficacy and safety of a testosterone patch for the treatment of hypoactive sexual desire disorder in surgically menopausal women: a randomized, placebo-controlled trial. *Menopause* **13:** 387–396.
66. Mehta, A. & G. Bachmann. 2008. Vulvovaginal complaints. *Clin. Obstet. Gynecol.* **51:** 549–555.
67. Loprinzi, C.L. *et al*. 1997. Phase III randomized double-blind study to evaluate the efficacy of a polycarbophil-based vaginal moisturizer in women with breast cancer. *J. Clin. Oncol.* **15:** 969–973.
68. Suckling, J., A. Lethaby & R. Kennedy. 2006. Local oestrogen for vaginal atrophy in postmenopausal women. *Cochrane Database Syst. Rev.* (4): CD001500.
69. Perrotta, C. *et al*. 2008. Oestrogens for preventing recurrent urinary tract infection in postmenopausal women. *Cochrane Database Syst. Rev.* (2): CD005131.
70. Simon, J. *et al*. 2008. Effective treatment of vaginal atrophy with an ultra-low-dose estradiol vaginal tablet. *Obstet. Gynecol.* **112:** 1053–1060.
71. Weisberg, E. *et al*. 2005. Endometrial and vaginal effects of low-dose estradiol delivered by vaginal ring or vaginal tablet. *Climacteric* **8:** 83–92.
72. Weiderpass, E. *et al*. 1999. Low-potency oestrogen and risk of endometrial cancer: a case-control study. *Lancet* **353:** 1824–1828.
73. Rosenberg, L.U. *et al*. 2006. Menopausal hormone therapy and other breast cancer risk factors in relation to the risk of different histological subtypes of breast cancer: a case-control study. *Breast Cancer Res.* **8:** R11.
74. Kendall, A., M. Dowsett, E. Folkerd & I. Smith. 2006. Caution: vaginal estradiol appears to be contraindicated in postmenopausal women on adjuvant aromatase inhibitors. *Ann. Oncol.* **17:** 584–587.

ANNALS OF THE NEW YORK ACADEMY OF SCIENCES
Issue: *Reproductive Aging*

Reproductive aging and its consequences for general health

Michael L. Traub and Nanette Santoro

Division of Reproductive Endocrinology and Infertility, Department of Obstetrics & Gynecology and Women's Health, Albert Einstein College of Medicine and Montefiore Medical Center, Bronx, New York

Address for correspondence: Dr. Nanette Santoro, Obstetrics and Gynecology, University of Colorado at Denver, 12631 East 17th Avenue, Mail Stop B-198, Academic Office 1, Room 4010, Aurora, Colorado 80045. glicktoro@aol.com

Reproductive aging coincides with endocrine changes that are not solely reproductive in nature and culminates in hypergonadotropic hypogonadism and amenorrhea. These changes are identifiable biochemically regardless of clinical manifestations. Changes in the hypothalamic–pituitary–ovarian axis are associated with changes in other hormonal axes, specifically the adrenal androgen and the somatotropic axis. A large body of literature indicates that reproductive aging is associated with a decline in the somatotropic axis. The interactions between reproductive aging and changes in the adrenal androgen axis are more complex and complicated by age-related declines in the adrenal axis early in the reproductive years. These changes may play an important role in overall health maintenance. Attempts to ameliorate hormonal declines with exogenous hormonal therapy have produced mixed results. Finally, the age-specific timing as well as the rapidity of the changes that occur with reproductive aging seems to have important consequences on metabolism, cardiovascular risk, cognition, bone density, and even mortality.

Keywords: reproductive aging; menopause; premature ovarian failure; dehydroepiandosterone sulfate (DHEAS); insulin-like growth factor-1

Reproductive aging

Reproductive aging refers to the biological processes in the hypothalamic–pituitary–ovarian (HPO) axis that culminate in menopause. Menopause is generally defined as the cessation of menses (amenorrhea) for 12 months. By that time, the HPO axis is in a state of hypergonadotropic hypogonadism, reflecting minimal ovarian estrogen production and a concomitant elevation in pituitary gonadotropins.[1] Most women have their final menstrual period between age 49–51 with some ethnic and geographic variation.[2] They become aware of reproductive aging by alterations in menstrual cyclicity or local/systemic symptoms of hypoestrogenism. Improvements in assisted reproductive technology, combined with serum, ultrasound, or dynamic testing of markers designed to estimate fertility potential and characterize the earliest stages of reproductive aging, have caused women and their doctors to become more aware of clinically undetectable declines in ovarian function (diminished ovarian reserve [DOR]).

Once menstrual cycles become irregular (the "early" menopause transition), follicle-stimulating hormone (FSH) is more consistently elevated, and cycles progressively lengthen until prolonged amenorrhea occurs (more than 3 months, indicative of the "late" phase of the menopausal transition). Once a woman has gone 12 months without a menstrual period, she is defined as menopausal. Women who have premature ovarian failure (POF) (i.e., 12 months amenorrhea prior to age 40) have a much higher rate of cycle recovery than do naturally menopausal women who experience their final menses after the age of 45. Almost 50% of karyotypically normal POF patients have sporadic follicle growth and ovulation.[3]

There have been concerns for many years that reproductive aging may herald other general medical consequences, especially now that women spend a greater proportion of their lives after the transition than in previous generations. Reproductive aging seems, at a minimum, to be associated with changes in other endocrine systems, and in some cases, may directly alter other systems. Although reproductive aging by definition refers to changes that ultimately affect the output of ovarian hormones, reproductive hormones, specifically estrogens, interact directly with hormones in other systems. For instance, estrogen has complex molecular interactions with insulin-like growth factor-1 (IGF-1), its receptors, and its binding proteins and the two molecules have profound influences on each other's pathways.[4] Because steroidogenesis involves conversion between various sex hormone end products, ovarian sex hormones, such as estradiol and adrenal hormonal output in the form of dehydroepiandosterone sulfate (DHEAS), specific to adrenal gland are interlinked by nature. So, therefore, a decline in one system may affect the amount of substrate available in another pathway. The adrenal corticosteroid pathway also undergoes some changes with aging. There seems to be an overall increase in cortisol production as well as a blunting of the evening cortisol nadir with age.[5] Although some gender-related differences have been observed, suggesting a possible role for sex hormones, meager data exist, and the significance of these findings are currently unknown. A combination of reproductive aging with its effects on these other systems may have discrete, and quantifiable, effects on long-term morbidity and mortality. Therefore, it is important to define these changes and understand their interrelationships, as there may be a rationale to intervene to prevent some of the hormone trajectories that occur with aging. What follows is a review of the interacting endocrine systems that have been investigated to date.

Aging of the neuroendocrine system

Most characterizations of reproductive aging are solely concerned with hormonal output at the ovarian level. They also provide an understanding of the quantity and timing of gonadotropin production at the pituitary level. However, they do not provide as much understanding of potential central nervous system (CNS) aging. Relatively little is known about age-related change in gonadotropin-releasing hormone (GnRH) pulsatility or the sensitivity of the pituitary to GnRH produced. Whether decreased ovarian hormone production ultimately alters hypothalamic and pituitary function or whether central alterations in GnRH and FSH/LH (luteinizing hormone) drive or contribute to the decrease in ovarian function is still not completely understood. The sequence of events may also differ between individual women.

Central feedback mechanisms

Of particular interest is how the hypothalamic axis responds to decreasing levels of ovarian steroid hormone output. Postmenopausal women had LH and FSH levels drawn at baseline and weekly for 4 weeks after the administration of exogenous estradiol in physiological doses. FSH and LH levels declined proportionally to duration of exposure, indicating appropriate negative feedback mechanisms. In addition, short-term supraphysiological doses resulted in an LH surge, indicating an appropriate positive feedback loop.[6] Despite chronological aging, women maintain an intact negative and positive feedback system. However, this positive feedback response may not be preserved in all women, or may be temporarily disrupted during the menopausal transition. In the Study of Women's Health Across the Nation (SWAN) cohort of women transitioning into menopause, LH responses to different patterns of endogenous estrogen were characterized. Some women had neither rises in estrogen or LH, some women demonstrated elevations in estrogen accompanied by an LH surge, and still others failed to mount an LH surge despite a seemingly adequate estrogen response. This finding suggests that some women experience an alteration in pituitary sensitivity to estrogen—which is likely a manifestation of hypothalamic insensitivity to rising estrogen levels during follicular growth.[7] Persistence of such a defect could lead to anovulation on a CNS basis, unrelated to ovarian follicle supply.

Hypothalamic signaling

Studying the free α subunits (FAS) of LH and FSH by serial serum sampling provides an indirect means of measuring GnRH pulsatility. Compared to younger postmenopausal women, older postmenopausal women have demonstrated slower FAS pulse frequencies and lower amplitudes.[8] Some evidence shows that the quality of LH pulses (mean 24-h LH secretion and LH pulse amplitude) appears decreased with chronological aging, but not reproductive aging alone.[9] Taken together, these data imply that hypothalamic processes that influence GnRH release may change with age and make CNS signaling less effective.

LH secretion

Late postmenopausal women (mean age 59 years) demonstrate reduced baseline LH and FSH compared to younger women with POF (idiopathic, normal karyotype, no autoimmune, or identified pathological cause) and early postmenopausal women.[9] This does not seem to arise from a defective negative feedback system. It may result from a change in GnRH pulsatility as mentioned earlier. No significant change in LH pulse frequency has been found in older compared to younger reproductive age women.[10]

Aging and the somatotropic axis

Role of endogenous estrogen in the functional integrity of the somatotropic axis

An abundance of medical literature supports overall differences in the somatotropic axis between men and women. The somatotropic axis refers to the growth hormone (GH) releasing hormone stimulated and somatostatin-inhibited pulsatile secretion of GH from the anterior pituitary, which in turn stimulates hepatic production of IGF-1. This hormonal axis influences local and systemic growth, glucose homeostasis, neurogenesis, and metabolism. The maturation of this axis in puberty with increases in GH pulsatile secretion and resulting increases in serum IGF-1 levels occur later in girls than in boys, although girls achieve peak IGF-1 levels earlier than boys.[11]

Evidence also points to differences in GH/IGF-1 in adults. Most data in young adult humans suggest that total GH secretion, measured by 24-h integrated GH concentration, is higher in women than in men.[12] When older men and women were compared, the difference in secretion rates disappeared when controlling for endogenous estradiol levels.[12] Thus, sex and possibly sex steroid based changes in GH and IGF secretion exist that predate age-related declines in endogenous estrogen.

Endogenous estrogen appears to positively modulate the somatotropic axis during the normal menstrual cycle (as opposed to the *inhibitory* effect of most *exogenous* estrogens on IGF-1 with release of negative feedback on GH secretion). In regularly cycling premenopausal women, IGF-1 levels are highest during the follicular phase, corresponding to elevated estradiol levels.[13] GH and IGF-1 secretion decrease over time in premenopausal women. Wilshire *et al.*[10] studied both younger and older (age 42–46) reproductive age women. The older reproductive age women demonstrated lower basal GH and IGF-1 levels despite the maintenance of estradiol levels.[10]

Decline of the somatotropic axis with aging

IGF levels decline with age in both men and women.[14] Ho *et al.* found that GH secretion patterns in young and old men and women demonstrated differences; GH pulse duration, amplitude, and total concentration were lower in older adults of both sexes compared to their younger counterparts.[12]

Women with POF provide a good means to distinguish between the effects of chronological and reproductive aging on changes in the somatotropic axis. In studies of three groups of women (POF, age-matched reproductive controls, and postmenopausal women), IGF-1 and insulin-like growth factor binding protein-3 (IGFBP-3) levels were clearly lower and comparable in the hypoestrogenic hypergonadotropic women (POF and postmenopausal) independent of age. No differences in single serum GH measurements were observed, although detailed sampling was not undertaken.[15] On the other hand, Lieman *et al.*,[16] replaced older postmenopausal women with estradiol and did not observe a return of GH to concentrations attained in younger POF women treated with estradiol. These findings argue that aging is related to irreversible declines in somatotropic axis function.

Several groups have found correlations between serum estradiol and IGF-1 and/or GH levels. GH secretion declines in older females were linearly related to estradiol levels. After adjusting for estradiol levels, the relationship between age and sex on IGF levels disappeared, providing strong data in support of estrogen's role.[17] Estradiol is probably one of several factors affecting GH changes.

Health implications of the age-related decline in IGF

Because growth factors have been linked to hyperplasia and cancer risk in many different settings,[18] reduced circulating growth factors in association with aging may in theory reduce the risk of neoplasia. In the Nurses' Health Study cohort,

IGF-1 levels were not related to breast cancer risk.[19] However, both high IGF-1 levels and reduced IGFBP-3 levels were independently associated with a higher relative risk of colorectal cancer. The significance of this relationship is uncertain, given the decline of IGF with reproductive aging.

Age-related declines in IGF-1 are associated with adverse cardiovascular/metabolic risk profiles. A Danish cohort of both male and female subjects with no heart disease at baseline found that participants with low IGF-1 levels at entry had twice the risk of ischemic heart disease compared to controls.[20] The risk was similar in both sexes. In a Swedish study, mean IGF-1 levels in women declined each year and correlated significantly and negatively with body mass index (BMI), blood pressure, cholesterol, and triglycerides.[21]

Finally, evidence exists for a role of IGF-1 in the maintenance of bone mineral density (BMD). In a Japanese study of both pre- and postmenopausal women, IGF-1 levels were decreased in the latter group and were significantly correlated with BMD in both groups, independent of age.[22] Low IGF-1 has also been associated with fracture risk in healthy postmenopausal women.[23]

Effect of exogenous estrogen on the somatotropic axis

Of recent interest is whether exogenous estrogen therapy can reverse the changes seen in the somatotropic axis with aging, and if so, by what mechanism. The Rancho Bernardo Study examined a cohort of current users (mean age about 70 years) of either estrogen or estrogen and progesterone. Basal IGF-1 levels were decreased in the estrogen only group compared to nonusers, but were intermediate in concentration to those taking progesterone as well, suggesting that estrogen specifically reduces peripheral IGF-1 levels.[24] Postmenopausal women treated with oral ethinyl-estradiol showed higher basal GH levels and 24-h GH responses after oral estrogen treatment than at baseline and after exercise than at rest.[25] These data suggest that exogenous oral estrogen suppresses peripheral IGF-1 levels leading to an increase in GH secretion, presumably from removal of IGF-1 negative feedback.

In contrast, transdermal estrogen treatment does not seem to produce the same effect. Postmenopausal women replaced with either oral ethinyl-estradiol or transdermal 17-β estradiol were compared to premenopausal controls. Oral estradiol decreased IGF-1 levels and increased both basal and 24-h GH secretion levels with increases in GH pulse amplitude without a change in frequency. However, women who received transdermal estradiol had slightly higher IGF-1 levels and no change in GH parameters.[14] It is not clear if higher doses of transdermal estradiol would have the same effects as oral estrogen.

Notable inconsistencies indicate that we do not yet completely understand these interactions. Although the majority of evidence supports a positive effect of exogenous estrogen on the restoration of the somatotropic axis, clinical benefits are at best unknown. Definitive conclusions are still difficult to make, due to the lack of standardization between types, route, and dosing levels of estrogen.

Adrenal axis

Changes in DHEAS with age

A large body of literature suggests that adrenal androgen production decreases with age. Adrenal androgens interact with and can be metabolized to other sex steroid hormones and can affect metabolism, menstruation, ovarian function, lipid parameters, and sexual development. DHEAS levels in postmenopausal women have been shown to be significantly lower than in premenopausal comparison groups, in parallel with estradiol, with preserved production of glucocorticoids.[26] The Melbourne Women's Midlife Health project showed that the decreases in DHEAS throughout the menopause transition occur despite maintenance of testosterone levels, clearly implicating decreased adrenal function rather than general steroidogenesis.[27] Although data are highly consistent, they are generally limited to cross-sectional studies.

Health implications of declining DHEAS

Declining DHEAS levels have been implicated in many disease states. An Italian cross-sectional study of men and women over age 65 found that low DHEAS levels were significantly related to lack of health, history of myocardial infarction, angina, arterial disease, hypertension, chronic obstructive lung disease, cancer, diabetes mellitus, depression, and osteoporosis.[28]

Declining DHEAS may have implications for bone health. DHEAS levels significantly correlate with BMD as measured by dual energy X-ray absorptiometry (DEXA).[29] DHEAS levels were lower in postmenopausal women with osteoporosis at the spine and/or hip when adjusted for age.[30] The effect of DHEAS on BMD may be related to the effects of estrogen.

DHEAS may also be related to cognitive functioning. Men and women >age 55 years in the Rotterdam Study who scored <26 on the Mini Mental Status Exam, consistent with cognitive impairment, had lower DHEAS levels compared to those with scores above the cutoff.[31] Alzheimer's dementia has been correlated with significantly decreased DHEAS levels.[32]

However, other groups have failed to confirm an association between cognition, physical functioning, and DHEAS. A prospective study of women over age 65 failed to find correlations between DHEAS levels and measures of mental status, depression, visual abilities, and motor performance.[33] Attempts to improve cognition with exogenous DHEA in both younger and older patient populations have failed to show benefit.[34] Although there is limited evidence that DHEAS is associated with isolated parameters of cognition and physical function in some studies, the bulk of the evidence does not even favor an association, and a causal role of DHEAS appears unlikely.

Because of the confounding nature of the data to date, one would hope that studying the model of premature menopause would clarify the relationships between DHEAS, cognition, and physical function. However, such studies have also yielded conflicting results. DHEAS was lower in some POF patients compared to age-matched controls,[35] but in other studies there were no differences.[36] One would expect a more consistent or impressive decline in DHEAS levels in women with POF if the relationship between reproductive aging and declining DHEAS were stronger.

Finally, within-woman changes in multiple hormone levels have been monitored in the longitudinal, observational SWAN study, through various stages of menopause. DHEAS levels declined with age when examined cross-sectionally. After adjusting for age, BMI, and smoking, a significant association was seen between DHEAS levels and the transition to late but not early menopause.[37] Overall, the data support a complex interaction between reproductive aging and adrenal androgen declines.

Overall effect of reproductive aging on long-term morbidity and mortality

The predominance of clinical evidence suggests that reproductive aging has overall adverse health consequences. The initiation of reproductive aging is variable and its timing in the life cycle is probably of significant importance. Likely, women with significant health problems are more likely to undergo earlier menopause and their data may skew the distribution such that otherwise healthy women who are averaged in with them appear less healthy. It is not currently known how applicable or clinically important the diagnosis of DOR truly is with respect to other organ systems. There is also reason to believe that the nature of the process, that is, gradual versus abrupt, is related to subsequent health. To separate these from the overall effects of chronological aging, three patient populations are particularly useful: women undergoing natural menopause, those with POF, and women undergoing surgical menopause.

Premature ovarian failure

POF and its contributions to the understanding of interactions between reproductive and chronological aging have been discussed earlier. The relationship of POF to general health is most reasonably drawn from the study of those patients with idiopathic disease, which removes many of the associated comorbidities that would confound general health assessments. Despite such exclusions, clinical data suggest some adverse health outcomes linked to POF compared to age-matched, normally cycling women, including a higher prevalence of depression, anxiety, stress, and sexual complaints.[38] Caution is needed in interpreting psychosocial data in POF women as they may be related to the adverse implications of this diagnosis for childbearing. Estrogen has known positive effects on BMD. Endogenous estrogen contributes significantly to the strengthening of bone and exogenous estrogen therapy is known to have clinical benefits in treating osteoporosis.[39] Estrogen also has been shown to have an overall beneficial effect on lipid parameters, usually raising high-density lipoprotein (HDL) and lowering low-density lipoprotein (LDL). Although evidence is mixed, estrogen seems to positively affect glucose metabolism and insulin resistance.[40] Together these data link alterations in estrogen with an overall effect on cardiovascular risk in women. Lipid, DEXA, oral glucose tolerance test, and hormone parameters have also been compared between POF women, age-matched controls, and women undergoing natural menopause. Similar and significant detrimental effects were observed for ovarian failure at any age on total cholesterol, LDL, HDL, fasting glucose, fasting insulin/glucose ratio, and BMD.[41]

Surgical menopause

Surgical menopause refers to bilateral oophorectomy (BSO, removal of both ovaries). The most common indication for BSO in the past was as an adjunct to hysterectomy, especially in patients over age 45. It had long been assumed that the elimination of risk of subsequent ovarian cancer justified removal of ovaries that had only a small number of years left to function. General recommendations regarding routine BSO have changed due to data illustrating overall detrimental consequences. In one study, inputs of the incidence of various comorbidities and their relationship to hormone status were used to simulate the effect of BSO at a variety of ages, with mortality as the primary endpoint. This model estimated that women undergoing BSO (without exogenous hormone therapy) prior to age 59 had almost a 4% excess mortality by age 80 and those who underwent BSO prior to age 55 had almost a 9% excess mortality. BSO prior to age 75 was still associated at each intermediate year with excess estimated mortality.[42] This empirical modeling led others to look clinically to answer this question. The Mayo Clinic Cohort Study of Oophorectomy and Aging in premenopausal women compared to age-matched controls found a significant increase in mortality in women who underwent BSO prior to age 45 without further exogenous estrogen therapy (hazard ratio 1.67, 95% CI 1.16–2.40). These women also had an unexplained, significant increase in the risk of estrogen-related cancers.[43]

This same cohort was also analyzed with respect to cognitive functioning. A significant linear relationship between BSO and both dementia and Parkinsonism was observed, which was magnified in women who underwent surgery at a younger age.[44] A pooled meta-analysis showed that oophorectomy before age 50 conferred a fivefold increase in cardiovascular disease.[45] These data support an overall detrimental effect of surgical menopause magnified by an earlier age at time of surgery. Small amounts of endogenous estradiol that are produced after menopause or even the premature loss of progesterone production that occurs after BSO may play a critical role in maintaining health that may or may not be duplicated by exogenous dosing.

The data earlier assume that the postmenopausal ovary continues some steroidogenic activity, which contributes positively to overall health. There is evidence for aromatase activity as well as production of androstenedione and testosterone postmenopausally.[46] However, Couzinet et al.[47] studied postmenopausal women with adrenal insufficiency relative to similar postmenopausal controls, with each group containing some women with prior BSO. Testosterone, free testosterone, and androstenedione were present in significant quantities in the control group but were all undetectable in patients with adrenal insufficiency, even after exogenous human chorionic gonadotropin injections. Local steroidogenic enzyme expression in homogenized ovarian tissue samples measured by immunostaining revealed little to no enzyme activity.[47] The ability of the postmenopausal ovary to produce significant amounts of androgens remains unclear and so does the identity of the substance(s) produced by the postmenopausal ovary that seem to maintain overall health.

Natural menopause

The median age at natural menopause is 51 years. The National Health and Examination Survey (NHANES) database of 3000+ women aged 50–86 indicated a trend toward increased mortality if menopause occurred prior to age 40 (RR 1.5, 95% CI 0.97–2.23).[48] A cohort of almost 20,000 Norwegian women showed a decrease in mortality of 1.6% per 3 year of increase in age at menopause.[49] However, late menopause is not inherently healthy. Because of its association with more exposure to estrogen, it has been linked to increased risk of gynecologic cancers.[50]

Summary

Reproductive aging can be detected and predicted long before clinical implications are evident. Neuroendocrine changes seem to occur independent of chronological aging but are certainly influenced in some ways by aging itself. Changes in the HPO axis seem to play an important role in changes that occur in the somatotropic axis. The influence of reproductive aging on the adrenal axis is less clear, but declining endogenous estrogen levels seem to herald changes in DHEAS secretion. More data are needed to clarify these relationships. In most women, earlier reproductive aging seems to be detrimental to overall health and even life expectancy. Whether this link is strictly a result of reproductive aging or a reflection of the effects of reproductive aging on other endocrine systems still requires more research. Clearly, reversing the effects of reproductive aging by counteracting hormone deficiencies to prevent long-term morbidity and mortality has not been a successful clinical approach. A prospective, randomized, double blinded, placebo controlled trial of DHEA in perimenopausal women failed to demonstrate benefit with respect to lipid profiles, mood, cognition, or perimenopausal symptoms.[51] The use of exogenous GH replacement has also produced mixed benefits, with substantial side effects in women.[52] Finally, results from the Women's Health Initiative have failed to find substantial cardiac[53] or cognitive benefit[54] with sex steroid hormone replacement. This overall lack of proven success may be due to the complex nature of the problem itself, differences in the formulations of exogenous treatments, or differences in characteristics of the patient populations studied. It is important to note that much of the work involving the somatotropic and adrenal axes has been done in postmenopausal women and more data are needed in younger women. In addition, most of the data exploring the timing of menopause and associated long-term risks focuses on early reproductive aging. Women with very late menopause have rarely been studied and represent an area where future research may provide particularly interesting answers. Likely, the general cardiovascular risk profile of individual women plays an important role in determining the baseline risk of adverse risks and the impact the differences in reproductive aging play in modifying risk. As our understanding of these relationships improves, we will be in a better position to counsel patients on surgical and/or medical therapy for common clinical diseases.

Conflicts of interest

The authors declare no conflicts of interest.

References

1. Nelson, H.D. 2008. Menopause. *Lancet* **371:** 760–770.
2. Melby, M.K., M. Lock & P. Kaufert. 2005. Culture and symptom reporting at menopause. *Hum. Reprod. Update* **11:** 495–512.
3. Welt, C.K., J.E. Hall, J.M. Adams & A.E. Taylor. 2005. Relationship of estradiol and inhibin to the follicle-stimulating hormone variability in hypergonadotropic hypogonadism or premature ovarian failure. *J. Clin. Endocrinol. Metab.* **90:** 826–830.
4. Mendez, P., F. Wandosell & L.M. Garcia-Segura. 2006. Crosstalk between estrogen receptors and insulin-like growth factor-I receptor in the brain: cellular and molecular mechanisms. *Front. Neuroendocrinol.* **27:** 391–403.
5. Van Cauter, E., R. Leproult & D.J. Kupfer. 1996. Effects of gender and age on the levels and circadian rhythmicity of plasma cortisol. *J. Clin. Endocrinol. Metab.* **81:** 2468–2473.
6. Nishi, T., S. Yagi & R. Nakano. 1987. Feedback of estrogen in postmenopausal women. *Acta Obstet. Gynecol. Scand.* **66:** 309–313.
7. Weiss, G. et al. 2004. Menopause and hypothalamic-pituitary sensitivity to estrogen. *JAMA* **292:** 2991–2996.
8. Hall, J.E., H.B. Lavoie, E.E. Marsh & K.A. Martin. 2000. Decrease in gonadotropin-releasing hormone (GnRH) pulse frequency with aging in postmenopausal women. *J. Clin. Endocrinol. Metab.* **85:** 1794–1800.
9. Santoro, N. et al. 1998. Effects of aging and gonadal failure on the hypothalamic-pituitary axis in women. *Am. J. Obstet. Gynecol.* **178:** 732–741.
10. Wilshire, G.B. et al. 1995. Diminished function of the somatotropic axis in older reproductive-aged women. *J. Clin. Endocrinol. Metab.* **80:** 608–613.
11. Juul, A. et al. 1994. Serum insulin-like growth factor-I in 1030 healthy children, adolescents, and adults: relation to age, sex, stage of puberty, testicular size, and body mass index. *J. Clin. Endocrinol. Metab.* **78:** 744–752.
12. Ho, K.Y. et al. 1987. Effects of sex and age on the 24-hour profile of growth hormone secretion in man: importance of endogenous estradiol concentrations. *J. Clin. Endocrinol. Metab.* **64:** 51–58.
13. Helle, S.I. et al. 1998. Alterations in the insulin-like growth factor system during the menstrual cycle in normal women. *Maturitas* **28:** 259–265.
14. Weissberger, A.J., K.K. Ho & L. Lazarus. 1991. Contrasting effects of oral and transdermal routes of estrogen replacement therapy on 24-hour growth hormone (GH) secretion, insulin-like growth factor I, and GH-binding protein in postmenopausal women. *J. Clin. Endocrinol. Metab.* **72:** 374–381.
15. Milewicz, T. et al. 2005. 17beta-estradiol regulation of human growth hormone (hGH), insulin-like growth factor-I (IGF-I) and insulin-like growth factor binding protein-3 (IGFBP-3) axis in hypoestrogenic, hypergonadotropic women. *Endokrynol. Pol.* **56:** 876–882.
16. Lieman, H.J. et al. 2001. Effects of aging and estradiol supplementation on GH axis dynamics in women. *J. Clin. Endocrinol. Metab.* **86:** 3918–3923.
17. Ho, K.Y. & A.J. Weissberger. 1990. Secretory patterns of growth hormone according to sex and age. *Horm. Res.* **33**(Suppl. 4): 7–11.
18. Bertoni, E. & M. Salvadori. 2009. Antineoplastic effect of proliferation signal inhibitors: from biology to clinical application. *J. Nephrol.* **22:** 457–462.

19. Hankinson, S.E. et al. 1998. Circulating concentrations of insulin-like growth factor-I and risk of breast cancer. *Lancet* **351:** 1393–1396.
20. Juul, A. et al. 2002. Low serum insulin-like growth factor I is associated with increased risk of ischemic heart disease: a population-based case-control study. *Circulation* **106:** 939–944.
21. Landin-Wilhelmsen, K. et al. 1994. Serum insulin-like growth factor I in a random population sample of men and women: relation to age, sex, smoking habits, coffee consumption and physical activity, blood pressure and concentrations of plasma lipids, fibrinogen, parathyroid hormone and osteocalcin. *Clin. Endocrinol. (Oxf)* **41:** 351–357.
22. Nasu, M. et al. 1997. Effect of natural menopause on serum levels of IGF-I and IGF-binding proteins: relationship with bone mineral density and lipid metabolism in perimenopausal women. *Eur. J. Endocrinol.* **136:** 608–616.
23. Garnero, P., E. Sornay-Rendu & P.D. Delmas. 2000. Low serum IGF-1 and occurrence of osteoporotic fractures in postmenopausal women. *Lancet* **355:** 898–899.
24. Goodman-Gruen, D. & E. Barrett-Connor. 1996. Effect of replacement estrogen on insulin-like growth factor-I in postmenopausal women: the Rancho Bernardo Study. *J. Clin. Endocrinol. Metab.* **81:** 4268–4271.
25. Dawson-Hughes, B., D. Stern, J. Goldman & S. Reichlin. 1986. Regulation of growth hormone and somatomedin-C secretion in postmenopausal women: effect of physiological estrogen replacement. *J. Clin. Endocrinol. Metab.* **63:** 424–432.
26. Parker, C.R. Jr. et al. 2000. Effects of aging on adrenal function in the human: responsiveness and sensitivity of adrenal androgens and cortisol to adrenocorticotropin in premenopausal and postmenopausal women. *J. Clin. Endocrinol. Metab.* **85:** 48–54.
27. Burger, H.G. et al. 2000. A prospective longitudinal study of serum testosterone, dehydroepiandrosterone sulfate, and sex hormone-binding globulin levels through the menopause transition. *J. Clin. Endocrinol. Metab.* **85:** 2832–2838.
28. Ravaglia, G. et al. 2002. Dehydroepiandrosterone-sulfate serum levels and common age-related diseases: results from a cross-sectional Italian study of a general elderly population. *Exp. Gerontol.* **37:** 701–712.
29. Tok, E.C. et al. 2004. The effect of circulating androgens on bone mineral density in postmenopausal women. *Maturitas* **48:** 235–242.
30. Haden, S.T. et al. 2000. Effects of age on serum dehydroepiandrosterone sulfate, IGF-I, and IL-6 levels in women. *Calcif. Tissue Int.* **66:** 414–418.
31. Kalmijn, S. et al. 1998. A prospective study on cortisol, dehydroepiandrosterone sulfate, and cognitive function in the elderly. *J. Clin. Endocrinol. Metab.* **83:** 3487–3492.
32. Näsman et al. 1991. Serum dehydroepiandrosterone sulfate in Alzheimer's disease and in multi-infarct dementia. *Biol. Psychiatry* **30:** 684–690.
33. Yaffe, K. et al. 1998. Neuropsychiatric function and dehydroepiandrosterone sulfate in elderly women: a prospective study. *Biol. Psychiatry* **43:** 694–700.
34. Wolf, O.T. et al. 1997. Effects of a two-week physiological dehydroepiandrosterone substitution on cognitive performance and well-being in healthy elderly women and men. *J. Clin. Endocrinol. Metab.* **82:** 2363–2367.
35. Doldi, N. et al. 1998. Premature ovarian failure: steroid synthesis and autoimmunity. *Gynecol. Endocrinol.* **12:** 23–28.
36. Elias, A.N., M.R. Pandian & F.J. Rojas. 1997. Serum levels of androstenedione, testosterone and dehydroepiandrosterone sulfate in patients with premature ovarian failure to age-matched menstruating controls. *Gynecol. Obstet. Invest.* **43:** 47–48.
37. Lasley, B.L. et al. 2002. The relationship of circulating dehydroepiandrosterone, testosterone, and estradiol to stages of the menopausal transition and ethnicity. *J. Clin. Endocrinol. Metab.* **87:** 3760–3767.
38. Van Der Stege, J.G. et al. 2008. Decreased androgen concentrations and diminished general and sexual well-being in women with premature ovarian failure. *Menopause* **15:** 23–31.
39. Venken, K., F. Callewaert, S. Boonen & D. Vanderschueren. 2008. Sex hormones, their receptors and bone health. *Osteoporos. Int.* **19:** 1517–1525.
40. Lobo, R.A. 2008. Metabolic syndrome after menopause and the role of hormones. *Maturitas* **60:** 10–18.
41. Senoz, S., B. Direm, B. Gülekli & O. Gökmen. 1996. Estrogen deprivation, rather than age, is responsible for the poor lipid profile and carbohydrate metabolism in women. *Maturitas* **25:** 107–114.
42. Parker, W.H. et al. 2005. Ovarian conservation at the time of hysterectomy for benign disease. *Obstet. Gynecol.* **106:** 219–226.
43. Rocca, W.A. et al. 2006. Survival patterns after oophorectomy in premenopausal women: a population-based cohort study. *Lancet Oncol.* **7:** 821–828.
44. Rocca, W.A., B.R. Grossardt & D.M. Maraganore. 2008. The long-term effects of oophorectomy on cognitive and motor aging are age dependent. *Neurodegener. Dis.* **5:** 257–260.
45. Atsma, F., M.L. Bartelink, D.E. Grobbee & Y.T. Van Der Schouw. 2006. Postmenopausal status and early menopause as independent risk factors for cardiovascular disease: a meta-analysis. *Menopause* **13:** 265–279.
46. Adashi, E.Y. 1994. The climacteric ovary as a functional gonadotropin-driven androgen-producing gland. *Fertil. Steril.* **62:** 20–27.
47. Couzinet, B. et al. 2001. The postmenopausal ovary is not a major androgen-producing gland. *J. Clin. Endocrinol. Metab.* **86:** 5060–5066.
48. Cooper, G.S. & D.P. Sandler. 1998. Age at natural menopause and mortality. *Ann. Epidemiol.* **8:** 229–235.
49. Jacobsen, B.K., I. Heuch & G. Kvale. 2003. Age at natural menopause and all-cause mortality: a 37-year follow-up of 19,731 Norwegian women. *Am. J. Epidemiol.* **157:** 923–929.
50. Ossewaarde, M.E. et al. 2005. Age at menopause, cause-specific mortality and total life expectancy. *Epidemiology* **16:** 556–562.
51. Barnhart, K.T. et al. 1999. The effect of dehydroepiandrosterone supplementation to symptomatic perimenopausal women on serum endocrine profiles, lipid parameters, and

health-related quality of life. *J. Clin. Endocrinol. Metab.* **84:** 3896–3902.
52. Harman, S.M. & M.R. Blackman. 2004. Use of growth hormone for prevention or treatment of effects of aging. *J. Gerontol. A Biol. Sci. Med. Sci.* **59:** 652–658.
53. Rossouw, J.E. *et al.* 2007. Postmenopausal hormone therapy and risk of cardiovascular disease by age and years since menopause. *JAMA* **297:** 1465–1477.
54. Espeland, M.A. *et al.* 2004. Conjugated equine estrogens and global cognitive function in postmenopausal women: Women's Health Initiative Memory Study. *JAMA* **291:** 2959–2968.

ANNALS OF THE NEW YORK ACADEMY OF SCIENCES
Issue: *Reproductive Aging*

Longitudinal, epidemiologic studies of female reproductive aging

Rebecca J. Ferrell[1] and MaryFran Sowers[2]

[1]National Institute on Aging, National Institutes of Health, Bethesda, Maryland. [2]Department of Epidemiology, School of Public Health, University of Michigan, Ann Arbor, Michigan

Address for correspondence: Rebecca J. Ferrell, Scientific Review Branch, National Institute on Aging, National Institutes of Health, 7201 Wisconsin Avenue, Room 2C212, Bethesda, Maryland 20892. rebecca.ferrell@nih.gov

Human female reproductive aging consists of multiple processes and interacts with other physiological systems in unique ways. Here we discuss eight recent longitudinal, epidemiologic studies of female reproductive aging that include endocrine data to highlight their contributions to our understanding of these various aging processes and their interactions. Specifically, we review data on ovarian and nonovarian reproductive aging processes and reproductive staging. We consider these data in the context of longitudinal research design and research goals, identify limitations of the studies but also ways in which existing longitudinal data can further our understanding of aging processes, and make recommendations for future studies of female reproductive aging.

Keywords: ovarian aging; hypothalamic-pituitary axis; menstrual cycle; epidemiologic

Introduction

Understanding the relationship between female reproductive aging and more general aging processes is uniquely challenging in modern humans. Few species and none of our primate relatives have postreproductive intervals as long, relative to total lifespan, as those observed in human females.[1] Reproductive aging in women appears markedly accelerated against the backdrop of the elongated human lifespan and the more gradual aging processes observed in other physiological systems.

Interest in female reproductive aging is not only evolutionarily based but reflects a desire to address health concerns that women face at various stages of reproductive and postreproductive life. Changes in fertility, menopausal symptoms, and health risks during and after the menopausal transition are primary concerns, and research advances continue to inform decisions about appropriate treatments and interventions.[2] Assessing the interaction among various aging processes is a necessary part of this research effort. Reproductive aging itself can include multiple processes that occur at different rates and whose interactions are not fully understood. Elevated follicle-stimulating hormone (FSH), for example, may be a function of final depletion of ovarian follicles (and associated steroid hormones) at midlife, which in turn releases gonadotropins from negative feedback, but also could be the result of gradual changes in hypothalamic-pituitary sensitivity to this negative feedback.[1,3–5] Also, it is not always clear whether a particular health risk or outcome is the result of reproductive changes, more generalized age-related processes, or both.[4]

Addressing these research challenges requires a broad approach. Epidemiologic studies represent an important component of the research effort because they inform us about female reproductive aging in population-based samples. The first large-scale studies of female reproductive function, in the mid-20th century, were epidemiologic in design and focused on describing the "normal" menstrual cycle changes with age.[6,7] With the development of modern assay methodologies that could be implemented in populations, researchers established normative hormone profiles in the menstrual cycle[8,9] as part of a larger effort to understand the proximate reproductive functions of the hypothalamic-pituitary-ovarian axis. Concurrently, clinical researchers were identifying conditions in which "normal" reproductive function was in some way compromised, in an effort to understand and treat infertility.[10] And observations of postmenopausal estrogen loss, associated health risks (e.g., osteoporosis, cardiovascular risk), and the potential for hormone replacement therapy to reduce those risks, became a primary focus area.[1,5]

These research efforts laid the groundwork for more recent longitudinal, epidemiologic studies to characterize female reproductive aging in community samples and to reflect the sources of variation that may be contributed by race/ethnicity, body size, or behavioral practices. The studies vary in terms of research goals and therefore in hormonal, menstrual, and covariate measures, but the projects together provide an important body of data. In this chapter, we focus on eight of these studies, specifically those that have assessed reproductive hormone changes with age. Several of these studies have been reviewed elsewhere as parts of broader discussions of female reproductive aging, including the relationship and interactions between reproductive aging and more generalized aging processes.[1,4,11,12] Here, we consider the findings of these studies in terms of ovarian and nonovarian aging processes, research goals and epidemiologic study design, and suggestions for future research efforts.

Study summaries

The following is a summary of longitudinal, epidemiologic studies of reproductive aging that have been established within the last 25 years, and have endocrine measures, reasonable retention, and sufficient duration of follow-up to describe important physiological processes. In most, but not all cases, the age range of participants reflects a focus on perimenopause. Many of the studies are based on cross-sectional studies in order to establish some degree of generalization to community samples.

There is a notable contrast across studies in hormone sampling, which reflects differences in overall research goals. Those studies that aim to develop a broad, global understanding of the menopausal transition and collect data on a wide range of variables, including social, behavioral, and physiological factors, tend to have annual or monthly hormone measures that can be analyzed in conjunction with many other variables over a number of years. Annual hormone measures are taken in the early to middle follicular phase, which is easily identifiable relative to the menstrual period. Such measures can not, however, provide information about ovulatory status or within-cycle hormone patterns, and may or may not be comparable to other studies that sampled on slightly different days of the follicular phase. Those studies that focus more on the physiological, mechanistic underpinnings of reproductive aging generally take more frequent (e.g., weekly or daily) samples which allow for analyses of changes in within- and across-cycle hormone dynamics (e.g., peak progesterone, luteinizing hormone [LH], or FSH occurring outside the follicular phase) and ovulatory status. The drawbacks of an intensive collection routine are financial and time constraints, participant burden, and (usually) lower participant numbers compared with annual-measure studies.

With these factors in mind, the studies are described here in the time order of their development, and key features are compared in Figure 1.

Massachusetts Women's Health Study (MWHS)—longitudinal component: 1986–1991

The goal of the MWHS was to describe how women respond to the menopausal transition, and to determine how social, behavioral, and physiological factors affect women's experiences of the transition.[13] Initially, white women ($n = 8050$) from the greater Boston area who were not postmenopausal (<11 months amenorrhea) were followed from 1981 to 1986 (ages 45–55 at baseline). Multiple questionnaires covering menstrual, health, lifestyle, and sociodemographic characteristics were administered.[13] A second component of the study (MWHS II; 1986–1991; $n = 427$ from original cohort) included annual collection of biomarker data.[14,15] These participants (ages 50–60 years) had an intact uterus, at least one ovary, and no more than 11 months of amenorrhea.[5] Annual blood samples (from day 5 to 7 of menstrual cycle) were assayed for estradiol (E2) and FSH. Data were excluded if women used exogenous hormones or were missing predictor or outcome measures.

Specific topics addressed in the MWHS included menopausal symptoms, sexual function, depression, healthcare utilization, weight, cardiovascular health, women's attitudes about menopause, and reproductive hormone patterns.[13] The study was particularly important in demonstrating for

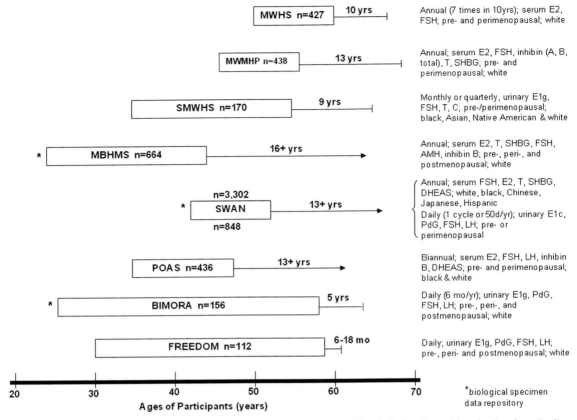

Figure 1. Longitudinal, epidemiologic studies of female reproductive aging that include substantial endocrine data. Studies are in order from least to most recent start date (top to bottom). Box width depicts the baseline age range of participants for each study. Number of years listed on line to the right of each box is the maximum number of years during which endocrine data were/are collected; arrow indicates that a study is ongoing. Information to the right includes sampling strategy, hormones measured, allowable menopausal stages at baseline, and ethnicity. All annual or monthly samples were taken during the early follicular phase of the menstrual cycle. Across all studies, women were excluded if they did not have at least one ovary, were pregnant or breastfeeding, or were taking exogenous hormones or other medications known to affect reproductive hormone values. MWHS, Massachusetts Women's Health Study; MWMHP, Melbourne Women's Midlife Health Project; SMWHS, Seattle Midlife Women's Health Study; MBHMS, Michigan Bone Health and Metabolism Study; SWAN, Study of Women's Health Across the Nation; POAS, Penn Ovarian Aging Study; BIMORA, Biodemographic Models of Reproductive Aging project; FREEDOM, Fertility Recognition Enabling Early Detection of Menopause study. Hormones: C, cortisol; AMH, anti-Müllerian hormone; E2, estradiol; E1c/E1g, estrogen conjugates; FSH, follicle-stimulating hormone; LH, luteinizing hormone; PdG, pregnanediol glucuronide; T, testosterone; SHBG, sex hormone-binding globulin; DHEAS, dehydroepiandrosterone sulfate.

the first time and in a community-based sample the generally positive attitudes of women in the menopausal transition, and that some negative experiences at midlife may not be menopause-specific but rather age-related psychosocial and physiological changes.[13] The annual biomarkers allowed for additional longitudinal analyses of associations among biological reproductive hormones and other biological and psychosocial factors, such as depression, weight, and menstrual cycle characteristics.[14,16]

Melbourne Women's Midlife Health Project (MWMHP)—longitudinal component: 1991–2004

The MWMHP was similar to the MWHS in addressing a variety of physiological, social, and behavioral factors associated with the menopausal transition. This study also focused more specifically on hormonal changes of the transition.[11] A cross-sectional sample of 2001 Australian-born white women (45–55 years of age) was the sampling frame for the study's longitudinal component.

The longitudinal study included 438 women who initially had menstruated within 3 months of examination and were without exogenous hormone use. Women participated in annual health, menstrual, and hormone data collection over a subsequent 13-year period.[17] Annual blood samples were collected (from days 4 to 8 of menstrual cycle or after 3 months of amenorrhea) and assayed for E2, FSH, inhibin A, inhibin B, immunoreactive total inhibin, testosterone (T), and sex hormone-binding globulin (SHBG).[17–19]

Factors addressed by the MWMHP included mood, sexual function, cholesterol levels, bone density, weight, hormone therapy, well-being and quality of life, menopausal symptoms, cardiovascular health, diet, and cognition.[17] The MWMHP was one of the first studies to show long-term changes in a number of reproductive hormones and in menstrual cycle characteristics, relative to the timing of the final menstrual period (FMP). It also identified associations between reproductive hormones and variables, such as menopausal symptoms, menstrual cycle characteristics, sexual function, and lipids and blood pressure.[11]

Seattle Midlife Women's Health Study (SMWHS)—longitudinal component: 1996–2005

The SMWHS, initiated in 1990, also addressed a wide range of menstrual and health factors potentially associated with the menopausal transition.[20] A multi-ethnic community sample (recruited by telephone) of women ($n = 367$, ages 35–55 years, median age 41 at baseline, 20% nonwhite [black, Asian, Native American]) was characterized with annual health and more intensive menstrual data by diary. Commencing in 1996, a subsample of women ($n = 170$) was selected for a more detailed evaluation of hormonal changes.[21] Women were excluded if they had not menstruated in the past 12 months, had both ovaries removed, or were unable to speak or understand English. Eligible women provided a single "day 6" urine specimen for eight to 12 menstrual cycles per year from 1996 to 2000 and quarterly from 2001 to 2005.[21] Urine samples were assayed for estrone glucuronide (E1g), follicle-stimulating hormone (uFSH), testosterone (uT), and cortisol (uC).

Specific topics addressed by the SMWHS included women's perceptions and expectations of menopause, mood and depression, memory, menopausal symptoms, hormone therapy, well-being, stress, and menopausal stages, for example, Refs. 22 and 23. Endocrine data have been assessed in relation to factors, such as menopausal symptoms and stages, and genetic polymorphisms in the estrogen pathway.[21,24] The wider age range and multi-ethnic nature of the participants in this study was also different from the earlier studies.

Michigan Bone Health and Metabolism Study (MBHMS)—longitudinal study: 1992–present

The MBHMS is a population-based longitudinal study with a primary focus on the relationships among reproductive hormones, musculoskeletal and metabolic diseases, and health risks.[25,26] A total of 664 white women (24–44 years at baseline) were recruited using two sampling frames to generate the population census from which 81% were successfully recruited. Participants have provided annual blood and urine specimens (days 2–7 of menstrual cycle, or on the anniversary of study enrollment for those participants with increasingly irregular cycles) and menstrual and health data. Among a range of other analytes, serum has been assayed for E2, T, SHBG, FSH, anti-Müllerian hormone (AMH), and inhibin B.[25–27]

The MBHMS has investigated potential relationships between reproductive hormones and factors, such as bone mineral density, osteoarthritis, physical activity, body composition, and metabolic biomarkers, for example, Refs. 27–29, and has also modeled changes in estrogen and FSH relative to the timing of menopause.[25,26] The study is also notable in having the youngest age range at baseline of all the studies discussed here, allowing it to more fully address the transition though reproductive aging into the postmenopause.

Study of Women's Health Across the Nation (SWAN)—longitudinal component: 1996–present

SWAN is investigating social and biological factors associated with aging and the menopausal transition. It is one of the most comprehensive of the studies described here, in the sense that it attempts to measure a large number of variables and potential confounding factors, and to collect both annual and daily endocrine data.[30] The initial longitudinal study was based on a cross-sectional survey of a multi-ethnic, community-based sample of women

($n = 16,063$; ages 40–55) from seven study sites.[31] From this survey, a subset of 3-302 women (ages 42–52 at baseline) was selected. To be enrolled in the study, women were still menstruating, had an intact uterus and at least one ovary, and could not be pregnant or taking exogenous hormones.[32] Enrollees self-identified as African-American (25%), Caucasian (50%), Chinese (8%), Hispanic (8%), and Japanese (9%). There have been assays of early follicular phase FSH, E2, T, SHBG, and dehydroepiandrosterone sulfate (DHEAS) in the annual serum collections.[33] Interview and diary data included information about each woman's menstrual cycle.

A multi-ethnic subset of 848 women are taking part in the longitudinal Daily Hormone Study (DHS) in which women collect daily first-morning urine samples during one menstrual cycle (or an interval of 50 days), annually; these urine specimens are analyzed for estrone conjugates (E1c), pregnanediol glucuronide (PdG), uFSH, and uLH.[32]

The topics investigated in SWAN are extensive and include analyses of health, psychosocial, and biomarker data.[34] The endocrine data for SWAN are unique because they include, for an unprecedented number of women in a multi-ethnic sample, both annual and daily (for one menstrual cycle per year) data. The annual data, for example, have contributed to our understanding of long-term changes in hormones with reproductive age.[33] The daily data allow for analyses of potential associations among health and psychosocial factors (e.g., hypothalamic-pituitary function, sleep, bone mineral density, cardiovascular risk) and within-cycle hormone patterns and ovulatory status, for example, Refs. 30, 32, 33, and 35–40.

The Penn Ovarian Aging Study (POAS)—longitudinal component: 1996–present

The POAS was designed to "identify hormonal, clinical, behavioral, and demographic factors associated with ovarian aging" (p. 544).[41] Participants were 436 randomly selected black and white women from Philadelphia County (35–47 years at baseline) who had had a regular menstrual cycle (22–35 days long) within the 3 months prior to enrollment. Single day hormone data (day 1–6 of menstrual cycle) were collected four times between 1996 and 2004, with approximately 2 years between collection times.[42] Women also kept a record of menstrual cycles.[42] Women had an intact uterus and at least one ovary and were absent health states that might affect ovarian or hormonal function.[41] Blood samples were assayed for E2, FSH, LH, inhibin B, and DHEAS.[43]

The POAS focuses on the relationship between endocrine changes related to ovarian aging (see further) and a variety of potentially associated factors, including menopausal symptoms, sleep, depression, weight gain, physical activity, menstrual bleeding patterns, and sexual function, for example, Refs. 43–46.

The Biodemographic Models of Reproductive Aging (BIMORA) project–longitudinal study: 1998–2002

The BIMORA project was developed to assess the hormonal and menstrual changes occurring with reproductive aging and to understand how these changes are related to the process of follicular depletion, whereby the ovary is ultimately depleted of ova and their associated steroid hormones. Enrollees were 156 white women (ages 25–58; mean 47.6 years at baseline) drawn from participants in the TREMIN Research Program on Women's Health.[47] Enrollees collected daily urine samples for 6 months (January–July) for 5 consecutive years (1998–2002) while continuing to provide menstrual diary data. At baseline, women had to have at least one ovary, and could not be pregnant, breastfeeding, or using exogenous hormones. The urine specimens were assayed for E1g, PdG, uFSH, and uLH.[47,48]

Similar to SMWHS, MBHMS, and POAS, the participant age range of the BIMORA study was wide, with a low minimum age at baseline, to capture any endocrine changes that might occur earlier in adult reproductive life. The BIMORA project was the first to collect daily data for multiple consecutive menstrual cycles from each participant. This type of data allowed for the modeling of individual 6-month trajectories of steroid and gonadotropin hormones over as many as 5 consecutive years.[47,48] It has also allowed for analyses of within-cycle hormone dynamics, including frequency of anovulation with age, and changes in progesterone, estrogen, and unopposed estrogen with age or menopausal stage.[49,50]

The Fertility Recognition Enabling Early Detection of Menopause (FREEDOM) study—longitudinal component: 1998–2000

The FREEDOM study is similar to the BIMORA study in that it was designed to understand changes in reproductive hormones and menstrual cycle dynamics at the daily level.[51–54] The sample was 112 white British women (ages 30–58; median 44 years at baseline) volunteers. They collected daily urine samples and menstrual bleeding data for 6–18 months, consecutively. Urine specimens were assayed for E1g, PdG, uFSH, and uLH. Women were excluded if they were pregnant or breastfeeding, using exogenous hormones, or had pituitary disorders.[51]

The FREEDOM study has demonstrated important relationships among the reproductive hormones and cycle characteristics within and across consecutive menstrual cycles, including the relationship between FSH, follicular phase length, and age[51] and the lengthening of the follicular phase associated with estrogen that remains low for a portion of the follicular phase.[53]

Hormonal changes with reproductive age

Epidemiologic studies provide an opportunity to evaluate hormonal changes at both the individual and population levels, and the eight studies just described have variously extended our understanding of some of the key hormonal changes that occur during female reproductive aging.

Declining inhibin B and AMH (products of the ovarian follicles) are emerging as some of the earliest indicators of approaching perimenopause. These declines, which occur during late premenopause when menstrual cycles are still regular, are thought to reflect the dwindling pool of ovarian follicles. Furthermore, concurrently increasing FSH may reflect a decrease in negative feedback control by these ovarian hormones.[11,12,33] The MWMHP was one of the first studies to describe this late premenopausal decline in inhibin B alongside a slight rise in FSH, in a population sample.[11] Findings from the POAS[44,55,56] and the MBHMS[57] were similar and the latter showed that AMH declines to values below detection 5 years before the FMP.[57]

When variation in menstrual cycle length starts to occur during early perimenopause, a combination of elevated FSH and maintained or elevated estrogen has been observed, which may indicate that remaining ovarian follicles are being hyperstimulated by elevated FSH.[58] In the MWHS, cohort E2 averages did not change significantly across three annual follow-up measures, even though the percentage of premenopausal women decreased through time.[14] The SMWHS and the FREEDOM study found that E1g remained stable during early perimenopause.[21,51] The MWMHP found that FSH increased up to and beyond the FMP and then leveled off, while E2 was sustained until 2 years before the FMP.[11,18] In the MBHMS, population-average E2 remained stable from −10 to −2 years before the FMP.[26] Elevated estrogen has been observed in the BIMORA study and in earlier work by some of the SWAN investigators[50,59] in which daily hormone data could be analyzed for complete menstrual cycles. Variability in estrogen within and across cycles, the age ranges of participants, and differences in sampling schemes may partly explain the different results across studies.

Rapid declines in estrogen occur during late perimenopause (when variable menstrual cycle length becomes more pronounced), largely in the 2 years before the FMP.[11,26] This decline is thought to reflect the imminent depletion of estrogen-producing ovarian follicles. Several of the epidemiologic studies have identified this estrogen decline at the population or individual levels.[11,21,26,33,44,47] Annual measures have the advantage of being able to characterize the drop in estrogen relative to the FMP, and can estimate the rates of the decline. Daily measures are informative in showing trajectories of declining estrogen for individual women.[47,50]

Some menopausal changes may extend beyond the process of follicular depletion. Increasing FSH may be in part the result of an aging hypothalamic-pituitary axis[60] in which gonadotropin pulsatility changes and the responsiveness to ovarian steroids decreases.[3,12] Also, indirect evidence suggests the possibility of reduced LH sensitivity to estrogen: although cycle-average estrogen levels are lower for anovulatory cycles,[40,49] some anovulatory cycles have follicular phase estrogen levels similar to ovulatory cycles[40] yet do not elicit an LH surge.

Other age-related changes to the oocytes and ovarian tissues may also be affecting cell function, including chromosomal mutations, decreasing telomere length, and accumulating metabolic debris.[61,62] It remains understudied as to how environmental factors affect declining oocyte quality or ovarian tissues, or how these changes interact with the process of follicular depletion. While direct measurements of such changes at the cellular and tissue levels are challenging in longitudinal epidemiologic studies, these studies may generate indirect evidence for relationships among oocyte quality, ovarian tissue function, and follicular depletion. For example, smoking is known to affect egg and ovary quality[63] and has also been shown (Sowers et al.[63a]) to be associated with earlier menopause and faster rates of AMH decline.

Defining reproductive stages

At late reproductive ages, generally menstrual cycles become shorter, and then cycle length variability increases during perimenopause.[6] On the basis of such changes, many of the epidemiologic studies have used series of menstrual bleed data (in some cases complemented by hormone values) to establish population-based reproductive staging paradigms.[20,25,26,50,54,64–72] Certain levels of change from the menstrual cycle "norm" of earlier reproductive ages are defined as marking entry into the next stage of reproductive aging. In this way, data can be grouped for women of similar biological age regardless of their chronological ages. The high level of variation in aging trajectories and age at menopause also makes staging systems particularly valuable for giving women an idea of their proximity to menopause. It should be noted, however, that the common staging systems in use today have been developed in large part from consensus discussions, not all women go through all stages or expected sequences of stages, and FMP can not be precisely predicted by any staging system.

The Stages of Reproductive Aging Workshop (STRAW) reproductive staging system[66] consists of seven stages. The mid- to late reproductive stages are indicated by regular cycles and lower FSH, followed by regular cycles and elevated FSH. In the perimenopausal stages, FSH remains elevated and cycles change from variable length to intermenstrual intervals greater than 60 days. Recent analyses[69,70] have applied the STRAW system to multiple longitudinal menstrual cycle diary data (MWMHP, TREMIN, SMWHS, SWAN) in order to test the utility of the stages for determining entry into early and late perimenopause, and for determining time to FMP. The use of cycle variability >6 days for defining early perimenopause, and the use of a first >60 day intermenstrual interval for entering late perimenopause, were supported, but additional research is needed to identify more precise markers of the early perimenopause.

Multiple hormone values or hormonal patterns across a menstrual cycle have also been proposed for refining reproductive stages, since it is clear that single hormone values (e.g., FSH) do not provide significantly different information about proximity to the FMP than menstrual cycle changes.[65,69] Miro et al.[54] proposed a five-stage classification based on FSH and menstrual cycle characteristics that differs from STRAW by defining an FSH threshold for each stage and by including in the definitions cycles with delayed follicular development and anovulation. This method may be difficult to apply in clinical settings where, for example, anovulatory cycles can not easily be identified. Longitudinal analyses of annual FSH[25] and E2[26] measures in the MBHMS study have shown systematic rates of change and acceleration of change in FSH and E2 that can be characterized in definable stages in the years around the FMP. It should be noted that even these multi-sample methods do not provide precise information about time to FMP.

Limitations and future research

Researchers are cognizant of several issues that limit the analyses that can be done with longitudinal data. First, age ranges of participants in most of the studies described were in the 40s and early 50s, and consideration of those early or mid-late reproductive events that entrain events of the late reproductive age were thereby excluded from study. Second, epidemiologic studies may be constrained to incorporate exclusion criteria that preclude certain population segments (e.g., ethnicity, hormone therapy users, early or late entrants into a particular reproductive stage), thus limiting generalizability of results. Finally, as discussed above, the hormone sampling strategies used in longitudinal studies vary, from large samples of annual measures to smaller samples of daily hormone measures, thus limiting the types of analyses that can be done in each type of study.

One of the goals in female reproductive aging research continues to be to understand attributes leading up to and following the FMP. For studies with annual hormone data, additional data can be pursued using archival specimens, including genetic data, for additional assays thought to be associated with reproductive aging. It is possible that a combination of hormonal and genetic markers from a single annual specimen, along with specific covariate and menstrual cycle data, could improve upon current reproductive staging efforts. To achieve this, however, specimens and suitable analytical techniques must be available.

For studies with daily hormone data, existing data on full and sequential menstrual cycles could be used to develop sampling strategies for predicting time to FMP and to identify subtle relationships among hormone changes across the various parts of the menstrual cycle. Indeed, a main advantage of existing daily hormone data is that they allow for analyses of sampling strategies or hormonal relationships outside of the early follicular phase. Because daily data span the menstrual cycle, however, external markers, apart from actual menstrual bleeding, would be required to relate to the timing of the various parts of their cycle (e.g., initial estrogen rise, ovulation, luteal phase). The daily hormone studies could also add genetic and additional hormone assay data from archival specimens, but these may require additional analytical work to understand how the urinary excretion levels are related to cellular and molecular attributes.

The combination of rich longitudinal data from the perimenopause and postmenopause could be invaluable for showing associations between perimenopause, health outcomes (e.g., osteoporosis, cardiovascular disease), and other aging processes. The heath-related outcomes will need to be common because the sample sizes of most studies of longitudinal aging preclude having sufficient health events to identify, for example, cancers, autoimmune disorders for which women are disproportionately affected, or neurological impairments that may disproportionately affect women.

Genetic markers are likely to become an increasingly common tool in longitudinal studies. Already, some longitudinal studies have examined relationships between specific polymorphisms in the estrogen pathway and later reproductive and health outcomes.[73] Markers for differences in follicular depletion rate and markers for specific disease states are likely to grow in number, and it will be important for existing studies with specimens to consider if and how additional genetic markers can inform their studies. More generally, the use of repositories for specimens and data sets will become increasingly important. The longitudinal studies are in various stages in terms of making their data available to others; most of those studies for which data are not yet available to others are still in the active analysis phase. Others may require additional resources to engage in broad data sharing.

Existing longitudinal studies can play an important role in informing future study designs in human female reproductive aging, through an appreciation of both their limitations and their accomplishments. Expanded age ranges, careful consideration of covariates that modify the rate of follicular depletion or other aging processes, tracking of women over long periods of time, and plans for specimen and data storage and sharing should all be considered. Although there are limits to what any single study can do, collaborative efforts across multiple study sites may have potential for identifying key linkages between reproductive aging and age- and disease-related health issues. In addition, future studies will likely be enhanced by ever-improving technological innovations in the areas of biological imaging, hormonal assays, blood sampling methods, genetic methods, and bioinformatics.

Acknowledgments

Dr. Ferrell contributed to this article in her personal capacity. The views expressed are her own and do not necessarily represent the views of the National Institutes of Health or the United States Government.

Conflicts of interest

The authors declare no conflicts of interest.

References

1. Bellino, F.L. 2007. Female reproductive aging and menopause. In *Physiological Basis of Aging and Geriatrics*. P. Timiras, Ed.: 159–184. Informa Healthcare. New York.
2. Utian, W.H. *et al.* 2008. Estrogen and progestogen use in postmenopausal women: July 2008 position statement of The North American Menopause Society. *Menopause* **15**: 584–602.
3. Downs, J.L. & P.M. Wise. 2009. The role of the brain in female reproductive aging. *Mol. Cell Endocrinol.* **299**: 32–38.
4. Santoro, N. & G. Neal-Perry. 2009. Normal aging and the menopausal transition: what to expect. In *The Menopausal Transition: Interface between Gynecology and Psychiatry. Key Issues in Mental Health*. C.N. Soare & M. Warren, Eds.: 1–17. Karger. Basel.

5. Avis, N. 1999. Women's health at midlife. In *Life in the Middle: Psychological and Social Development in Middle Age*. S. Willis & J. Reid, Eds.: 105–147. Academic Press. San Diego.
6. Treloar, A.E., R.E. Boynton, B.G. Behn & B.W. Brown. 1967. Variation of the human menstrual cycle through reproductive life. *Int. J. Fertil.* **12:** 77–126.
7. Vollman, R.F. 1977. The menstrual cycle. *Major Probl. Obstet. Gynecol.* **7:** 1–193.
8. Sherman, B.M. & S.G. Korenman. 1975. Hormonal characteristics of the human menstrual cycle throughout reproductive life. *J. Clin. Invest.* **55:** 699–706.
9. Metcalf, M.G., R.A. Donald & J.H. Livesey. 1981. Classification of menstrual cycles in pre- and perimenopausal women. *J. Endocrinol.* **91:** 1–10.
10. Broekmans, F.J. *et al.* 2006. A systematic review of tests predicting ovarian reserve and IVF outcome. *Hum. Reprod. Update* **12:** 685–718.
11. Burger, H.G., G.E. Hale, D.M. Robertson, & L. Dennerstein. 2007. A review of hormonal changes during the menopausal transition: focus on findings from the Melbourne Women's Midlife Health Project. *Hum. Reprod. Update.* **13:** 559–565.
12. Santoro, N. 2005. The menopausal transition. *Am. J. Med.* **118**(Suppl. 12B): 8–13.
13. Avis, N.E. & S.M. McKinlay. 1995. The Massachusetts Women's Health Study: an epidemiologic investigation of the menopause. *J. Am. Med. Womens Assoc.* **50:** 45–49, 63.
14. Avis, N.E., S. Crawford, R. Stellato, & C. Longcope. 2001. Longitudinal study of hormone levels and depression among women transitioning through menopause. *Climacteric* **4:** 243–249.
15. Stellato, R.K., S.L. Crawford, S.M. McKinlay & C. Longcope. 1998. Can follicle-stimulating hormone be used to define menopausal status? *Endocr. Pract.* **4:** 137–141.
16. Crawford, S.L., V.A. Casey, N.E. Avis, & S.M. McKinlay. 2000. A longitudinal study of weight and the menopause transition: results from the Massachusetts Women's Health Study. *Menopause* **7:** 96–104.
17. Guthrie, J.R. *et al.* 2004. The menopausal transition: a 9-year prospective population-based study. The Melbourne Women's Midlife Health Project. *Climacteric* **7:** 375–389.
18. Burger, H.G. *et al.* 1998. Serum inhibins A and B fall differentially as FSH rises in perimenopausal women. *Clin. Endocrinol. (Oxf.)* **48:** 809–813.
19. Burger, H.G. *et al.* 2000. A prospective longitudinal study of serum testosterone, dehydroepiandrosterone sulfate, and sex hormone-binding globulin levels through the menopause transition. *J. Clin. Endocrinol. Metab.* **85:** 2832–2838.
20. Mitchell, E.S., N.F. Woods & A. Mariella. 2000. Three stages of the menopausal transition from the Seattle Midlife Women's Health Study: toward a more precise definition. *Menopause* **7:** 334–349.
21. Woods, N.F. *et al.* 2007. Symptoms during the menopausal transition and early postmenopause and their relation to endocrine levels over time: observations from the Seattle Midlife Women's Health Study. *J. Womens Health (Larchmt.)* **16:** 667–677.
22. Woods, N.F., A. Mariella & E.S. Mitchell. 2002. Patterns of depressed mood across the menopausal transition: approaches to studying patterns in longitudinal data. *Acta Obstet. Gynecol. Scand.* **81:** 623–632.
23. Woods, N.F. & E.S. Mitchell. 2005. Symptoms during the perimenopause: prevalence, severity, trajectory, and significance in women's lives. *Am. J. Med.* **118**(Suppl. 12B): 14–24.
24. Mitchell, E.S. *et al.* 2008. Association of estrogen-related polymorphisms with age at menarche, age at final menstrual period, and stages of the menopausal transition. *Menopause* **15:** 105–111.
25. Sowers, M.R. *et al.* 2008. Follicle stimulating hormone and its rate of change in defining menopause transition stages. *J. Clin. Endocrinol. Metab.* **93:** 3958–3964.
26. Sowers, M.R. *et al.* 2008. Estradiol rates of change in relation to the final menstrual period in a population-based cohort of women. *J. Clin. Endocrinol. Metab.* **93:** 3847–3852.
27. Sowers, M.R. *et al.* 2008. Change in adipocytokines and ghrelin with menopause. *Maturitas* **59:** 149–157.
28. Sowers, M., M.L. Jannausch, W. Liang & M. Willing. 2004. Estrogen receptor genotypes and their association with the 10-year changes in bone mineral density and osteocalcin concentrations. *J. Clin. Endocrinol. Metab.* **89:** 733–739.
29. Sowers, M.F. *et al.* 1996. Association of bone mineral density and sex hormone levels with osteoarthritis of the hand and knee in premenopausal women. *Am. J. Epidemiol.* **143:** 38–47.
30. Santoro, N. *et al.* 2004. Body size and ethnicity are associated with menstrual cycle alterations in women in the early menopausal transition: The Study of Women's Health across the Nation (SWAN) Daily Hormone Study. *J. Clin. Endocrinol. Metab.* **89:** 2622–2631.
31. Sowers, M. *et al.* 2001. The association of menopause and physical functioning in women at midlife. *J. Am. Geriatr. Soc.* **49:** 1485–1492.
32. Meyer, P.M. *et al.* 2007. Characterizing daily urinary hormone profiles for women at midlife using functional data analysis. *Am. J. Epidemiol.* **165:** 936–945.
33. Randolph, J.F. *et al.* 2004. Change in estradiol and follicle-stimulating hormone across the early menopausal transition: effects of ethnicity and age. *J. Clin. Endocrinol. Metab.* **89:** 1555–1561.
34. SWAN. *SWAN: Study of Women's Health Across the Nation, Publications List*. 2009. http://www.edc.gsph.pitt.edu/swan/public/Documents/PublicationsPresentations/Publication_List.doc (accessed April 8, 2009).
35. Santoro, N. *et al.* 2003. Assessing menstrual cycles with urinary hormone assays. *Am. J. Physiol. Endocrinol. Metab.* **284:** E521–E530.
36. Santoro, N. *et al.* 2008. Factors related to declining luteal function in women during the menopausal transition. *J. Clin. Endocrinol. Metab.* **93:** 1711–1721.
37. Matthews, K.A. *et al.* 2006. Relation of cardiovascular risk factors in women approaching menopause to menstrual cycle characteristics and reproductive hormones in the follicular and luteal phases. *J. Clin. Endocrinol. Metab.* **91:** 1789–1795.
38. Grewal, J. *et al.* 2006. Low bone mineral density in the early menopausal transition: role for ovulatory function. *J. Clin. Endocrinol. Metab.* **91:** 3780–3785.

39. Kravitz, H.M. *et al.* 2005. Relationship of day-to-day reproductive hormone levels to sleep in midlife women. *Arch. Intern. Med.* **165:** 2370–2376.
40. Weiss, G. *et al.* 2004. Menopause and hypothalamic-pituitary sensitivity to estrogen. *JAMA* **292:** 2991–2996.
41. Nelson, D.B. *et al.* 2004. Predicting participation in prospective studies of ovarian aging. *Menopause* **11:** 543–548.
42. Schmitz, K.H. *et al.* 2007. Association of physical activity with reproductive hormones: the Penn Ovarian Aging Study. *Cancer Epidemiol. Biomarkers Prev.* **16:** 2042–2047.
43. Gracia, C.R. *et al.* 2007. Hormones and sexuality during transition to menopause. *Obstet. Gynecol.* **109:** 831–840.
44. Freeman, E.W. *et al.* 2008. Symptoms in the menopausal transition: hormone and behavioral correlates. *Obstet. Gynecol.* **111:** 127–136.
45. Morrison, M.F. *et al.* 2001. DHEA-S levels and depressive symptoms in a cohort of African American and Caucasian women in the late reproductive years. *Biol. Psychiatry* **50:** 705–711.
46. Freeman, E.W. *et al.* 2004. Hormones and menopausal status as predictors of depression in women in transition to menopause. *Arch. Gen. Psychiatry* **61:** 62–70.
47. Ferrell, R.J. *et al.* 2005. Monitoring reproductive aging in a 5-year prospective study: aggregate and individual changes in steroid hormones and menstrual cycle lengths with age. *Menopause* **12:** 567–577.
48. Ferrell, R.J. *et al.* 2007. Monitoring reproductive aging in a 5-year prospective study: aggregate and individual changes in luteinizing hormone and follicle-stimulating hormone with age. *Menopause* **14:** 29–37.
49. O'Connor, K.A. *et al.* 2009. Progesterone and ovulation across stages of the transition to menopause. *Menopause* **16:** 1178–1187.
50. O'Connor, K.A. *et al.* 2009. Total and unopposed estrogen exposure across stages of the transition to menopause. *Cancer Epidemiol. Biomarkers Prev.* **18:** 828–836.
51. Miro, F. *et al.* 2004. Relationship between follicle-stimulating hormone levels at the beginning of the human menstrual cycle, length of the follicular phase and excreted estrogens: the FREEDOM study. *J. Clin. Endocrinol Metab.* **89:** 3270–3275.
52. Miro, F. & L.J. Aspinall. 2005. The onset of the initial rise in follicle-stimulating hormone during the human menstrual cycle. *Hum. Reprod.* **20:** 96–100.
53. Miro, F. *et al.* 2004. Origins and consequences of the elongation of the human menstrual cycle during the menopausal transition: the FREEDOM Study. *J. Clin. Endocrinol. Metab.* **89:** 4910–4915.
54. Miro, F. *et al.* 2005. Sequential classification of endocrine stages during reproductive aging in women: the FREEDOM study. *Menopause* **12:** 281–290.
55. Freeman, E.W. *et al.* 2005. Follicular phase hormone levels and menstrual bleeding status in the approach to menopause. *Fertil. Steril.* **83:** 383–392.
56. Freeman, E.W. *et al.* 2007. Symptoms associated with menopausal transition and reproductive hormones in midlife women. *Obstet. Gynecol.* **110:** 230–240.
57. Sowers, M.R. *et al.* 2008. Anti-mullerian hormone and inhibin B in the definition of ovarian aging and the menopause transition. *J. Clin. Endocrinol. Metab.* **93:** 3478–3483.
58. Klein, N.A. *et al.* 1996. Reproductive aging: accelerated ovarian follicular development associated with a monotropic follicle-stimulating hormone rise in normal older women. *J. Clin. Endocrinol. Metab.* **81:** 1038–1045.
59. Santoro, N., J.R. Brown, T. Adel, & J.H. Skurnick. 1996. Characterization of reproductive hormonal dynamics in the perimenopause. *J. Clin. Endocrinol. Metab.* **81:** 1495–1501.
60. Wise, P.M. *et al.* 1994. Neuroendocrine concomitants of reproductive aging. *Exp. Gerontol.* **29:** 275–283.
61. Kirkwood, T.B. 2008. Understanding ageing from an evolutionary perspective. *J. Intern. Med.* **263:** 117–127.
62. Santoro, N. & D. Tortoriello. 1999. Endocrinology of the climacteric. In *Menopause: Endocrinology and Management*. D. Seifer & E. Kennard, Eds.: 21–34. Humana Press. Totowa, NJ.
63. Motejlek, K., F. Palluch, J. Neulen & R. Grummer. 2006. Smoking impairs angiogenesis during maturation of human oocytes. *Fertil. Steril.* **86:** 186–191.
63a. Sowers, M.R. *et al.* 2010. Relating smoking, obesity, insulin resistance, and ovarian biomarker changes to the final menstrual period. *Ann. N.Y. Acad. Sci.* **1204:** 95–103.
64. Dudley, E.C. *et al.* 1998. Using longitudinal data to define the perimenopause by menstrual cycle characteristics. *Climacteric* **1:** 18–25.
65. Randolph, J.F., Jr. *et al.* 2006. The value of follicle-stimulating hormone concentration and clinical findings as markers of the late menopausal transition. *J. Clin. Endocrinol. Metab.* **91:** 3034–3040.
66. Soules, M.R. *et al.* 2001. Executive summary: Stages of Reproductive Aging Workshop (STRAW). *Menopause* **8:** 402–407.
67. Brambilla, D.J., S.M. McKinlay & C.B. Johannes. 1994. Defining the perimenopause for application in epidemiologic investigations. *Am. J. Epidemiol.* **140:** 1091–1095.
68. Gracia, C.R. *et al.* 2005. Defining menopause status: creation of a new definition to identify the early changes of the menopausal transition. *Menopause* **12:** 128–135.
69. Harlow, S.D. *et al.* 2006. Evaluation of four proposed bleeding criteria for the onset of late menopausal transition. *J. Clin. Endocrinol. Metab.* **91:** 3432–3438.
70. Harlow, S.D. *et al.* 2008. The ReSTAGE Collaboration: defining optimal bleeding criteria for onset of early menopausal transition. *Fertil. Steril.* **89:** 129–140.
71. Lisabeth, L.D. *et al.* 2004. Staging reproductive aging: a comparison of proposed bleeding criteria for the menopausal transition. *Menopause* **11:** 186–197.
72. Robertson, D.M. *et al.* 2008. A proposed classification system for menstrual cycles in the menopause transition based on changes in serum hormone profiles. *Menopause* **15:** 1139–1144.
73. Sowers, M.R., A.L. Wilson, C.A. Karvonen-Gutierrez & S.R. Kardia. 2006. Sex steroid hormone pathway genes and health-related measures in women of 4 races/ethnicities: the Study of Women's Health Across the Nation (SWAN). *Am. J. Med.* **119:** S103–S110.

Corrigendum for Ann. N. Y. Acad. Sci. 1173: 865–873

Abraham Klepfish, Lugassy Gilles, Kotsianidis Ioannis, Rachmilewitz Eliezer, and Schattner Ami. 2009. Enhancing the Action of Rituximab in Chronic Lymphocytic Leukemia by Adding Fresh Frozen Plasma: Complement/Rituximab Interactions & Clinical Results in Refractory CLL. *Ann. N. Y. Acad. Sci.* **1173**: 865–873.

In the above article, the names of the last two authors are incorrect. The author list should read as follows: Abraham Klepfish, Lugassy Gilles, Kotsianidis Ioannis, Eliezer A. Rachmilewitz, and Ami Schattner.

The authors apologize for this error.